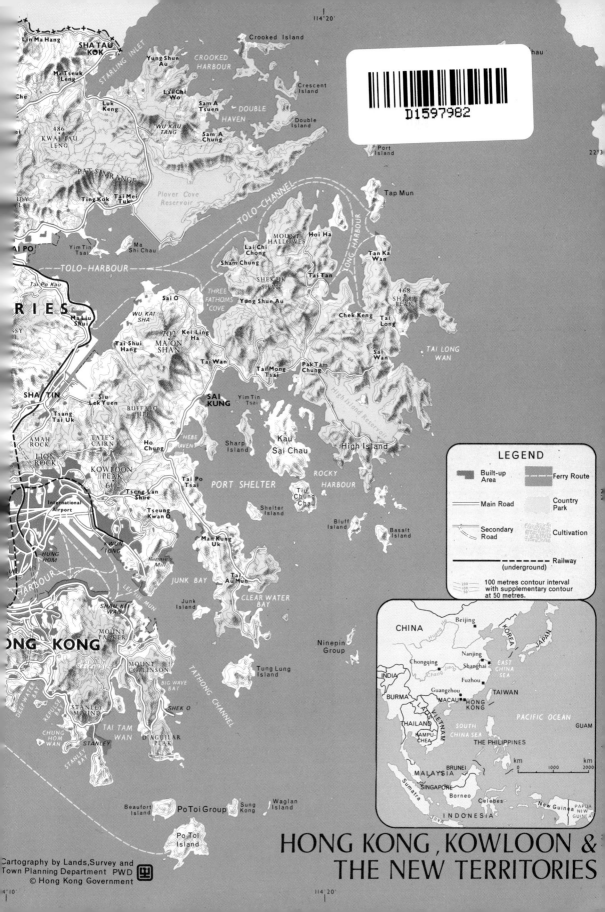

Mary E. Rice

THE SEA SHORE ECOLOGY OF HONG KONG

THE SEA SHORE ECOLOGY OF HONG KONG

Brian Morton
Professor of Zoology, University of Hong Kong

John Morton
Professor of Zoology, University of Auckland

With all good wishes
Brian Morton
Hong Kong   June 1988

HONG KONG UNIVERSITY PRESS

© Hong Kong University Press, 1983

ISBN 962-209-027-3 (Paperback)
      962-209-028-1 (Casebound)

Printed in Hong Kong by Shanghai Printing & Binding Co.

With respect and affection to

    Sir Charles Maurice Yonge, Kt., C.B.E., D.Sc., F.R.S.,

friend, master and counsellor of us both, as of so many others, in all our studies of molluscs and the sea shore.

Contents

Preface — xiii

PART ONE
An Introduction to Hong Kong's Sea Shore — 1

1. Hong Kong: Its Climate and Hydrography — 3
   Climate — 5
   Hydrography — 6
   Biological Seasons — 12

2. The Tidal Shore and its Zoning — 15
   Life on the Shore — 15
   The Kinds of Shores — 16
   Tides, Waves and Zonation — 18

3. Classification of Plants and Animals — 25

PART TWO
Hard Shores — 29

4. Exposed Rocky Shores — 31
   The Moulding of the Hard Shore — 31
   Wave-beaten Shores — 33

5. Sheltered Rocky Shores — 56
   Rock Surfaces — 56
   Algae — 58
   Mollusca — 61
   High Shelter with Low Salinity — 66
   Moderate Exposure with Low Salinity — 69
   From Exposure to Shelter — 69

6. Wharf Piles - 75
    Hydroids and Polyzoa - 78
    Wood-borers - 83

7. Boulder Shores - 86
    A Sheltered Boulder Beach: Tai Tam Harbour - 87
    Crabs and their Habitats - 91
    Molluscs - 98
    Sponges - 106
    Special Habitats - 108

PART THREE

Soft Shores 127

8. Mobile Beaches - 129
    Origin and Moulding - 129
    Donax Beaches - 137
    The Range of Soft and Mobile Habitats - 144

9. Protected Flats - 146
    Meretrix-umbonium Flats: Under Slight Protection - 147
    Crustacea - 151
    Gastropoda - 158
    Bivalves - 162
    Coelenterates - 170
    Burrowing Worms - 170

10. Sheltered Flats - 178
    The Study of Soft Flats - 179
    Tai Tam Harbour: Uca-Anomalocardia Flats - 180
    Flora and Fauna of Tai Tam - 181
    Cerianthus-Atrina Flats - 189
    Hemichordates and Cephalochordates - 194
    Fishes and Cephalopods - 196
    Commensals and Parasites - 200

11. Enclosed Flats - 206
    Salt-Meadows with Zoysia Grass - 208
    Mangrove Swamps - 220
    The Deep Bay Marshes - 230

Contents

  Mud-flats with Oyster Beds — — — — — — — — — — — — — 242
  Sea Grasses — — — — — — — — — — — — — — — — — — — 249

PART FOUR
Eastern Coral Shores   253

12. The Coral Sub-littoral — — — — — — — — — — — — — — — — 255
  Introduction — — — — — — — — — — — — — — — — — — — 255
  Mirs Bay and the Eastern Islands — — — — — — — — — — — 257
  The Coral Sub-tidal — — — — — — — — — — — — — — — — 258
  The Families of Stony Corals — — — — — — — — — — — — 259
  Coral Relatives — — — — — — — — — — — — — — — — — — 264
  Interstitial Communities of Coral — — — — — — — — — — 268
  Molluscs of the Coral Sub-littoral — — — — — — — — — — 278
  Fishes of the Coral Sub-littoral — — — — — — — — — — — 291
  The Terraces and Pools of Ping Chau — — — — — — — — — 295

PART FIVE
The Sea Shore's Future   301

13. Epilogue: Pollution and Conservation — — — — — — — — — — 303
  Reclamation and Sub-division — — — — — — — — — — — — 304
  Pollution — 305
  Carrying Capacity — — — — — — — — — — — — — — — — 307
  Towards Coastal Conservation — — — — — — — — — — — 308

Glossary — 313

Taxonomic Guide — — — — — — — — — — — — — — — — — — 327

Index — 333

Plates

colour   facing page
1. Rocky shore fauna — — — — — — — — — — — — — — — — — 18
2. Birds of rocky coastlines — — — — — — — — — — — — — — 19

black-and-white *facing page*

3. Cape D'Aguilar ------------------------- 34
4. Beaches at Stanley and Tai Tam ------------------ 35

colour

5. Rocky shore algae ------------------------ 66
6. Boulder shore fauna ----------------------- 67

black-and-white

7. Open sand beach at Tong Fuk ------------------ 82
8. Protected flats at Shiu Hau and Tai Tam Bay ----------- 83

colour

9. Sandy shore fauna ------------------------ 114
10. Sheltered sand flats in Tolo Harbour -------------- 115

black-and-white

11. Droppers of the mangrove *Kandelia candel* and fringe of trees of the outer mangrove ------------------------- 130
12. Mai Po marshes -------------------------- 131

colour

13. Effect of impoundment upon river mouths ------------ 162
14. Contrasting coastal vegetation ----------------- 163
15. Sandy shore fauna ------------------------ 178
16. Mangrove plants ------------------------- 179

black-and-white

17. Mud flats of Deep Bay and the *Kei Wais* in the Mai Po marshes -- 210
18. Rocky coasts of Ping Chau and Wang Chau ---------- 211

colour

19. The fauna of mangroves --------------------- 226
20. Birds of soft shores and marshes ---------------- 227

black-and-white

21. Some Hong Kong corals --------------------- 258
22. Some Hong Kong corals --------------------- 259

Contents

colour

facing page

23. Some Hong Kong corals – – – – – – – – – – – – – – – – – – – 274
24. Coral associated fauna – 275

black-and-white

25. Some Hong Kong corals – – – – – – – – – – – – – – – – – – – 290
26. Some Hong Kong corals – – – – – – – – – – – – – – – – – – – 291

colour

27. Hong Kong nudibranch molluscs – – – – – – – – – – – – – – 306
28. Some aspects of coastal pollution in Hong Kong – – – – – – – – 307

Preface

We began this book as an unpretentious short guide to the intertidal ecology of Hong Kong, with enough illustrations to make it useful to the beginner. We were soon to find how much pretension this in fact involved. The marine flora and fauna of Hong Kong is extremely diverse. Previously little had been discovered about Hong Kong shore communities and almost nothing published. Though many first-class systematic books have been written in Japan, we could find little guidance for a work on the local ecology. The different taxonomic groups are very unevenly known for Hong Kong, and—even with Japanese and Chinese help—we could be criticized for the scant treatment we have given to some of these. Our heavy leaning towards molluscs and crustaceans can be justified not only because they are the best known but by their wider importance in the pattern of communities. Certain major groups, most conspicuously the algae, but also the polychaete worms, were well outside our primary expertise. We hope to have atoned for brevity by showing the opportune ways for others to take up the study of these organisms.

We have tried to ascribe no specific names that were not within our special knowledge or expertly confirmed by others. We wish to avoid any nomenclatorial slips which would be perpetuated by this book. With so little known, however, some will be inevitable.

With so much apology, we still believe that there is a wide need in Hong Kong for such a book as this. We hope it will encourage the study of marine biology in the schools and universities of Hong Kong. Its real success would be to supersede itself by leading others to publish more definitive studies. We hope that some of our enjoyment of the shores of Hong Kong may come through to the students who read the book and use it in the field.

We gratefully acknowledge the help of a number of friends in field studies, but this book is an especial appreciation of the undergraduate and postgraduate students of the Department of Zoology, University of Hong Kong, who have over the years brought to light much of what is written here.

Mr D. Brandt and the late Mr T. Garland have made an invaluable contribution by their help on the pulmonates and the birds respectively. 'Ms P. J. B. Scott has similarly been of considerable help to our understanding of the Hong Kong coral community.

Professor Brian Morton is indebted for the identification of many of the animal and plant species of the Hong Kong shore to museum specialists and friends throughout the world; particularly to Dr John Taylor, Solene Morris and Ailsa Clark of the British Museum, Dr T. Habe, Dr M. Imajima, and Dr S. Mawatori of the National Museum of Tokyo, and Dr Raymond Manning of the Smithsonian Institution. Dr Soh Cheng Lam, who visited Hong Kong from Singapore in 1975, identified many of the crabs discussed here whilst Dr William Chan, formerly of the Fisheries Research Station, Aberdeen, Hong Kong, has been of continuing help with Hong Kong fishes. Scleractinian corals have been identified by Dr J. E. N. Veron of the Australian Institute of Marine Science.

Professor John Morton was the holder of a Royal Society Visiting Professorship to the University of Hong Kong. He wishes also to acknowledge the continuing hospitality during his visit with his family in 1975 of the University of Hong Kong, and of many friends, especially Mrs Margaret Richards.

The researches of Brian Morton and his students, on which this book rests, has resulted from the support and encouragement of Professor Brian Lofts, Head of the Zoology Department at the University of Hong Kong.

BRIAN MORTON, JOHN MORTON

PART ONE
An Introduction to Hong Kong's Sea Shore

PART ONE

An Introduction to Hong Kong's Sea Shore

CHAPTER 1

Hong Kong: Its Climate and Hydrography

The islands and mainland territories of Hong Kong lie 320 kilometres south of the Tropic of Cancer on the southern coast of China. They mark the end of an eroded continental mountain chain. The whole of Hong Kong is 1036 sq. km in area, and has a surprisingly long coast-line, one-fifth the total coast length of England and Wales.

Huge aggregates of population in the twin cities of Kowloon on the mainland and Victoria on Hong Kong Island border the magnificent harbour that first brought the colony into being. The countryside of Hong Kong and its New Territories is largely separated from the cities by high ridges. But today's population is breaking out of its urban confines; and the 'Nine Dragon' hills of Kowloon have been largely levelled and removed to the foreshore for large-scale reclamations.

The population expansion, seen in the new centres of Shatin, Tsuen Wan, Tuen Mun, Tai Po, Aberdeen and Little Sai Wan, threatens Hong Kong's natural environment. The village community, practising conservation with its traditional *fung shui* grove of trees, and using a wide range of plants and animals as food, is being replaced by an urban population with little feeling for nature.

There is much to lose, for Hong Kong has today a splendid range of habitats, most of all in the diversity of its sea shore.

Hong Kong owes its flora and fauna to its strategic geographical position (Fig. 1.1), at the junction of the vast temperate Palaearctic Japonic zoogeographical region with the tropical Oriental region itself but a small component of the huge Indo-Pacific province. With elements of both regions, Hong Kong is particularly rich in tropical species that manage to survive the rather cold winters. Nowhere is this juxtaposition of two faunas better seen than on the sea shore, where it is responsible for much of the ecological appeal. Thus, temperate algae are found in abundance, but only at the winter period of low temperatures. Mangroves flourish with near tropical diversity, but are dwarfs in

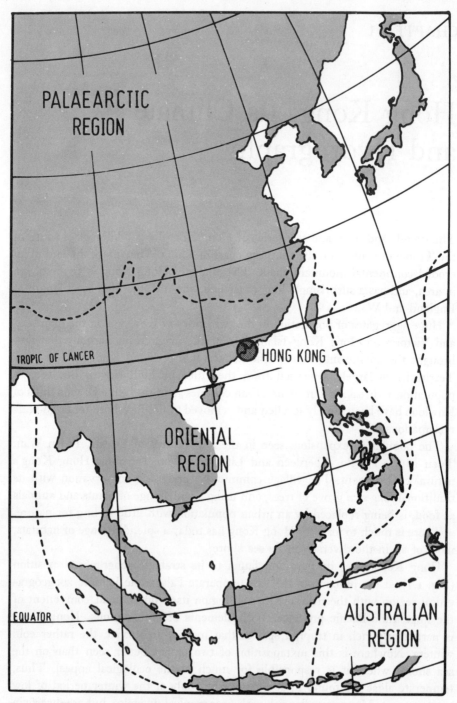

FIG. 1.1. The geographical position of Hong Kong.

comparison with their Malaysian relatives. Corals have an almost tropical range and variety, but with few exceptions are restricted by the low winter air temperatures to the sub-tidal zone.

CLIMATE

The climate of Hong Kong is dominated by the two monsoons, the warm, rain-bearing south-easterly in summer, and the cool dry north-easterly in winter. The cool season generally lasts from October until mid-March. The hot season typically begins in April and continues until September (Fig. 1.2). Spring and autumn—so far as they can be said to occur—are times of fluctuation, when one day may be warm and humid, as the south-easterly monsoon asserts itself, and the next cold and dry.

Summer is the wet season, bringing about 80% of the 217 cm average annual rainfall. Much of this arrives with the typhoons which are prevalent in summer, with gale-force winds, lashing rain and mountainous seas. Summer is generally humid and bright, averaging as much as seven hours of continuous, tropical sun. Air temperatures average 28°C in July, though extremes of 36°C have been recorded.

Winter is often cold and dry and the temperatures on mountain peaks may fall to zero, with resultant frosts. A usual winter minimum is 15°C in February. Winter is typically less humid, and the daily mean of sunshine falls to three hours in March. Greater average wind speeds are recorded in winter, (despite the periodic high force of the summer typhoons), and the sea shore at this time is cold and gusty (Table 1.1).

TABLE 1.1   CLIMATE OF HONG KONG

	Mean air temperature (°C)	Mean Rainfall (mm)	Mean wind speed (knots)
January	15.4	31.7	7.7
February	15.2	46.9	8.9
March	17.5	72.2	9.4
April	21.3	135.8	8.7
May	25.2	292.7	8.3
June	27.3	401.2	7.6
July	27.9	371.7	6.8
August	27.7	370.8	6.5
September	27.1	278.8	7.8
October	24.6	99.2	8.5
November	20.9	43.1	7.8
December	17.3	24.9	7.2

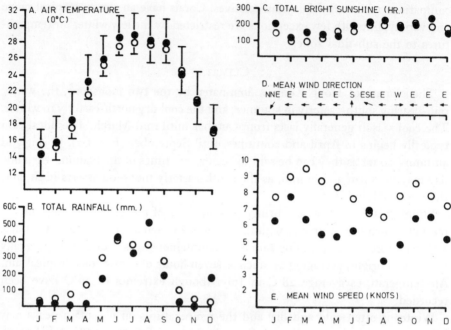

FIG. 1.2. The climate of Hong Kong (●—mean values; ○—usual values).

Hong Kong is thus a climatic compromise, with the weather exerting a dominating influence on coastal waters. The inter-play of the monsoons determines the relative time of arrival of the different water masses that warm or cool the seas; and in extreme cases, the monsoons can have direct effects upon the plants and animals on the shore.

HYDROGRAPHY

The hydrography of Hong Kong's territorial waters above the continental shelf is created by several water masses originating from different sources. The predominating influence throughout the year, but particularly in summer, is the Hainan Current (South China Sea Water) which is characterized by high salinity ($34.4^0/_{00}$–$34.6^0/_{00}$) and variable temperature (29°C at the surface and 12°C at 300 mm depth). It is replaced in winter by the Kuroshio current of high salinity ($34.4^0/_{00}$–$35^0/_{00}$) and high temperature (26°C–29°C), originating in the Pacific and invading the South China Sea via the Luzon Straits. It is this water mass that keeps the coastal waters of Hong Kong relatively warm in winter months, and maintains the sub-tropical and sub-littoral coral communities (Fig. 1.3). The Taiwan current (North China Sea Water) from the East China Sea, characterized by moderate water temperature (19°C–23°C) and reduced salinity ($31^0/_{00}$–$33^0/_{00}$), also invades Hong Kong waters in winter. Figure 1.4 shows

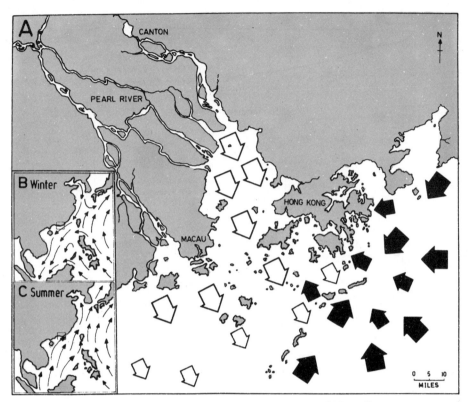

FIG. 1.3. Water masses and ocean currents affecting the coasts of Hong Kong.
A. The influence of the Pearl River, with fresh water run-off (white arrows) and the oceanic water of full salinity (black arrows). B and C. The influence of winter and summer water movements in the South China Sea.

profiles through the South China Sea, to the south of Hong Kong, in summer and winter. The differential effects of the various water masses upon the coasts of Hong Kong can be explained by the interaction of waters of different qualities. A good account of the hydrography of Hong Kong's waters is given by Watts[1] which contains references also to other works on South China Sea hydrography.

At the local level, the inshore hydrography is affected by fresh water arriving from two sources. The heavy rainfall (averaging 217 cm/year) associated with the monsoons prevailing from May to September appreciably dilutes the coastal waters. Moreover, a huge accession of fresh water issuing from the mouth of the Pearl River affects the western coast of Hong Kong and the Portuguese colony of Macau, about 64 km to the west of Hong Kong on the river's opposite bank. The influence of the Pearl River is greatest in summer when the rainfall is highest

[1] Watts, J. C. D. 1973. Further observations on the hydrology of the Hong Kong territorial waters. *Hong Kong Fisheries Bulletin* 3: 9–35.

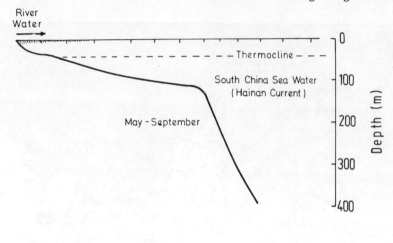

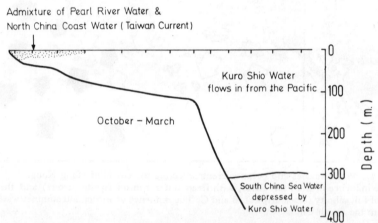

FIG. 1.4. Profiles through the South China Sea to the south of Hong Kong showing the changing influences of a number of oceanic and coastal water masses (after Williamson, G. R., 1970. The hydrography and weather of the Hong Kong fishing grounds. *Hong Kong Fisheries Bulletin* 1: 43–49).

in southern China; and surface salinities to the west of Hong Kong may then, at times, fall to as little as $1^0/_{00}$–$2^0/_{00}$. This spate of turbid, fresh water is identified by its murky colour, often cutting a distinct brown swathe through the blue oceanic water to the south of Lantau Island. Chau[2] has shown by water quality analysis at stations across the Pearl River estuary, that the river water is diverged towards the west, Macau thus being more influenced than Hong Kong (Fig. 1.5). Its waters, as a consequence, are always dilute and sediment laden.

The fresh water selectively reduces the salinity of the western and north-

[2] Chau, Y. K. 1961. The influence of the outflow of the Pearl River on the waters of the South China Sea, with special reference to the phosphate and nitrate content. (M.Sc. thesis, University of Hong Kong).

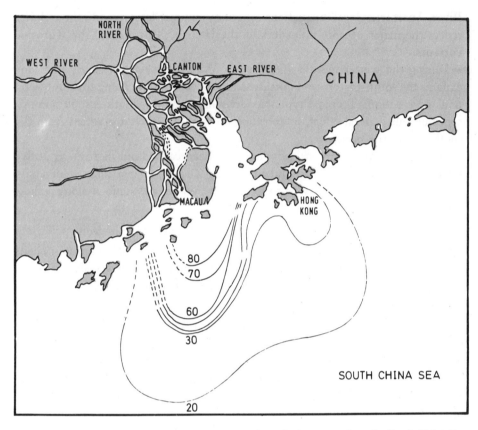

FIG. 1.5. A diagram showing how the discharge of Pearl River water into the South China Sea is diverted to the west, away from Hong Kong. Numbers represent % dilution by fresh water.

western reaches of Hong Kong. Estuarine conditions prevail, with a reduced fauna and flora of euryhaline species centred around the mud flats and mangrove swamps. In the same way, coral growth is precluded. In Deep Bay to the north-west, the shallowness causes a uniform temperature in the surface and bottom waters, and also a more rapid warming and cooling than elsewhere.

The effect of the Pearl River is progressively reduced towards the south-east. The farthest outpost of Waglan Island has high oceanic salinity, with uniform surface and bottom values caused by the mixing effects of heavy swell.

At intermediate stations, for example at Tai Lam Chung, stratification occurs in summer, with waters of low salinity from the Pearl River flowing south-eastwards over the denser, more saline oceanic water mass arriving from the south-east.

By contrast with the west, the south-eastern, eastern and north-eastern coasts of Hong Kong are virtually unaffected by the fresh-water drainage of the Pearl

River. Apart from local fluctuations caused by summer land run-off, these shores reflect the major oceanic influences of the Hainan, Taiwan and the Kuroshio currents.

During the winter months (October to March) the north-easterly monsoon exposes the south-eastern and north-eastern coasts of Hong Kong to strong wind and wave action. Tropical typhoons arriving from the south-east in summer continue this general effect, establishing an exposure/shelter gradient from the south-east to the north-west.

Fortnightly hydrographical sampling has been carried out in 1972–73 at five selected stations by research students in the Zoology Department, University of Hong Kong. On a north-west to south-east transection, stations along it show the following characteristics:

1. Waglan Island: high exposure, high salinity and freedom from pollution.
2. Stonecutter's Island: moderate exposure, in the fast-flowing harbour channel, and high pollution, especially with sewage and oil from vessels.
3. Aberdeen Harbour: little circulation, and high pollution, from oil leakage and sewage discharge.
4. Tai Lam Chung: moderate exposure with strong turbulence generated from currents. Pollution relatively slight.
5. Tsim Bei Tsui: highly sheltered and close to the Pearl River mouth; low and fluctuating salinities, much turbidity from sediments, and with high organic pollution by river-borne sewage from Canton, and from the pig and duck farms of the Yuen Long Plain.

The graphs in Fig. 1.6 provide comparative hydrographical information from three of the stations:

Sea temperature was in general similar at all sites, except for more extreme summer and winter values at the highly enclosed Tsim Bei Tsui. Some contrast with Tsim Bei Tsui is shown by the lower, more uniform temperatures at Waglan Island.

Salinity shows a progressive increase from west to east. The highest values are in winter and early spring, when fresh-water influx is minimal, followed by a progressive decrease, most marked at Tsim Bei Tsui, and least apparent at Waglan Island. There is a summer stratification, with differences in surface and bottom values, most obvious at Tai Lam Chung, and least at Tsim Bei Tsui (low salinity by dilution) and at Waglan Island (high salinity by turbulent mixing).

Dissolved oxygen levels are generally high in winter and spring but low in summer, possibly an effect of animal respiration. The highest values were from off Waglan Island, the next at Tai Lam Chung, Tsim Bei Tsui was also relatively high. The

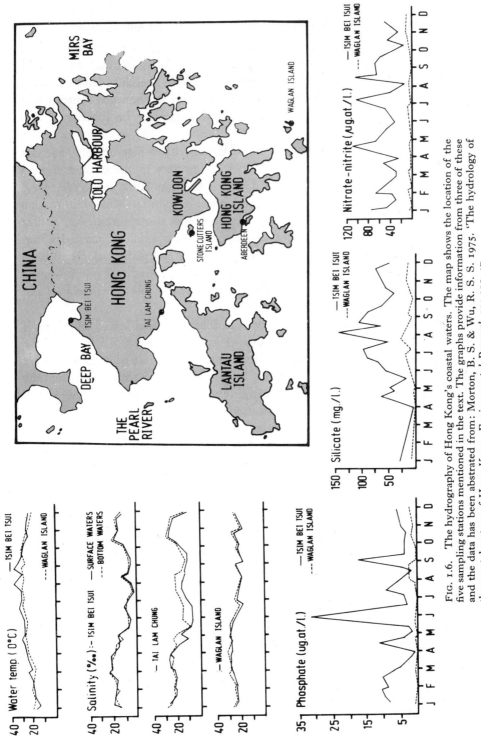

FIG. 1.6. The hydrography of Hong Kong's coastal waters. The map shows the location of the five sampling stations mentioned in the text. The graphs provide information from three of these and the data has been abstracted from: Morton, B. S. & Wu, R. S. S. 1975. 'The hydrology of the coastal waters of Hong Kong. *Environmental Research* 10: 319–47.

lowest were at the polluted sites of Stonecutters Island and Aberdeen.

Phosphate levels were highest at Tsim Bei Tsui, followed by Stonecutters Island and Tai Lam Chung. Low values were obtained at Aberdeen, and lowest of all at Waglan Island. At all stations except Tsim Bei Tsui, values fell in summer from a high winter level. *Nitrite-nitrate* trends show less seasonal contrast than for phosphate. The highest *silicate* levels are in summer, with invasion of land run-off.

Sewage and industrial effluents increase the activity of heterotrophic bacteria and deplete the waters of oxygen. Values fell to zero in the effluent from a chemical dye-works at Sha Tin in Tolo Harbour. Low levels for oxygen were also recorded in Victoria Harbour. The highest oxygen levels were at Waglan Island, next at Tai Lam Chung, by agitation from water turbulence, and at Tsim Bei Tsui, from high photosynthesis in shallow, well-oxygenated water. Low readings were obtained at Aberdeen (enclosed and with small circulation) and at Stonecutters Island which is near several sewage discharge pipes. Marine pollution, which is severe in Hong Kong is discussed in more detail in Part 4 of this book.

From the above, however, it can be seen that, at a local level Hong Kong's territorial waters (Fig. 1.7) can be divided into three zones. To the west the Pearl River creates an estuarine environment. To the east oceanic waters are dominant with the Pearl River exerting little influence and with runoff from the land and smaller streams only somewhat diluting the sea inshore. The central region of Hong Kong is a zone of transition in which, in summer, surface waters of reduced salinity, higher temperature and rich in dissolved oxygen flow over the more saline and cooler oceanic waters which are low in dissolved oxygen. In winter, with a reduced flow from the Pearl River, the waters of this region are vertically more uniform (Fig. 1.8). The limits of this zonation scheme are not strict, the transition zone in particular fluxing from winter to summer with the changing climate brought about by the interplay of the two monsoons and with the varying effect of the Pearl River.

The importance of this longitudinal zoning system is that Hong Kong's geologically diverse shores are further diversified by this hydrographic transition, so that within the very small area of Hong Kong a wide range of inter-tidal and sub-littoral habitats and communities are to be found.

BIOLOGICAL SEASONS

The climate and hydrography of Hong Kong have produced a mixed sub-tropical fauna and flora. The northern and southern water masses impinging at different times on these shores bring larvae of varying origins. There are many truly tropical species, such as the corals and their associated fauna of the east that remain sub-tidal, avoiding the effects of the winter atmosphere and being

Hong Kong: Its Climate and Hydrography

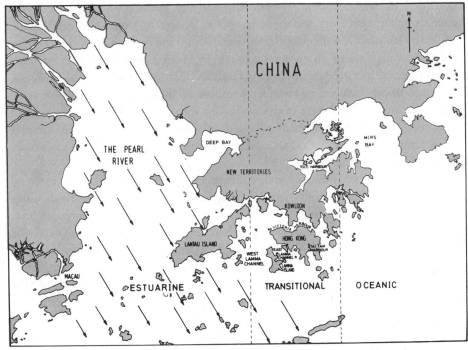

FIG. 1.7 A map of Hong Kong showing the impact of the Pearl River and how the area can be divided into three hydrographic zones.

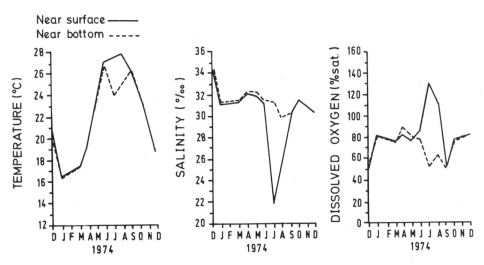

FIG. 1.8. Temperature, salinity and dissolved oxygen figures for Victoria Harbour waters for 1974.

sustained by warm Pacific waters. Alternatively the Taiwan Current brings animals and plants from the more temperate shores of Japan and east China; while the Hainan Current contributes species from the South China Sea. During the winter maxima of shore algal growth in Hong Kong, a number of more temperate herbivores come in-shore to feed. As the algae die off towards summer, the more tropical species of herbivores appear in-shore, often only temporarily.

The oceanic water masses to the south-east introduce a high standing crop of plankton. The in-shore animals of the south-eastern coasts breed in summer when temperatures are high, and a single peak production of their larvae occurs in the plankton at this time.

To the north-west, however, many factors combine to alter the basic seasonal pattern. The reduced salinity from the high mid-summer influx of fresh water appears to inhibit breeding, in spite of the favourable temperatures. A first spawning thus takes place in spring, when the temperature is rising and the salinity is still high, followed by a second in autumn when the salinity again rises and the temperature is falling. Such a breeding pattern has been identified—from falls of settling larvae—for shore barnacles by Dr Rudolph Wu[3], for ship-worms (Teredinidae) by Miss L. F. Fung[4], and for the commercial oyster *(Crassostrea gigas)* grown in Deep Bay by Mr T. K. Mok[5].

---

[3] Wu, R. S. S. 1975. The distribution of littoral barnacles in Hong Kong. In *Proceedings of the Pacific Science Association Special Symposium on Marine Sciences, Hong Kong 1973*, pp. 146–53.

[4] Fung, L. F. 1975. The distribution of marine wood-borers in the coastal waters of Hong Kong. In *Proceedings of the Pacific Science Association Special Symposium on Marine Sciences, Hong Kong, 1973*, pp. 136–45.

[5] Mok, T. K. 1973. Studies on spawning and setting of the oyster in relation to seasonal environmental changes in Deep Bay, Hong Kong. *Hong Kong Fisheries Bulletin* 3: 89–101.

CHAPTER 2

The Tidal Shore and its Zoning

LIFE ON THE SHORE

The shore between tides has always had the strongest attraction for the naturalist. For the past generation—in the temperate and later in the tropics—much research has gone into its species and community patterns. But it has only recently been realized how well the shore lends itself to experimental study.

The inter-tidal is probably the most elaborate and complex of all habitats, rivalled in this—or perhaps outdone—only by the many-layered tropical rain forest. As compared with forest, though, the shore is spatially compact, and very easily accessible from top to bottom. For a major habitat it is a uniquely narrow one. It runs in a thin girdle round the sea and land margin, the strip over which the tide ebbs and flows twice daily. In most temperate regions, there is a regular semi-diurnal tidal rhythm; but many tropical countries present tidal anomalies, and these are shown in some measure in Hong Kong (Fig. 2.1).

Whatever their detailed pattern, the tides produce a shortspan alternation of two sorts of regime, aerial and aquatic, that is unique to the sea shore. This changing balance of wet and dry, from high to low water, leads to a steeply graded transition known as an 'ecotone', with its successive zones often very clear-cut. The sequence of different life forms from periwinkles on supra-littoral bare rock, to rich low tidal algae, in a metre or so of vertical space, is a dramatic one—comparable with the stratification on a mountain, and parallelled nowhere else in so short a space.

The inter-tidal shore is rich beyond comparison in animal life, and has representatives of more of the great groups or 'phyla' than anywhere else. Only the insects and land arthropods, and the higher vertebrates are seriously under-represented. Though the plants of the shore are simple and primitive, (with the full achievement of the plant kingdom reserved for the land), yet it is in the inter-tidal that the algae attain their zenith and gianthood.

The most important animals of the shore are sessile or permanently fixed, often colonial and generally forming a permanent ground cover. By far the predominant mode of feeding is by collecting planktonic food, with every variety and ingenuity of filtering device. Most sessile animals feed by cilia with mucus, using up little energy in the pursuit of food. The barnacles, however, use a muscular-powered sweeping net of cirri. The plankton filterers are preyed upon in turn by animals that can be called by the special term 'grazing carnivores'. Food is thus easy to get, not only from diffusely floating plankton, but also from the living animal protein that—like plant tissues—is found encrusting the ground. Such a diet is constantly abundant, needing no foraging to secure.

If not food, it is settlement and living space that is in small supply. This is a short resource, frugally parcelled out and intricately organized. Microclimates are so variable and competition so intense, that within a few centimetres from its zonal boundaries a species may not be able to sustain life. Even the mobile animals, molluscs, crustaceans and echinoderms, are generally slow-moving and territorial.

On a sloping or vertical tidal surface, the spatial geometry is apt to be simple and to lend itself excellently to experiment. All those parameters that control the living patterns, such as the ratio of emersion/immersion, wave action, current speed, illumination, nature and texture of substratum, warming and desiccating power of the atmosphere, can be altered, simulated or manipulated at need.

Not only are the habitat and most of its animals miniaturized; whole communities are often complete and comprehensible within a few metres space. Of convenient proportions for study is also the length of life, both for species growth and community build-up. An ecological succession can be completed and observed over a year or two's working span, in contrast with the brief few days of a micro-organism colony, or the centuries needed for a forest.

In this book, the chapters and detailed treatment of shores are set out in ecological order according to communities. But first, for the introduction of the general reader, a discussion on the types of shore terrain, developed between tides the world over, and on the divisions of the shores is also given.

THE KINDS OF SHORES

Wherever shore biology is studied, a primary distinction must be made between the two kinds of shore, *hard* and *soft*. In geological terms, these are respectively the shores of *erosion* and *deposition*.

Hard shores are places where the exposed bedrock confronts the action of waves refracted on to capes and promontories, or impacting directly on open coasts. In Hong Kong the bedrock is chiefly igneous or metamorphic. But in many parts of the world hard shores are built of sedimentary rocks, and it is chiefly these softer materials that lend themselves to the cutting of broad inter-tidal platforms.

The shores exposed to the highest wave action tend to be steep, and if platforms exist they are high and narrow. Under the strongest wave attack, the slope can be almost vertical.

The Tidal Shore and its Zoning

In Chapter 4 these outer *exposed shores* of the open, south-facing coasts, have been described from extreme examples such as the tip of Cape D'Aguilar.

Sheltered and *semi-sheltered* hard shores tend to be less steep, and come under lighter wave attack in places away from direct oceanic fetch. Examples are shown in this book from Deep Water Bay and at the mouth of Tai Tam Bay, protected from high surge, but with still enough water movement to wash them clean of loose debris. Their fauna and flora have generally a large proportion of species different from those of exposed hard shores.

Some of the richest habitats are formed by the extension of these hard shores into still further shelter. The shore profile becomes low-pitched, sometimes almost flat, and large fields of boulders accumulate. The complexity of communities under boulders increases greatly; the majority of the crabs and gastropod molluscs live here, protected from strong light, and under a uniformly low temperature and high humidity that would never be presented on the exposed surface.

Boulder beaches differ from each other according to relative wave action or shelter. In Chapter 7 rounded and wave exposed boulders, mobile in stormy weather are distinguished, then more stable beaches of slabs and cobbles and gravel, and finally boulders resting on anaerobic muddy sites. A further type of shaded and secluded habitat, rather akin to boulders is that of *crevices* between shales and other laminar rocks. A description is given in Chapter 6 of *wharf piles*, a special case of hard shores, where the fauna is in a similar way protected against the harsh effects of light and desiccation.

The second great category is that of *soft shores*, where the subdued wave action, or the angle of slope is insufficient to carry all the soft burden away. A net accumulation of particles is left behind. The most wave-exposed of soft shores are appreciably sloped, and their particles well-sorted and mobile. They form sand beaches of no great inter-tidal extent and are regularly swept by sheets of surf. Such beaches, with direct ocean aspects, are not numerous in Hong Kong, but good examples are found along the south-east facing coasts. They are scenically beautiful, and their clean almost fluid sand is characterized by the small bivalve *Donax*.

After the most open beaches, the next division of soft shores enjoys moderate shelter from wave fetch, lying in the wide mouths of bays, or under some protection by the adjacent headlands. Still of clean and rather mobile sand, these beaches have a much wider inter-tidal extent than the *Donax* shores. They are also more level, and a mantle of organic detritus is available as food. The array of burrowing and surface-moving animals is very large, and among the rich fauna, the most characteristic species are the bivalve *Meretrix* and the gastropod *Umbonium*. It is at Shiu Hau, on Lantau Island, that this kind of beach is best represented.

Passing further into the shelter of harbours are *protected sand-flats* with finer sand, and an enriching organic layer. The sediments become increasingly stable, with a grey anaerobic layer at about 10 cm below the surface. They are still however composed of silty sand, rather than mud. The great inter-tidal expanses of Tai Tam, Hebe Haven, Starling Inlet and Three Fathoms Cove are typical of this class of flats, with the cockle *Anomalocardia* and the tiny crabs *Scopimera* the most characteristic indicator species. On these and on the *Meretrix-Umbonium* flats, the species tally reaches its greatest for all soft shores.

Finally are the soft shores that are virtually land-locked, forming the broad shelving inter-tidal flats of the deepest bays and estuaries. Unlike the 'protected shores', these *enclosed shores* are built predominantly of the clay and silt mixtures that form soft and unconsolidated muds. Animals burrow less deeply, and the anaerobic black layer rises to a short distance below the surface.

Mud-flats have in many places undergone much alteration by man, particularly those in Hong Kong. Extensive oyster cultivation for example transforms the flats of Deep Bay, and the large areas of cultch introduce on soft shores an important holding ground for a hard shore fauna. Large areas of reclaimed mud-flats are given over to fish ponds and paddy fields in the Mai Po marshes.

On sheltered soft shores, communities of vegetation are a conspicuous feature both inter-tidally and above normal high water. Hong Kong has rather few of the sea grasses normally characteristic of protected inter-tidal flats. But two important formations are *salt-meadows* and *mangrove swamps*. Salt-meadows are developed normally above the high water level, or are only exceptionally inundated at high spring tides. The commonest Hong Kong species, amongst which terrestrially tolerant molluscs and crabs abound, is *Zoysia* grass. There are a number of succulent angiosperm species, though the flora of these 'saltings' is rather poorer than in some other parts of the world.

Mangroves belong to no single taxonomic family but form an inter-tidal scrub or forest, developed in the tropics and subtropics where conditions are rich and moist enough. Though Hong Kong lies near the northern limits of the mangrove zone, its mangrove flora is rich, with eight species properly belonging to the inter-tidal 'mangal' but with numerous other associated trees, creepers, grasses, herbs and shrubs. Common animals associated with the mangal are the *Uca* and other burrowing crabs, and numerous infaunal bivalves. The bird life of the Deep Bay marshes and ponds is justly renowned internationally.

TIDES, WAVES AND ZONATION

Tides result from a vertical movement of water caused by the response of the sea to the gravitational pull of the moon and sun upon the earth. During the earths rotation, the tidal bulges (one facing the moon, the other, opposite, and resulting from a centrifugal counter force) remain stationary relative to the moon and at any point on the earths surface there is a diurnal cycle of alternate high

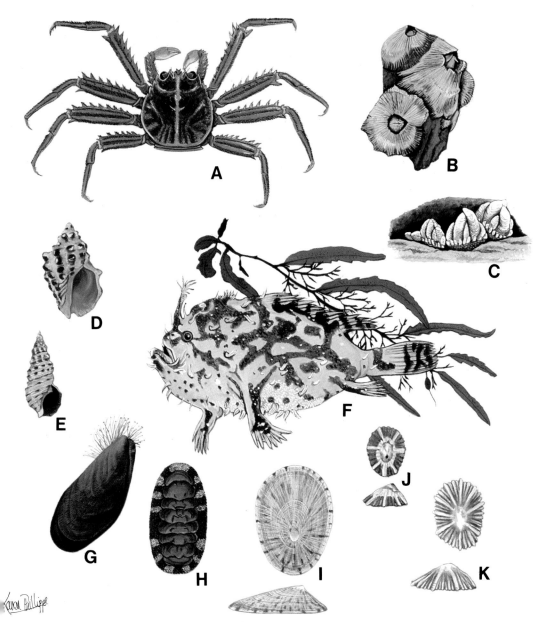

PLATE 1. Rocky shore fauna: A. *Percnon affinis*; B. *Tetraclita squamosa*; C. *Pollicipes mitella*; D. *Thais clavigera*; E. *Batillaria sordida*; F. The Sargassum fish, *Histrio histrix*; G. *Septifer bilocularis*; H. *Liolophura japonica*; I. *Cellana toreuma*; J. *Patelloida saccharina*; K. *Siphonaria sirius*.

PLATE 2. Birds of rocky coastlines: A. Reef Egret, *Egretta sacra*; B. Herring Gull, *Larus argentarus*; C. Cormorant, *Phalacrocorax carbo*; D. White-bellied Sea Eagle, *Haliaetus leucogaster*; E. Common Tern, *Sterna hirundo*.

tide, low tide, high tide and low tide within the period of a lunar day i.e. 24 hours 50 minutes. Though the lunar generated tides vary from day to day with changes in the angle of declination of the moon and its position in its elliptical orbit, most seas experience this semi-diurnal tidal effect. The earth revolves once every 24 hours. The combined effect is thus to produce two tide-generating forces every 24 hours 21 minutes, so that a single tide occurs every 12 hours 12 minutes.

The effect of the sun on the tides is much less than the moon's because of its greater distance from the earth, but does exert a considerable modifying influence.

When the moon, earth and sun are in line, i.e. when the moon is full or new (which happens approximately twice each lunar month or every $14\frac{1}{2}$ days) the combined gravitational pull on the earth is greatest. The high tides are then highest, and the low tides at their lowest. Such extremes are called 'spring tides'. When, however, the moon and sun are at right angles to each other with respect to the earth, high tide levels are lowered and low tides raised. Occurring in the first or last quarter, these are termed 'neap tides'.

The effect of the sun also varies with its angle of declination, being greatest at the equinoxes and least at the solstices. At these times spring tides have their maximum and minimum ranges respectively and the neaps their minimum and maximum ranges, again respectively.

Complexities in tidal behaviour arise because the oceans are broken up by land, each ocean possessing its own unique tidal oscillation pattern. The oscillations interract with each other and are deflected by the earth's rotation causing them to swing around a centre known as the amphidromic point.

As a result of Hong Kong's compromised geographical position, the basic pattern of two tides a day is altered (Fig. 2.1). There is a regular transition from semi-diurnal tides (two tides a day) at the time of spring tides to diurnal tides (tides once a day) at the time of neaps.

There is also a diurnal inequality in the tides. From October to March and in the evening from April to September, the early morning low tide is lower than that at the middle of the day.

In the summer months, then, sensitive inter-tidal animals, especially those living low on the shore, are protected from the heat of the mid-day tropical sun. Moreover, in winter, the same community is protected from exposure to the cold air temperatures at night. Hong Kong's tidal pattern thus augments its geographical advantage as a border between the tropical and temperate zones; and facilitates an unusual mixture of tropical and temperate species.

The flowing (incoming) tide comes predominantly from south-east to north-west (Fig. 2.2) bringing with it oceanic water. This water influences the patterns of communities that can establish on the shores, and larvae from a wide surround-

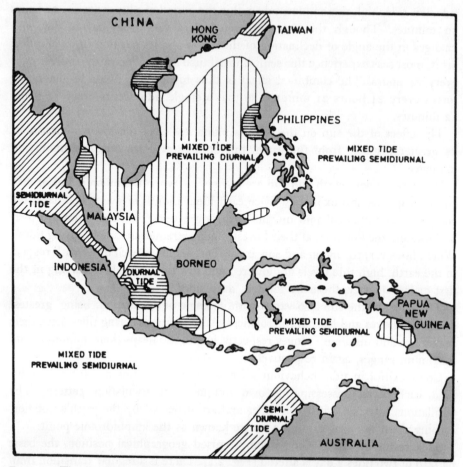

FIG. 2.1. The distribution of tidal types in S. E. Asia (after, Wyrtki, K. 1962. Physical Occanography of the South-east Asia waters. *Naga Report*, 155–63).

ing region arrive here. The ebbing (out-going) tide moves in an opposite direction, causing an influx of fresh water by enhancing the outflow from the Pearl River. Thus, over a complete tidal cycle, sea-shore animals, especially in the middle ('transitional') area of Hong Kong, are exposed to a complex of hydrographical changes, not only in salinity but also in temperature and nutrient level. In the extreme north-west 'euryhaline' species, with wide salinity tolerance, predominate; while the south-east has more 'stenohaline' species, tolerating only a narrow range of salinity.

The major effect of the tides, by their regular rise and fall, is to determine the vertical distribution of the animals and plants. Each shore species tends to colonize an optimum level on the shore, giving rise to the zoned distribution

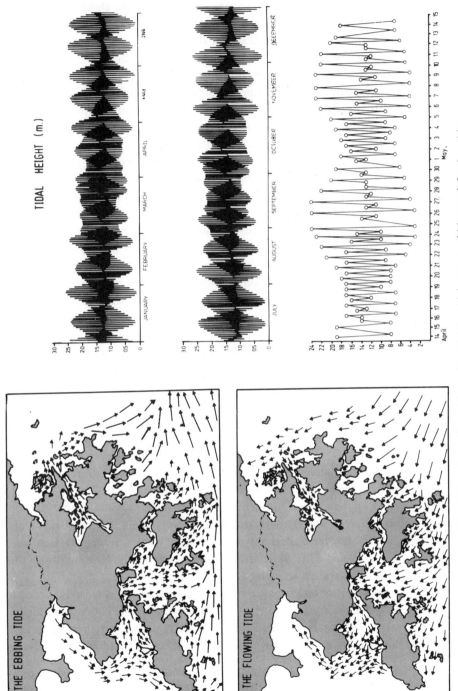

Fig. 2.2. The tidal cycle at Hong Kong. (*Left*) The tidal stream, ebbing and flowing tide. (*Right*) The amplitude of spring and neap tides over a twelve-month period and in greater detail for April and May.

so prominent on rocky shores. Three major levels of a hard shore can generally be recognized:

(i) *Lichen* and *Littorine Zone* (Littoral fringe).
(ii) *Limpet, Barnacle* and *Bivalve Zone* (Eulittoral Zone).
(iii) *Brown* and *Red Algal Zone* (Sub-littoral fringe) (Fig. 2.3).

Such a basic pattern, with the kinds of organisms in each zone, is common to virtually all the rocky shores of the world, though the component species vary greatly from place to place, and from tropics to temperate and colder zones[1]. The basic themes are, however, remarkably alike. It is as if a relatively few plays had been written; and the great diversity is in the different casts that fill comparable roles in the various parts of the world.

On soft shores of sand or mud, a similar pattern of zonation could be suggested, but no universally applicable scheme is at present accepted. With such large differences of substrate, features common to hard and soft shores could only be delineated with high generality and much sacrifice of detail. But on soft shores as on hard, the tides have a dominant role in establishing the patterns of life.

Waves are generated by the action of wind upon the open water surface and their length and amplitude are dependent upon the two major factors, wind-speed and the distance over which the wind operates (i.e. 'fetch'). Wind-produced waves travel outwards from a storm area like ripples on a pool. They rarely affect the deeper waters, effective wave depth being equivalent to no more than half the wave length. On reaching shallow water the wave crest is elevated to produce a steep wave that ultimately topples over to form a concave front, and then break. The water now surges up the shore in a powerful sheet known as 'swash', receding as 'backwash' largely within the sand before the next ensuing wave arrives.

Waves fall into two categories: *low waves* with a long wavelength and constructive action, carrying material up the shore, and *high waves*, with a short wavelength and destructive effects, removing material from the beach.

Long waves approaching a headland meet with shallow water sooner than in a bay. With lowered speed they tend to converge upon the headland while diverging within the bay: receiving concentrated wave impact, the headland forms a steeply eroded cliff; while reduced wave-force in the bays builds up the bay-head beaches common in Hong Kong.

When waves approach a long sandy bay, their leading edges are slowed down as they enter shallow water, and they tend to approach the shore at an angle. Their swash sends material up the shore obliquely, which may then be carried down by backwash at a right angle to the shore. Sediments are thus moved along

---

[1] The recent work by T. A. and Ann Stephenson, *Life between Tide Marks on Rocky Shores* (San Francisco, W.H. Freeman and Co., 1972), summarizes their lifetimes' research of the world zonation of coast-lines and is an important work for any student of hard shore zonation patterns.

The Tidal Shore and its Zoning

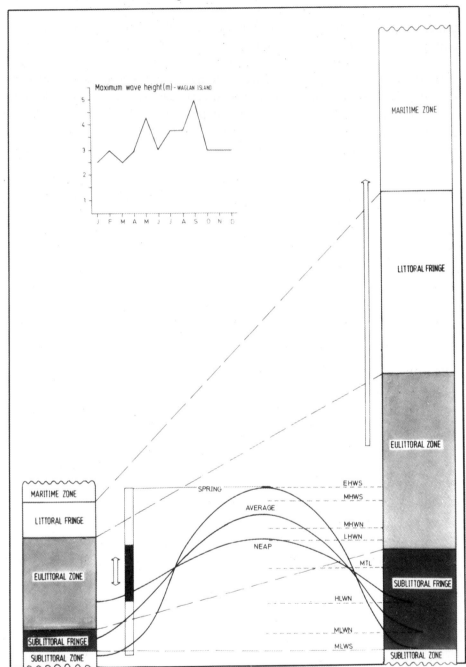

FIG. 2.3. The tidal curves and conventional levels for spring, neap and 'average' tides, with vertical scale in metres. The columns *(left* and *right)* show the variations in the limits of the zones, from the extremes of shelter to exposure. Vertical arrows indicate splash reach at mid-tide. Inset *(above)* shows the seasonal variation in maximum wave height (metres) for Waglan Island

the shore, with the phenomenon of 'long shore drift'. A good example of a beach built up by such drift is Shiu Hau (south Lantau), at the western end of a long open sand beach.

On the same coast there is a related building of sand bars across estuaries by coastal drift. The bars grow from west to east, because the ebbing tide runs mainly in this direction, bringing with it deposits from the Pearl River. These may fully block the stream mouths in winter, but they are cleared by summer rains with resulting flooding of the stream.

CHAPTER 3

Classification of Plants and Animals

Naming and classification are disciplines all biologists must learn early in their education. Living organisms have evolved and speciated, adapting to every variety of niche space. No serious student would be impatient of this lavish diversity, or wish biology to be as tidily concise as the physico-chemical sciences.

The detailed classification of the animals and plants in this book is set out on p. 327. While this would be unsuitable to learn straight off, the reader will constantly find the main entities turning up through these pages. To help those coming new to taxonomy—the science of classification—an outline introduction is given here to the major categories of the PLANT and ANIMAL KINGDOMS. In both, the first largest sub-divisions each involving a distinctive structural plan is the PHYLUM (plural PHYLA). The phyla are in turn divided into large groups called CLASSES, and the grouping at the next level, to which there are frequent references, are ORDERS.

For the PLANT kingdom, no one would go to the sea shore for a full conspectus. It is only the most primitive phyla, those embracing the separate groups of algae, that are widely represented. As well as microscopic forms, these include many that are multi-cellular, with complex constructed thalli, firmly attached to the ground. The largest of these, and probably of all shore organisms, are to be found within the phylum PHYCOPHYTA. Amongst these, the Phaeophyceae, or brown algae, are the most important but are never as large in the tropics as on colder temperate shores. The other main shore groups are the green algae CHLOROPHYCEAE and the red algae RHODOPHYCEAE. Still classed as algae, but much more primitive than all the rest of the plants, are the small and simply structured blue-green algae of the class CYANOPHYCEAE (or MYXOPHYCEAE). In the plankton, hardly considered at all in a book about the shore, are the unicellular plants, including silica-cased diatoms, BACILLARIOPHYCEAE, the cellulose-armoured dinoflagellates, DINOPHYCEAE, and the naked microscopic flagellates,

CHRYSOPHYCEAE (and some also among the CHLOROPHYCEAE).

The higher plants are well exemplified on the shore only by a few metaphytes or seed-plants, phylum SPERMATOPHYTA. Predominantly land-based groups have sent back salients adapted to the inter-tidal. In Hong Kong, the most characteristic are those of the tidal and supra-tidal swamp known as the 'mangal', with representatives of many families; and the succulent and halophytic plants of the salt-meadow.

Of the ANIMAL kingdom, for the present purposes the lowermost phylum and sub-kingdom, the unicelluluar PROTOZOA may be largely ignored. The larger sub-kingdom, the METAZOA, is sometimes for working purposes divided into two very unequal groups, 'chordates' and 'non-chordates'. The first group consists only of the single phylum CHORDATA, albeit the one most familiar, containing the vertebrate animals including man.

As with the highest plants, the most advanced CHORDATA are securely land-based. Those most often venturing on to the shore are the birds (class AVES), especially the waders and other marsh and swamp birds; but there are many on beaches and stony shores as well. Included in this book are all those birds important or noticeable on the shores of Hong Kong, but no attempt has been made to treat them exhaustively.

Almost as abbreviated has been the coverage of the bony fishes (class ACTINOPTERYGII, subclass TELEOSTEI); but in the section on coral reefs some attention is given to those fishes: wrasses, cardinals, angel fish, trigger fish, and the rest, that are as colourful and characteristic of reefs as the corals themselves.

Also part of the CHORDATA is the most primitive section, not 'vertebrates' in any sense, but including those solitary or colonial forms known as sea-squirts (class TUNICATA). Their real link with higher chordates is revealed only by the 'tadpole-like' free-swimming larva.

NON-CHORDATA: The phylum MOLLUSCA is the largest invertebrate group after the arthropods, and is probably the most numerous phylum on the shore. By far the commonest classes are GASTROPODA and BIVALVIA. Most of the first, and all the second, are enclosed in a hard calcareous shell secreted by the mantle. Gastropods are most typical of rocky shores, and (at least primitively) have a single-piece spirally coiled shell. The head carries a proboscis, paired tentacles and eyes. The mantle cavity—in marine forms—contains a gill (ctenidium), but in land gastropods is converted to an air-filled lung. The foot forms a broad creeping surface. The most highly modified of gastropods are however naked, with a colourful integument, forming the 'sea-slugs' of the sub-class OPISTHOBRANCHIA.

The BIVALVIA have a shell of two hinged pieces. The majority burrow with the compressed, spade-like foot in soft sediments, and communicate with the surface by siphons. Bivalves are specialized fine-particle feeders, having lost the head and mouth-parts, and developed the gill as a filter. An important section of the bivalves, numerous in Hong Kong, attach to hard surfaces by a beard or 'byssus' or cement directly to the rock.

In overall numbers the ARTHROPODA, or jointed-limbed animals, are the largest of all phyla. But their greatest group, the Insects, are hardly found at all between tides. Of marine arthropods in Hong Kong the large archaic 'spider relative', *Tachypleus,* the king crab, is a notable living fossil.

The rest of the shore arthropods are CRUSTACEA, with segmented body, jointed appendages and hard external skeleton, shed periodically by moulting. The more primitive Crustacea are light-bodied shrimps, either swimming freely, or burrowing, as in *Penaeus* and the snapping shrimps *(Alpheus)*. Larger Crustacea have become heavy and bottom-dwelling, as with rock lobsters and the immensely numerous crabs, highly successful shore forms modified for backward or sideways progression, and protected by pincers or chelae. The most individually numerous of Crustacea are the sessile barnacles, settling on rock and building a conical shell. They use feathered limbs as a sweeping plankton net.

The ECHINODERMATA have a five-sided radial symmetry and appear to subvert every normal rule of animal design. Named from their spiny skin they also have in common a 'water vascular system' operating a hydraulic system of tube-feet. Their five-sided geometry has a distinctive pattern in each of the five classes. ECHINOIDEA or sea urchins have hollow globular hard shells ('tests') that are modified to heart-shaped or biscuit-shaped in urchins burrowing on soft shores. The starfishes ASTEROIDEA are flattened stars with five or more points. The brittle stars, OPHIUROIDEA, have a small central disc and five flexibly jointed slender arms. The sea cucumbers (HOLOTHUROIDEA) are naked echinoderms, drawn out into a muscular cylinder, with the mouth and anus at opposite ends. Finally the CRINOIDEA or feather stars are attached to the ground, with delicate pinnate arms collecting particles from the plankton.

Below these three 'higher' phyla of the invertebrates, are several phyla of worm-like animals, each separate but once associated together under the loose term 'VERMES'.

The first and the highest of these worm phyla is the ANNELIDA, whose chief marine group is the bristle-worms or POLYCHAETA. These are named from the bristles of 'setae' on their paired, paddle-like limbs. They are divided into uniform segments, each with limbs, and unlike the arthropods and molluscs they have retained a true coelomic body cavity. There are two traditional divisions: the SEDENTARIA, mainly tube-dwelling or attached, and collecting food with special head appendages; and the ERRANTIA, moving about freely in sediments or on hard ground.

Of uncertain affinity, near the base of the ANNELIDA, are two small phyla of unsegmented worms, both commonly exemplified in Hong Kong: the ECHIURA and the Sipuncula. They burrow peristaltically, and are both limbless and un-segmented at least in the adult.

Lower worms without segmentation comprise two phyla: the ribbon worms, Nemertea, are long and flattened, often bright-coloured. They have no limbs, but a long, eversible proboscis usually used to capture prey.

The flat-worms Platyhelminthes have two large parasitic classes (flukes and tape-worms) but the Turbellaria are free-living, with many species on the shore. These are broad and leaf-like, progressing by cilia, or crawling (even swimming) by a muscular wave. The head carries numerous small eyes, and the mouth lies far back, upon the ventral surface.

The important phyla that remain have a sessile habit, forming encrusting colonies organized as aggregates of small individuals.

They typify three very different levels of organization. The highest, being tiny triploblastic animals with a true coelom, are the Bryozoa (Ectoprocta). The individual organisms (polypides) live in chitinized or calcified boxes, assembled into colonies of various shape and form. Food is collected by a ciliated tentacle crown, the 'lophophore'. This organ links the bryozoans with two small phyla of non-colonial invertebrates, the Brachiopoda and Phoronida, typified in Hong Kong by the sand-burrowing *Lingula* and the worm-like *Phoronis* respectively.

At a much lower level, the bottom of the Metazoa, are the Coelenterata. Those found on the shore have a 'polyp' life form, though one whole class, the Scyphozoa, and some Hydrozoa are medusoid, with a jellyfish habit. In colonial polypoid coelenterates, Hong Kong is very rich; it just reaches the tropical province where corals play a major part.

A polyp is of simple, two-layered construction, with a mouth surrounded by a circlet of tentacles, leading to a simple 'coelenteron' (body-space) where food is digested. All coelenterates are united in the possession of stinging cells or 'cnidoblasts'. They are all carnivorous.

Under stones and on wharves, colonies of hydrozoan polyps are abundant. But the largest polypoid coelenterates belong to the class Anthozoa containing the naked anemones, and the Scleractinia or stony corals. The latter have polyps, secreting a calcareous cup (calice) and organized into a colony or 'corallum'. The soft corals (Alcyonacea) by contrast have the polyps embedded in a fleshy common matrix or 'coenenchyme'.

Even below the simplest Metazoa are the earliest of the multi-cellular animals, the sponges, of the phylum Porifera. These are aggregations of simple cells into colonies, sometimes of irregular shape, that lack, however, the metazoan attributes of distinct nervous system, muscles and gut. The sponge body forms a unity only in its regular orientation of inhalant and exhalant canals, lined by flagellated and absorbing choanocyte cells. The body is strengthened by skeleton fibres of silica, or in some sponges of calcite.

PART TWO
Hard Shores

CHAPTER 4

Exposed Rocky Shores

THE MOULDING OF THE HARD SHORE

Every habitat study must begin with geology. Pitman and Peck[1] have produced a readable account of the coastal geomorphology of Hong Kong, and for a good, second, general account the reader is referred to Allen and Stephens[2]. Hong Kong has a diversified geomorphology (Fig. 4.1), its land-mass being for the most part high-built and constructed of hard igneous (plutonic and volcanic) rocks, which

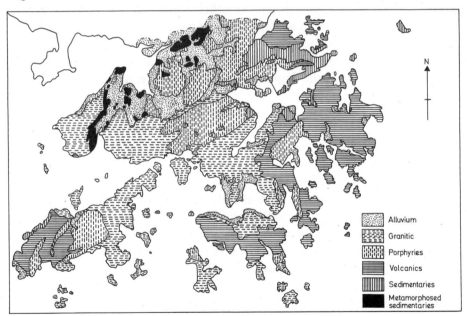

FIG. 4.1. Geological map of Hong Kong and its territories.

[1] Pitman, J. and Peck, R. 1979. The physical nature of the Hong Kong seashore. In *The Future of Hong Kong Seashore*, ed. B. S. Morton. Hong Kong, Oxford University Press.

[2] Allen, P. M. and Stephens, E. A. 1971. *Report on the Geological Survey of Hong Kong*. Hong Kong, Government Printer.

stand out rough and angular. Unlike softer sedimentary rocks, they are slow to weather at the coasts, and only in a few places do they exhibit broad wave-cut platforms. Hong Kong thus has a 'primary coast' in the sense that the original features presented by its rock mass are little altered, except by coastal drowning, with the resulting complexity of islands and inlets produced by the recent rise in sea level, following the retreat of the last ice age.

The oldest rocks of the present Hong Kong land-mass go back to the Permian, only in evidence in a few places today, for example, the sedimentary Tolo Harbour Formation of orange-red hued silt-stones and shales. After marine deposition they were uplifted and weathered, then submerged again. They are now unconformably overlain by later marine deposits from the Jurassic, and a stratified succession of sedimentary and volcanic rocks. The greatest and most important of these is the Repulse Bay Formation, consisting of acid rhyolitic lavas and associated sedimentary rocks. These Jurassic volcanics and sedimentaries later became folded along a NE–SW axis, slightly metamorphosed and intruded by various granites. At Repulse Bay they form a sequence 1350 metres thick.

At the end of the Jurassic, around 130 million years ago, came the major events of Hong Kong's geological history. Acid intrusions of granite and granodiorite began, giving the present massive structure of the north of Hong Kong Island, Lamma Island, the east of Lantau, and much of the New Territories. Subsequent coastal deposition of alluvium—with significant accessions by human reclamation—has given the final shape to a primary, largely drowned igneous coast-line.

In the shaping and moulding of the coast to its present form, marine waves have been the prime weathering agents. The coast forms finally produced by waves and other weathering depend greatly on the response of the various rock types. On cliffs of hard granite or rhyolite, the slap-up force of waves and the suction caused by their retreat produce powerful effects. The blocks of jointed bed-rock are loosened, torn out and dislodged. Rolling back and forth, they themselves contribute to further attrition. Where the texture and bedding of the rocks is suitable, wave-cut benches appear between tides. Though uncommon in Hong Kong, these are best displayed in the 6–10 metre wide platforms of the islands of Crooked Harbour and at Ping Chau in Mirs Bay, across the dip of the sedimentaries.

But most of the outer igneous coasts show only slight development of a foreshore zone. Their ruggedness or gentleness of contour are directly related to wave exposure and to the degree and rate at which erosion is still proceeding. Differences in the geology of the bed-rock and its joining and weathering pattern have thus had great influence on coastal forms.

Some familiar examples can be given. On north-east Lantau, at Num Shui Wan, the granite is invaded by dykes of porphyritic lava, which are fine-grained and far more slowly weathered than the granite. The latter is thus worn back in embayments between cliffs and promontories formed by the outcropping lava. At the heads of these bays, small crescentic beaches form, largely from wave-eroded material containing much coarse-grained quartz.

On the northern (Kwa Mun) shore of High Island, the rhyolite has weathered in a characteristic prismatic or columnar structure, forming vertically standing 'organ pipes' that are highly resistant to erosion, and well developed cliffs that receive direct, slap-on wave assault.

The resulting coast forms are much controlled by rock jointing. Joint planes of local faulting give rise to intensified chemical weathering along the fracture zones. These can be attacked and excavated by the sea to form clusters of stacks and islands. In Rocky Harbour and inner Port Shelter the rhyolite mass breaks up to give a spectacular topography: with parallel-sided re-entrants and narrow passages, small enclaves and stacks, and, in some places, sea arches through steep islands ('Hole' island).

Such erosion will always be greatest and fastest with the highest wave exposure. The very high cliffs looking south and south-east face the maximum wave assault. The highest cliffs are found on Basalt Island and Bluff Island, the latter reaching over 140 metres. There is never an appreciable foreshore at the base of these cliffs, and where there are beaches at all, as on High Island, these will consist of ramps of rounded, wave-turned pebbles and small boulders.

Sand beaches, even the cleanest and most mobile, develop only within the shelter of bays. Different rock types produce sands with different grain sizes. Where a material such as granite has been weathered and partly broken down to boulders and matrix on an exposed face, a large extent of material may be removed before a coherent cliff is formed. A wide bench will thus be produced, which in turn serves to inhibit continued wave erosion at the inner end of its profile.

Where a cliff is eroded by undercutting or gulleying, its corestones will be deposited on the foreshore to serve as a damper or buffer against the full force of the waves. Examples are seen in the boulder beaches at north Lamma, especially near Boulder Point, showing well-developed granitic cliffs with a characteristic red iron-stained hue from 'Zone II' weathering material. On more sheltered, north-facing coasts this wave-cliffing is restricted and shores will have gently inclined slopes, with the boulders, if present, angular. Wave action is insufficient to round them, and hence to promote the formation of sand beaches.

Wave-beaten Shores

Life between tide marks is zoned from high to low, often in complex ways, under the general influence of tides and waves. The interwoven causes of this zonation and its varying patterns from exposure to shelter, are still only partly understood. Their study is one of the continuing themes of shore ecology.

But the fact of zonation is plain; and it can be illustrated first by describing a habitat where it is simple and clean-cut. Such shores are not always the most accessible nor are they the safest to work, being generally cliffs and promontories under the maximal wave exposure. Here 'exposure' is to be understood as not the uncovering of a shore to the atmosphere between tides, but its subjection to high wave action, with the successive effects of surge, splash and spray reaching far up the shore. Even at low tide, such shores are never for any considerable period 'emersed' or left dry. The ranges of species that would normally be confined to the inter-tidal are pushed up far above the nominal high water by splash and spray, and during low tides the shore is regularly awash with heavy surge over at least the lower half of the tidal zone.

The tidal rhythm itself plays, then, a diminished part on exposed shores. Its role is increasingly taken over by waves, submerging the shore with a far greater frequency than the diurnal tides. In heavy seas a wave may have moreover a greater amplitude than the tidal range itself. In addition to their continual wetting of the rock surface, waves have great mechanical power: the battering and pounding from direct impact and the wrenching and dislodging effects of wave run-off. The vertical limits of species distribution are related only very broadly and inexactly to any given levels of the tides. From the scale of tidal heights placed along-side the introductory diagram (Fig. 4.2) it can be seen that the limits of wave-elevated (inter-tidal) zonation far over-reach the notional level of high spring tides.

Good examples of exposed shores can be found on any of the south-easterly directed capes and headlands, typically of granite, on Hong Kong, Lamma or east Lantau, or on the outlying volcanic islands to the east. The illustration is based on the steep granite coast of Cape D'Aguilar or Thai Long Head (Fig. 4.3), rugged and irregularly stepped, both with vertical faces up to a metre high, and with narrow shelves, often dipping back to retain temporary splash pools. Over the lower half of the shore strong surf breaks continually, so that—except in the calmest weather—low-water level can only be reached with care and some difficulty even at low tide.

It is customary to recognize three major belts of the tidal shore; these have much constancy the world over, and are well demarcated in Hong Kong. The first and highest is the LITTORAL FRINGE. This is seldom fully immersed even by the highest tides, but it is regularly visited by splash and spray. The granite bed-rock may have its natural grey-white colour at this level, but at least at lower levels it can carry a dull blackish film of primitive 'blue-green' algae (Myxophyceae). Small concavities and crevices lodge fine organic remains, including diatoms brought by splash. These are grazed by the two small herbivorous gastropods constantly found here, which are representatives of the littorine or

PLATE 3. Cape D'Aguilar. The low tidal shore is never continuously emersed. The sequence of wave break and surge rise and run-off is shown for a single wave.

PLATE 4. *(Above)* A boulder beach at Stanley and *(Below)* a cobble beach at Tai Tam.

Exposed Rocky Shores

FIG. 4.2. Cape D'Aguilar with exposed shore zonation on rocky headlands. The zoning pattern stretches above the inter-tidal, through *Verrucaria* (dark-hatched) over bare rock or grey-green lichen up to the cap of scrub and bush. *(Below left)* Sequence of zonation with vertical height in metres. The tidal range is shown for spring and neap tides and the broken curves represent maximal wave height.

ZONES AND SUBDIVISIONS:

A. Maritime vegetation—*Scaevola* and *Ipomoea*; B. Bare rock with grey-green lichen; C. *Verrucaria maura*; D. *Nodilittorina* species; E. *Bangia fusco-purpurea* and *Porphyra suborbiculata*; F. *Tetraclita squamosa*; G. *Balanus tintinnabulum volcano*; H. Pink *Lithophyllum* crust with *Corallina* turf; I. *Sargassum* belt.

SPECIES:

1. *Chthamalus malayensis*; 2. *Nodilittorina pyramidalis*; 3. *Nodilittorina millegrana*; 4. *Porphyra suborbiculata*; 5. *Pollicepes mitella*; 6. *Tetraclita squamosa*; 7. *Gelidium amansii*; 8. *Patella flexuosa*; 9. *Ophidiaster cribrarius*; 10. *Anthocidaris crassispina*; 11. *Balanus tintinnabulum volcano*.

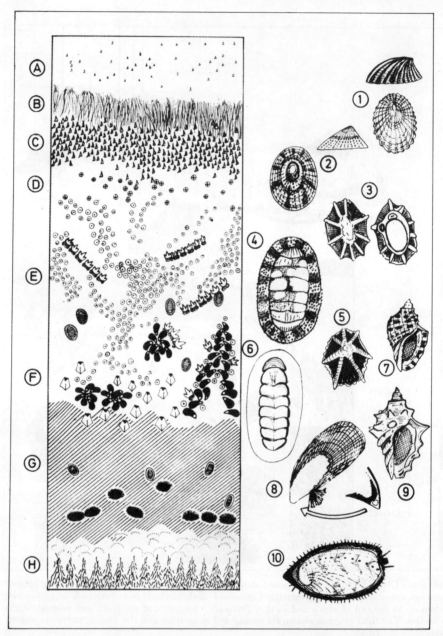

FIG. 4.3. Zonation of animals and plants on an exposed shore at Cape D'Aguilar.

ZONES:
A. *Nodilittorina* species; B. *Bangia fusco-purpurea*; C. *Porphyra suborbiculata*; D. Scattered *Chthamalus malayensis*; E. *Tetraclita squamosa*; F. *Balanus tintinnabulum volcano* with *Septifer bilocularis* and *Liolophura japonica*; G. Pink *Lithophyllum* crust with *Anthocidaris crassispina* and *Onithochiton hirasei*; H. *Corallina* turf and beginnings of *Sargassum*.

1. *Collisella dorsuosa*; 2. *Cellana grata*; 3. *Siphonaria sirius*; 4. *Liolophura japonica*; 5. *Patelloida saccharina*; 6. *Onithochiton hirasei*; 7. *Morula musiva*; 8. *Septifer bilocularis* with detail of septum and insertion of anterior adductor muscle; 9. *Thais luteostoma*; 10. *Sanhaliotis planata*.

periwinkle family; these typify the littoral fringe in almost every shore of the world. The conical and large noduled *Nodilittorina pyramidalis* has the higher range, often outdistancing by several metres the plumper, striped and fine to coarsely granulose *N. millegrana*. Rosewater[3] has reviewed the littorine snails of the Indo-Pacific shores and from Hong Kong records a third species—*N. exigua*—living in high exposure at Cape D'Aguilar. This author notes, however, that this species is 'closely related' to *N. millegrana* but is more strongly sculptured, lacks colour striping and has a stronger siphonal trough. Clearly, however, both species are very variable and we cannot separate populations to our satisfaction. The separation of these species, or forms, is thus an exercise worthy of further research. Though well-adapted air-breathers and able to resist desiccation by closing tight for long periods, periwinkles are held to the shore-line by a dependence on water for liberating the eggs, which hatch into free-swimming planktonic larvae.

The lichens of the upper shore call for special mention. Especially characteristic of open shores under a high regime of splash and spray, they are good indicators of a clean atmosphere, being seldom found in harbours or to the windward of polluted cities. The lowest and most nearly inter-tidal is the dull black film of *Verrucaria*, found in the upper part of the littorine range, and often—under spray—reaching much higher. It should not be confused with the more intense black streaks of blue-green algal film, developed on rocks in places of water seepage. Above *Verrucaria* are, first, an orange crustose lichen *Caloplaca*, and, still higher, varied cover of grey and grey-green species (*Lecanora, Thelidiopsis, Lecidia, Physcia*). These lichens have still to be properly studied in Hong Kong, as is true for many other shore-lines.

For the lichens above *Verrucaria*, together with the accompanying seed-plants, a separate MARITIME ZONE is generally designated. Common species are the screw-pine, *Pandanus tectorius,* the scrambling shrub, *Scaevola frutescens,* and the convolvulus *Ipomoea pes-caprae*. Other less common species include the creeping *Clerodendron inerme, Hibiscus tiliaceus* and *Vitex rotundifolia*. Smaller cyperacean sedges include *Cladium chinense, Cyperus javanicus, Eleocharis afflata* and *Fimbristylis cymosa* and gramineacean grasses include *Sporobolus virginicus and Zoysia sinica*. Under strong action of spray, the grey-green lichen cover can be raised high, sometimes more than 30 metres, over the outcrops and cliff-faces of pale granite.

The lower part of the littoral fringe, especially during late autumn and winter, has typically two primitive red algae (Rhodophyceae). At a higher level, the first, *Bangia fusco-purpurea,* forms thin, pink or mauve filaments, drying out in air and closely pressed against the rock. Immediately below comes the membranous

---

[3] Rosewater, J. 1974. The family Littorinidae in the Indo-Pacific. *Indo-Pacific Mollusca* 2: 417–534.

Porphyra suborbiculata, shiny and reddish black when wet, at other times stretched tight over the rock face.

One of the few highly active animals in the littoral fringe is the swift, crevice-haunting isopod *Ligia exotica,* which is closely related to woodlice, and is the first of its line to develop terrestrial adaptations.

Zoning Animals

The EULITTORAL ZONE is a broad stretch comprising most of the tidal shore. It is generally covered and uncovered by the tides twice daily. Even during the shortest neap tides on quiet shores, its upper levels are at least wave-splashed. On exposed shores the waves constantly transcend the tide effects. The sublittoral zone is a habitat of gregarious sessile animals; and in most parts of the world its upper part is strongly dominated by barnacles. These are highly successful Crustacea that begin as planktonic larvae, first a nauplius and then a cyprid, then settle to become entirely transformed by the secretion of shelly plates round the body. In acorn barnacles these form a conical tent, with varying numbers of valves (*see* Fig. 5.6). In stalked barnacles they are arranged as protecting shields at the end of a fleshy peduncle. In either case there are two pairs of movable valves to allow the exit of six pairs of curled limbs, the cirri, feathered with fine setae. These can be drawn rhythmically through the water to strain plankton; or with some barnacles in turbulent water (such as *Pollicipes mitella*) they are held open passively to intercept particles carried in the currents.

Four sorts of barnacle can be found on exposed shores. The commonest and the first noticeable is the large acorn barnacle *Tetraclita squamosa,* which is dome-shaped and grey, and sculptured with fine radial ridges. This barnacle is well protected both from desiccation by its small, almost pore-like aperture, securely closing between tides, and from insolation (heating up by the sun) by its large ratio of volume to surface area and the insulatihg character of the thick shell wall, with air-filled cellular spaces. At least two sub-species of *Tetraclita squamosa* occur on Hong Kong shores, *T. squamosa japonica* typical of exposed shores and *T. squamosa squamosa* in more shelter[4].

Below *Tetraclita,* and in places of constant surge, a still larger acorn barnacle *Balanus tintinnabulum volcano* predominates. Thinner shelled and with straight sides, it has a much wider aperture and may reach as much as 5 cm high. Less aggregated than *Tetraclita,* it is the largest barnacle in Hong Kong, and is confined strictly to the outermost exposed shores.

The third barnacle of exposed shores, *Pollicipes mitella,* like the oceanic goose-barnacle *Lepas anatifera* (Fig. 8.9), belongs to the stalked or pedunculate group. It is found high on the shore, in bands wedged into crevices or inserted

---

[4] It is probable that a number of other sub-species of *T. squamosa* occur on Hong Kong shores, but the distribution of these has never been studied.

in rows into the high-built crusts of *Tetraclita*. On digging a specimen out with a knife, the grey or yellow shell valves can be seen like flat claws on the golden brown peduncle, which is itself strewn all over with minute platelets. Inside the mantle cavity of *Pollicipes* lives a commensal nemertine worm, *Nemertopsis gracilis*, yellow with longitudinal black bands and which feeds upon particles collected by the barnacle.

None of these three species reaches to the top of the designated 'barnacle zone'. Thirty centimetres or more of vertical height are left largely bare, where, on warm temperate shores, a *Chthamalus* species would colour the rocks grey. Such a barnacle in fact does sparsely spatter exposed Hong Kong shores. An autumn settlement of *Chthamalus malayensis* takes place in scattered patches above *Tetraclita* but can fail to survive the winter. Small and flat *Chthamalus* has a circlet of six plates, and is recognized by the wavy sutures between terga and scuta. The more euryhaline *Euraphia withersi* will be encountered on shores with an input of fresh water.

The lower reaches of the eulittoral zone are generally the preserve of bivalve molluscs. On exposed shores in Hong Kong the finely ribbed, black mussel *Septifer bilocularis* forms sheets on horizontal rock surfaces awash with surge, or extends higher up narrow runnels where water courses through. In some sites a large proportion of the lower eulittoral is mussel-occupied. Like all mussels, *Septifer* attaches by the tough, tanned-protein fibres of its secreted byssus. These can be periodically broken by the leverage of the narrow muscular foot, as the mussel changes its site and secretes new attaching fibres. Co-occurring with the mussel and filling sediment-accumulating crevices is the anemone *Anthopleura pacifica*.

Low-tidal Algae

The lowest horizontal belt of the tidal shore, the SUB-LITTORAL FRINGE, can conveniently be defined by the beginning of the pink veneer of calcareous red algae. Often collectively called 'Lithophyllum' this crust also contains the basal 'paint' of *Corallina*, an early growth stage devoid of upright fronds that may continue long on open coasts without forming a turf. This pink formation is a constant part of all exposed shores, especially in the tropics and sub-tropics; it may cover the entire low-tidal surface, and carry numerous other algae, some of them delicate and seasonal, growing upon it.

The sub-littoral fringe is generally recognized from its algae, but in comparison with most temperate shores, the Hong Kong algae are for a great part of the year disappointingly impoverished. E. Yale Dawson[5] has provided a useful general guide to the seaweeds which the Hong Kong student will find invaluable. However, the existing literature on local marine algae is sparse. W. A. Setchell

---

[5] Dawson, E. Y. 1956. *How to Know the Seaweeds*. Dubuque, Iowa, Wm. C. Brown, Co.

relates how in 1926 his colleague F. A. McClure visited Lantau with his collector at the end of July and early August and to his disappointment found no seaweeds, although they journeyed halfway round the island. 'The villagers and fishermen were unanimous in their statement that there are no salt-water algae to be had there during July, August, September, October, November and December, but that beginning with January or February they come in increasing abundance'. On the Stanley Peninsula the seaweeds began to appear on the beach around January or February, being abundant by March and gradually diminishing in abundance until the end of August when the season is practically over.

The seasonal appearance of the majority of the larger Hong Kong algae is presumably related to the wide fluctuation in water temperature. The annual rise, fruiting and decline of algal species, including those in shallow water offshore, should be the most pressing concern of Hong Kong's marine botanists. Knowledge of algae is in such a rudimentary state that regular weekly or fortnightly observations could be valuable for almost any species.

The high-level growth of *Bangia* and *Porphyra* under the influence of heavy splash and spray comes on earliest, and *Bangia* is the first to arrive and decline, having disappeared by January. *Porphyra* is already well established as small plants in early December and reaches a maximum in January. It is collected and dried for the markets in January and February. Important near the high-tide mark are the shade growths of the two small red algae *Bostrychia tenella* and *B. binderi*.

In the sub-littoral fringe, the encrusting and turfing calcareous algae continue all year round. *Corallina* and *Jania* species form a close-set turf on the open surface, and grow thicker in pools. Coloured pink, mauve or bleached by the sun, these algae have a limy skeleton of small movable segments surrounded by a thin layer of protoplasm. The commonest species is *Corallina pilulifera*, pinnately branched and with cylindrical segments. Its turfs, up to 3 cm tall, intercept and accumulate fine sediments. There is a rich interstitial fauna especially in pools. A second species, *C. sessilis*, confined to exposed shores, forms shorter tufts in rock pools. The segments are flattened and somewhat hexagonal, imbricating or over-lapping as the fronds lie close-packed. A third calcareous turf species, *Jania undulata*, occurs either pure or mixed with *Corallina*, being distinguished by its dichotomous branching and spreading fans of small terminal segments (*see* Fig. 4.8). Found also with *Corallina*, and with long, dichotomously branching segments, but not fanning at the tips, is the densely tufted *Amphiroa ephedraea*. *Asparagopsis taxiformis* is a red alga epiphytic with *Corallina* turf.

Towards the end of December, the pink or grey colour of the coralline turf is enriched by a golden expanse of *Sargassum* species, locally the largest brown algae. Springing from a tuft of basal leaves, *Sargassum* grows in a long axis with more slender leaves, and, in the axils, bladders formed of modified leaves and reproductive organs.

A widespread tropical and sub-tropical genus, *Sargassum* has numerous species, based upon the varying characters of the leaves. Sargassoid algae have replaced all the other large sub-littoral brown algae of temperate shores. Kwok-Yan Lee[6] characterizes the three common local *Sargassum* species as follows:

A   With cylindrical vesicles: *S. horneri* (*see* Fig. 5.2(4))
     (stem twisted and furrowed, with sharp spinules; leaves linear and alternately pinnatisected).

AA Vesicles elliptical:
  B   *S. patens* (*see* Fig. 5.2(5))
       (basal leaves simple or divided, others with elevated midrib, margin coarsely serrated, and vesicle with a terminal serrate leaflet)
  BB *S. hemiphyllum* (*see* Fig. 5.2(3))
       (leaves without mid-rib, lower ones oblong-elliptical, upper with inner margin entire, and outer edge coarsely dentate).

S. A. Setchell[7] has, however, described very many more species of *Sargassum*.

A modified *Sargassum* relative, *Turbinaria trialata,* growing in wave-protected pools, has dense tufts of coarsely serrated peltate leaves.

Along with *Sargassum* grow the twining thongs of *Hypnea musciformis,* golden-brown in colour, though belonging to the red algae (Rhodophyceae). The tendrils, up to 10 cm long, have a cartilaginous texture, and encircle and entangle the *Sargassum* by their hook-like branchlets. *Hypnea* may also grow alone. Two other species of *Hypnea, H. cervicornis* and *H. japonica,* occur in Hong Kong. Another red alga found intertwining with *Sargassum* is *Spyridia filamentosa*.

An important group of late winter red algae are the *Grateloupia* species, often gathered as food. Purplish-brown in colour, they have a mucilaginous or cartilaginous feel. *G. imbricata* forms a rosette of short, wide, overlapping branches close to the rock surface. *G. filicina* and *G. livida* have narrower, linear branches radiating from a centre, the former pinnately branched and 18–25 cm long, and the latter dichotomous, 10–15 cm long.

A number of smaller red algae grow prostrate, spreading over the rocks at low tidal surge level or in shallow pools. The purplish red *Gymnogongrus flabelliformis,* used in agar-making in Japan, forms fan-shaped tufts of compressed dichotomous branches. *Gigartina intermedia* forms a tangle of recurved, irregularly pinnate branchlets. *Gloiopeltis furcata* has a similar habit. The *Laurencia* species are notably firmer and gelatinous, *L. japonica* forming thick tufts up to

[6] Lee, K. Y. 1964. Some studies on the marine algae of Hong Kong, I. *New Asia College Academic Annual* 6: 27–79.

[7] Setchell, S. A. 1931–36. Hong Kong seaweeds, Parts I–V. *The Hong Kong Naturalist* 2: 39–60 and in supplements 2, 4 and 5.

20 cm tall, and *L. undulata* spreading in rosettes from a disc-like holdfast.

Among the still smaller red algae, *Gelidiopsis variabilis* springs up, with rather coarse, dull greenish cylindrical branches, from amongst corallines. Fine tufts of *Polysiphonia* spp. intermingle with larger red algae on surf-beaten rocks. *Caulacanthus okamurai* forms a flat, red mat low on the shore. The small, cylindrical and cross-partitioned branches of *Champia parvula* are often found epiphytic upon *Sargassum*. Again Kwok-Yan Lee[8] has well described many of the commoner red algae.

The green algae (Chlorophyceae) are generally less important on open shores than the reds and browns; but are prominent by their bright splashes of colour on any low tidal surface. The sea-lettuce, *Ulva lactuca*, grows as broad, pleated sheets, common upon nearly all shores. Smaller green algae, especially prominent at Cape D'Aguilar and Shek O, include *Chaetomorpha antennina* with thick filaments made up of large brittle cells, and the delicate, feathered tufts of *Bryopsis plumosa*.

Mobile Animals

On exposed shores, the few rapidly moving animals are also able to grip the rocks firmly as the waves rise and fall. The grapsid crabs well illustrate this alternately mobile–prehensile habit; their fastest species are thoroughly adapted for life on surf coasts in all warm and tropical seas. At the top of the shore, among the barnacles and even higher, lives the greenish, white lined *Grapsus albolineatus*. The carapace is relatively small and roughly circular, but straight across the front. The legs are long and flattened, the joints tinted with orange, and the chelae mauve or pink. With strong spinose claws that cling to the rock, these crabs remain firmly poised against wave attack. At other times they run about noisily and can spring from one level to a lower by flexing the limbs. When disturbed they dart back to take impregnable refuge in narrow spaces. A second grapsid crab, *Plagusia depressa tuberculata,* plays the same role at low levels as *Grapsus* higher up. Pinkish-brown with depressed body and incised carapace front, it takes sure hold of the rock, or darts swiftly sidaways amidst carpets of algae. A second species, *P. dentipes,* is less common. A smaller low-tidal grapsid *Percnon affinis* is even flatter and faster, with long-spined, angled legs like a swift spider. Orange-brown and marbled with dull green, it is chiefly seen in the open or in shallow splash pools. The dull red porcelain-crab *Petrolisthes coccineus* inhabits crevices near mid-tide.

In addition to the grapsids, crevices and recesses harbour the more retiring crabs of the family Xanthidae. The larger of those illustrated (Fig. 4.4) is the coarsely spinose red-eyed crab *Eriphia laevimana smithii*. The smaller pilumnid

[8] Lee K. Y. 1965. Some studies on the marine algae of Hong Kong, II. Rhodophyta (Part 1). *New Asia College Academic Annual* 7: 63–110.

Exposed Rocky Shores

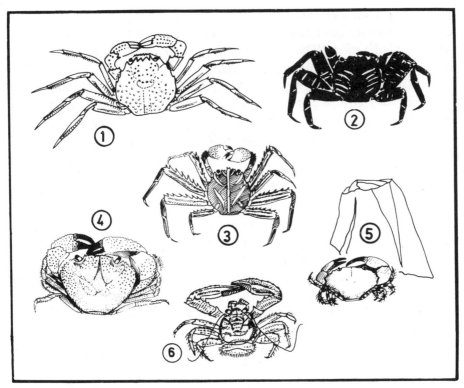

FIG. 4.4. Crabs of wave-exposed shores:
1. *Plagusia depressa tuberculata*; 2. *Grapsus albolineatus*; 3. *Percnon affinis*; 4. *Eriphia laevimana smithii*; 5. *Pilumnus* cf. *minutus* in *Balanus tintinnabulum volcano* shell; 6. *Petrolisthes coccineus*.

crabs, e.g. *Pilumnus minutus,* are short-legged and hairy. They are very sedentary, often living within the protection of empty *Balanus tintinnabulum volcano* shells.

Most surf shores have characteristic, highly prehensile echinoderms. In temperate regions there is generally a large mussel-feeding starfish, *Pisaster* or *Stichaster,* as well as a firmly attached urchin. On Hong Kong shores the only predatory starfish, the small tan *Ophidiaster cribrarius* is probably feeding on mussels. This species is easily recognized by the inequality of the five arms and the radial rows of tubercules along the arms. A second species is *Coscinasterias acutispina* which normally has seven or eight arms but which often splits in half and regenerates three of four arms. On surf coasts sea urchins are always important. The Hong Kong species is the jet black, stout-spined *Anthocidaris crassispina*. *Anthocidaris* is an indiscriminate grazer on corallines, more succulent algae or even upon dead organisms. Its depressed circular test is drawn close against the rock surfaces by the sucking cups of the tube-feet. The spines are usually splayed downwards against the rock during low tides, but become erect and bristling when wave-covered. *Anthocidaris* generally inhabits depressions or recesses where a dozen or more frequently lodge in a row. They are subject to an extensive local fishery. Their depletion from a shore may have a profound influence on the extent of low-zoned algae.

The large low-tidal anemone *Epiactis japonica* is scarcely mobile although it can periodically change position. Attached by its basal disc it is often found clustering in surge channels, and sometimes scattered with coarse sand. The outspread oral disc, with five marginal circlets of tentacles, is grey or greenish-brown, but is full concealed as the anemone closes and retracts for protection. The smaller, dark red *Actinia equina* can be found in rock pools. Both anemones are the prey of a small (1.5 cm long) wentletrap, *Epitonium* sp. The beautiful white shell is decorated with upraised vertical regularly spaced varices. They hide in crevices or in gravel at the bases of the anemones, thrusting their proboscis into the anemone's column and sucking the body fluids.

The gastropod molluscs of exposed shores must now be described. Each species is mobile but, like the periwinkles, faithful enough to its own shore level to form a basic part of the zonation pattern. There are broadly two types of shore gastropods, grazing herbivores and carnivores. The first are the most numerous in species, and along with them can conveniently be considered the chitons, or eight-valved coat-of-mail shells. These are not gastropods, but ecologically closely convergent with them.

The herbivorous species, differing widely in their vertical ranges, are further sub-divided by their shell form into limpets and the more mobile, spirally coiled snails. On shores under high wave attack, the herbivores at lower levels than the two periwinkles are all limpet-like, stream-lined and fast-attaching. There are three common species of true limpets (Patellacea) and two species of *Siphonaria* which are highly modified pulmonates and have secondarily returned to a limpet habitat on the shore. Their mantle cavity is modified as a lung, and opens by an air-breathing pore, protected by a slight expansion of the shell edge at the right side. Siphonariids can be distinguished by this small asymmetry and by their rather weak attachment to the rock, also by their habit of depositing gelatinous spawn crescents in shaded places, whereas common limpets shed eggs and sperm freely into the sea where fertilization occurs and a mobile planktonic larva develops that ultimately settles and metamorphoses into a miniature adult.

The highest level limpet is *Collisella dorsuosa,* with the shell elevated, radially ribbed and arched forwards. Found in shaded places on apparently bare rock, it ranges well above the *Porphyra* level. Lower down, amongst *Tetraclita,* lives the high conical *Cellana grata,* of about the same size as *Collisella* but with finer granulated radial ribs. Much smaller and flatter than either is another acmeid limpet, *Patelloida saccharina,* which is bluish black with a stellate pattern of seven strong white ribs. It should not be confused with the very similar black and white pulmonate, *Siphonaria sirius,* which is smaller, and nine-ribbed. A second siphonariid, *S. japonica,* is taller and greyish-brown ribbed. A much rarer, patellid limpet, found sub-tidally grazing on *Lithophyllum* crusts is *Patella flexuosa.* Flattened, strongly ribbed and with a deeply incised margin, it is easily recognized by its large size, often exceeding 7 cm in length. J. Christiaens[9] has provided a useful guide to the Hong Kong limpets.

Limpet-like in form, but with a small flattened spire is the ormer *Sanhaliotis planata,* immediately distinguished by the row of small exhalant holes. The foot is strong and mobile, extending somewhat beyond the shell and fringed with slender tentacles. Still found occasionally in the pink *Lithophyllum* zone the ormer or abalone is so relished as food locally that it seldom attains a large size close to Hong Kong.

There are two species of chiton or coat-of-mail shell on surf shores. The high-level species *Liolophura japonica* lives along with *Cellana grata,* grazing in the *Tetraclita* zone. The shell is roughly eroded with a transversely striped girdle, covered with small granules. *Onithochiton hirasei* which has smooth valves and a scale-less brown or orange girdle occurs in the pink *Lithophyllum* zone.

There remain the carnivorous gastropods belonging to the important family Thaidae, with representatives upon almost every intertidal shore. Their special habit is to feed upon barnacles or attached bivalves, gaining entry to mussels (or oysters) by boring a hole with the radula, aided by chemical action, and inserting the attenuate proboscis. Alternatively, smaller barnacle prey is opened by dislodging the opercular valves with the suction power of the foot. *Thais clavigera* (*see* Fig. 5.3) has the higher range and feeds generally on barnacles, and *T. luteostoma* is commonest at *Balanus* and *Lithophyllum* level where it is chiefly a mussel predator. A third, *Morula musiva* is widely distributed on more sheltered rocky shores. These gastropods, like the limpets, are extremely variable in shell dimension and thickness; the colour of the body is often helpful in identification.

The small fishes of wave-exposed shores deserve mention. They fall into two broad habit-groups, those adapted for quick spasmodic movements with the tail, but taking firm attachment by a ventral sucker during wave-fall, and those modified for a slow-moving existence, camouflaged against the fronds of brown algae (Fig. 4.5). The first group appears most on wave-exposed, mobile boulder shores. The second can often be found amongst sub-tidal *Sargassum. Histrio histrix,* the Sargassum fish, is plumply rounded, and rich brown and gold in colour like the seaweed; it is also camouflaged with filmy processes all over the body, and prominent head appendages. Swimming slowly and unadventuresome,

[9] Christiaens, J. 1980. The limpets of Hong Kong with descriptions of seven new species and subspecies. In *The Malacofauna of Hong Kong and Southern China,* ed. B. S. Morton, pp. 61–84. Hong Kong, Hong Kong University Press.

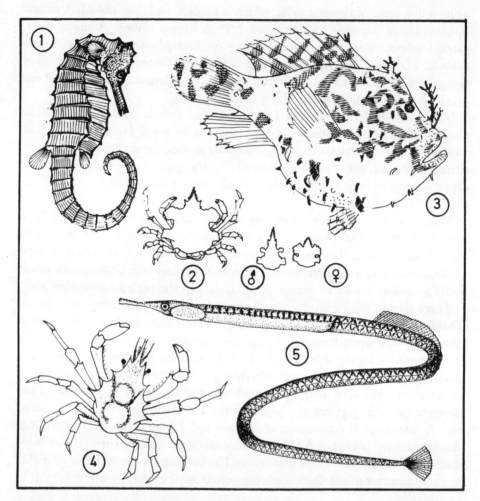

FIG. 4.5. The animals camouflaged in the *Sargassum* and other weeds: 1. *Hippocampus* cf. *kuda*; 2. *Huenia proteus* with carapace outlines of male and female; 3. *Histrio histrix*; 4. *Menaethius monoceros*; 5. *Sygnathus acus*.

Histrio relies for protection on the enveloping algal fronds.

The sea-horse *Hippocampus* cf. *kuda* is even more extremely modified. Compressed sideways, it has a rigid body-frame and a tapering, prehensile tail. Feeble swimming is performed with the reduced dorsal fin and the pectoral fins behind the cheeks. Normal movement is vertically up or down. Its food is small zooplankton, taken individually with the pipette-like narrow snout. The female is distinguished by a small anal fin, its place being occupied in the male by an abdominal brood pouch, into which the female transfers the eggs. Related to *Hippocampus* is the pipe-fish *Sygnathus acus* which has an elongated, snake-like form, a trumpet-shaped mouth and feeds on smaller animals of the interstitial algal fauna.

Huenia proteus, a small maiid-crab clinging to *Sargassum,* is highly camouflaged by its olive or brown colour, and the indentured expansions from the side of the carapace. The male and female have distinctive outlines. Also illustrated is the brown *Menaethius monoceros* found intertidally amongst algae (not necessarily *Sargassum*). A number of other maiid 'decorator' crabs occur in Hong Kong. *Micippa philyra* is found subtidally amongst rocks and covered with strips of weed and other organic debris, retrieved from its surrounding and carefully stuck to the body and arms. These crabs are exceptionally difficult to find, so closely do they match their habitat.

Algal Zonation: Mid-March

Any study of the open shore algae of Hong Kong should be continued from month to month. From the beginning of winter growth onwards, no one set of data would be valid even a few weeks later. The algal sequence of Hong Kong coasts is brief and dynamic. Its full expression can best be seen in the first half of March; as on the wave-exposed shores in Fig. 4.6 near Clear Water Bay, at the steep and dissected eastern reaches of the New Territories.

Less exposed than the shores at Cape D'Aguilar, Clear Water Bay lacks the surf coast barnacle *Balanus tintinnabulum* in significant number; the rocks are not granitic, but of rhyolite, breaking into vertical prismatic blocks and leaving cavities to form tide-pools which are not generally common on steep coasts. The algal patterns are essentially similar in March at Cape D'Aguilar and Clear Water Bay. Both shores are now dramatically different from their depauperate state in November.

The December *Bangia* strip of the littoral fringe has disappeared, and the *Porphyra* belt below it (beginning in late November as small pink plants only one centimetre long, thickening in January to a rich, dull red cover) has now begun to disintegrate. From a distance it stands out as a wide strip of yellowish-green; the plants have passed the reproductive phase and are already bleached and about to break up.

Below *Porphyra* comes a bare gap which is constant the year round, with scattered *Pollicipes mitella* in crevices, and the high-level limpets. This is bounded below by the rather straight edge of the *Tetraclita* barnacle zone. Thirty centimetres or more further down, deep brown swards of algae stand out from a distance with a rather level upper margin. One of the highest species is *Gelidium amansii,* in scattered patches 4–5 centimetres across. At the same level are the small tree-like or bushy tufts of *Dermonema frappieri,* not in close formation, but freely invading the *Tetraclita* zone. Mingling with *Dermonema* and to a great extent supplanting it below is a continuous belt of the tan or

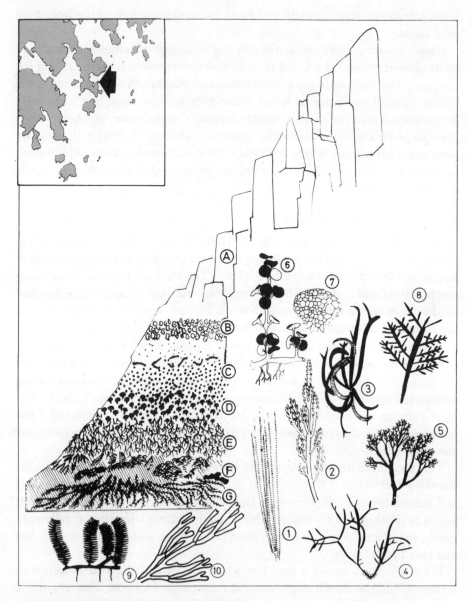

FIG. 4.6. Moderately exposed shore at Clear Water Bay. The steep rocky slopes of rhyolite blocks is shown with its zonation pattern (March).

ZONES:
A. Bare rock; B. *Porphyra suborbiculata* (regressing); C. *Tetraclita squamosa* with *Pollicipes mitella*; D. *Dermonema frappieri*; E. *Gloiopeltis furcata*; F. Zone of *Laurencia-Chaetomorpha-Gigartina*; G. *Sargassum*.

SEAWEEDS:
1. *Chaetomorpha antennina*; 2. *Laurencia japonica*; 3. *Gigartina intermedia*; 4. *Gloiopeltis furcata*; 5. *Dermonema frappieri*; 6. *Caulerpa peltata*; 7. *Dictyosphaeria cavernosa*; 8. *Gelidium amansii*; 9. *Caulerpa scalpelliformis*; 10. *Galaxaura fastigiata*.

yellowish-brown alga *Gloiopeltis furcata,* spreading over the emersed rocks. Somewhat slippery with mucilage, tubular and dichotomously branching, these plants dominate all moderately exposed middle shores in March and April.

Coralline algal turf is uncommon at Clear Water Bay, but the lower part of the eulittoral zone is diversified with a number of red and green species. *Laurencia undulata* and *L. japonica* grow in long tufts. Clumps of coarsely, filamentary and brittle *Chaetomorpha antennina* stand out bright green. In moist or shaded places grow the dark green, almost black, turfs of the resilient and springy *Gigartina intermedia. Gelidium amansii* has small dark red pinnate fronds. *Galaxaura fastigiata* forms short tufts of close set cartilaginous branchlets. *Sargassum hemiphyllum,* on a ground base of pink *Lithophyllum* is at its longest and densest in March and April. The clusters of small basal leaves, first noticed in late November, have now produced streamers up to 2 metres long, swaying in the surge and studded with narrow reproductive branchlets.

Rock Pools

The dislodgement of rhyolite blocks and prisms gives a broken contour to the shore-line, forming small, irregular pools, regularly breached and re-circulated by the surge. Pools are rich habitats, and their beauty and interest increase with close study. In early March, the prevailing colour in deeper pools is the golden-brown of *Colpomenia sinuosa,* whose fleshy and wrinkled bladders cover the entire bottom. In shallower pools there is a blend of pastel shades, olive-green, brown and, most pervasive of all, pink or mauve coralline paint.

The molluscs in pools are far more mobile and diverse than on the open surface (Fig. 4.7). Several gastropod families are represented here, such as the Turbinidae, Trochidae and Neritidae, that are found on open rocks only in far more sheltered places. The conical turbinid *Astraea rhodostoma* is, however, mostly confined to open coasts, and is often covered with coralline paint. It grazes on coralline algae and on the smaller succulent red species. The sculpture is of heavy oblique ridges, forming nodules at the keeled periphery. The base has a mauve streak, and the convex calcareous operculum is prettily marked in pink. *Batillaria sordida,* representative of a family (Potamididae) more characteristic of soft shores, is recognized by the stocky, knobbly shell and a distinctive sculpture of the two spiral rows of black nodes on each whorl. Common also amongst corallines is the smaller, carnivorous snail, *Pyrene punctata,* generally pink-encrusted. The hemispherical black and white snail, *Nerita albicilla,* is found in large numbers in every pool. Even commoner in pools than on the open rock are the broadly expanding anemones, *Anemonia sulcata* and the smaller *Anthopleura pacifica,* also the jet-black urchin *Anthocidaris crassispina.*

The largest molluscs feeding on succulent algae are the aplysioids or sea-hares. Normally sub-tidal in habit, they appear commonly in pools at the breeding time

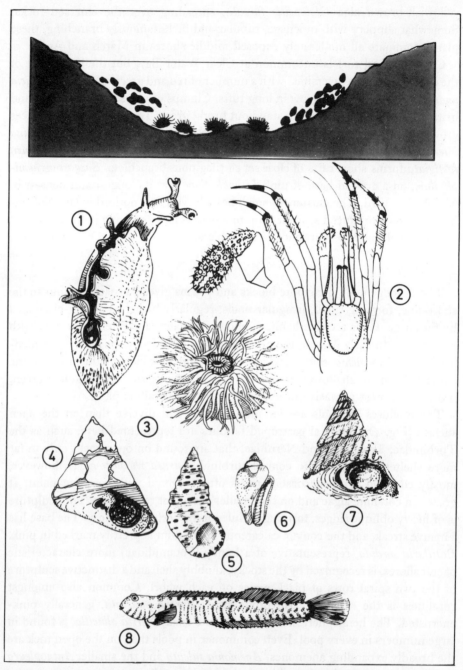

FIG. 4.7. A rock pool at Clear Water Bay with *Anthocidaris* and *Colpomenia*.
1. *Stylocheilus longicauda*; 2. *Clibanarius bimaculatus*; 3. *Anthopleura japonica*; 4. *Chlorostoma rustica*; 5. *Batillaria sordida*; 6. *Pyrene punctata*; 7. *Astraea rhodostoma*; 8. *Entomacrodus stellifer*.

Exposed Rocky Shores

from March onwards. Copulation takes place in chains, each hermaphrodite individual acting as female to the one in front of it and male to the one behind. The eggs are laid in springy coils of gelatinous yarn, loosely attached to rock or weed. The body of an aplysioid is soft and slug-like with the shell reduced to a small horny shield, within which the tissue forms the roof of the mantle cavity. The animal is approximately flask-shaped, with a long narrow sole. The slender head bears two pairs of tentacles, the anterior ones forming rolled tubes, containing the olfactory receptors. As the body expands behind, two muscular folds, the parapodia, grow up to enclose the mantle cavity from the sides, their margins being mobile and wavy.

Stylocheilus longicauda (Fig. 4.7) is speckled with opaque white over the sides of the foot. Other species common in spring are *Aplysia sagamiana* and the large darkly mottled *A. juliana*. A related aplysioid, *Bursatella leachii* appears in large numbers at the same time of year on protected soft shores (Fig. 10.6).

The small blennies *Istiblennius tanagashimae* and *Entomacrodus stellifer* are also to be found in the rock pools, curled up in and amongst empty barnacles or hidden deep in crevices.

In pools high up the shore, replenished only by splash or rain, mosquito larvae of the two species *Aedes togoi* and *Culex sitiens* and the non-biting midge *Chironomus sp.* abound. Such pools are often densely occupied by the filamentous almost hair-like, fronds of the green alga *Ulothrix* sp. which seems to prefer such pools with a fresh-water input.

Fauna of Coralline Turf

The turf expanses of *Corallina* and *Jania* or their fringes around pools are some of the richest habitats, in species and individuals, on the inter-tidal shore (Fig. 4.8). These calcareous algae are dense enough to protect their occupants against desiccation and wave action, yet open enough for small animals to move through them freely. They are well oxygenated by clean surge and only at the base (where specially modified animals live) do fine sediments collect that would smother the gills or mouth-parts of animals. The species are almost all small or semi-microscopic, and mostly active and mobile.

The molluscs of coralline algae are juvenile gastropods, especially *Monodonta, Planaxis* and *Lunella,* or bivalves. Small mussels *(Septifer bilocularis, Perna viridis* and *Modiolus)* settle here, freely breaking off, crawling about and then renewing their byssal attachment. Small ark-shells *(Barbatia virescens)* with *Cardita leana* attach near the bases of the thalli.

Many species of small gastropod, belonging chiefly to the Rissoidae and the Triphoridae, reach their adult life in corallines. The first are minute and very numerous, and by no means all their species have yet been determined. The shells are conical to rounded, and smooth or ribbed. The long narrow sole has

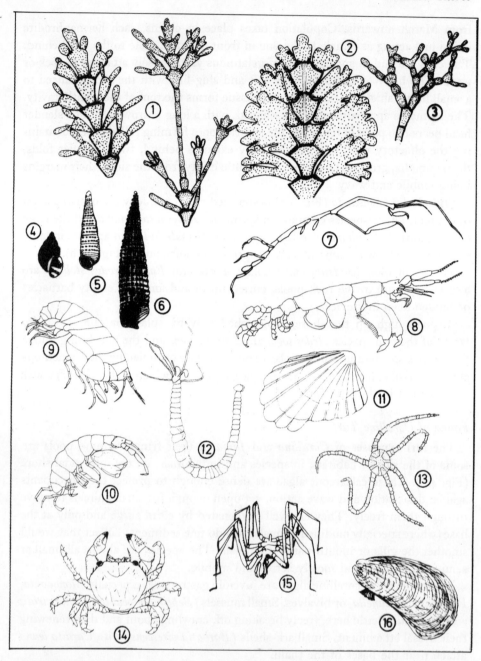

FIG. 4.8. *Corallina* algae and their interstitial fauna.
1. *Corallina pilulifera* compact and slender growth forms; 2. *Corallina sessilis*; 3. *Jania undulata*; 4. A rissoid snail; 5. *Notosinister granulata*; 6. *Viriola tricincta*; 7. *Caprella simia*; 8. *Caprella acutifrons*; 9. *Stegocephalus inflatus*; 10. *Orchestia platensis*; 11. *Cardita leana*; 12. A syllid-worm; 13. A brittle-star (*Amphiura* sp.); 14. *Pilumnus minutus* (juvenile); 15. A sea-spider *Achelis superba*; 16. *Modiolus agripetus*.

a single posterior tentacle and a mucous gland secreting a string by which the snail can hang suspended. While rissoids generally browse upon succulent algae, the Triphoridae feed upon sponges. They are long-spired, and easy to recognize: as well as fine beaded sculpture, they often have the shell coiled sinistrally, i.e. with the aperture to the left. The longer, dark brown species is *Viriola tricincta,* and the shorter, light brown one *Notosinister granulata* (Fig. 4.8).

Isopods and amphipods abound in coralline algae. Small and diverse, these crustaceans are not yet thoroughly described or identified in Hong Kong. They are, however, easy to collect merely by placing algal fronds in sea-water with a trace of formalin added, when the species will swim free and show up clearly in a white or enamelled dish.

The largest group of the isopods found here are the Flabellifera, with the Sphaeromidae the principal family. The body is broad and dorsally arched and it is able to curl up protectively into a ball. A distinctive character of this family is the large tail fan, with the inner uropod branch fixed against the telson. The first five abdominal somites are fused together, and the sixth coalesced with the telson.

A common amphipod with a rather caricatured shape is *Stegocephalus inflatus,* a species of the family Stegocephalidae. It is dark brown, flatly compressed and almost circular; the thorax is greatly deepened by the enlargement of its side plates or coxae.

The most bizarrely specialized of the algal amphipods include some of the commonest species, such as the Caprellidae or skeleton shrimps. Up to a centimetre long (some species are longer) they are found seething in coralline and also in more succulent red and brown algae. With thin, knobbled bodies, caprellids progress like looper caterpillars, taking hold in front by a pair of enlarged gnathopods, with chelae, and at the back by three pairs of clawed thoracic limbs or pereiopods. These give anchorage while the gnathopods are extended in front to seize copepods or other small swimming prey. The abdomen is a tiny vestigial knob, with neither limbs nor separate segments.

The Pycnogonida or 'sea-spiders' are an isolated group of arthropods closer to true spiders than Crustacea. Especially at home in coralline algae, they live partly suctorially and partly by seizing solid prey. The small body is slung from four pairs of thin, attenuated limbs, firmly attached by claws to the substrate. The narrow, segmented thorax carries behind it a short abdomen. The stomach gives off digestive diverticula that enter the long legs. Many pycnogonid species seize small hydroids by the pedipalps; and other species, living on coelenterates, holothurians or ascidians, may take body fluids by the insertion of their sharp mouth-parts (chelicerae).

The polychaete worms have two important families in coralline algae, both carnivores. Of larger size are the Nereidae, with several species winding freely between the branches or concealed in the sediments at the base and living on smaller worms or crustacea. Details of their head-parts and proboscis are illustrated in Fig. 9.12. Much smaller and individually more numerous are the family Syllidae. They are seldom longer than 2 cm, and a good lens or a binocular microscope is needed to see their distinguishing marks. The head bears four eyes, three tentacles and a pair of long unjointed palps. The second segment (peristomium) has two pairs of cirri, while the cirri of the parapodia on all succeeding segments in some species form long tendrils. The tubular pharynx has sharp denticles and is shot out to impale or grasp small prey.

Some of the commonest scavengers and deposit feeders in the coralline turf are the small brittle-stars, *Amphiura,* generally no more than 1 cm across. They are built of a central disc with five serpentine flexibly jointed arms, moving through the turf by flexing two leading arms, drawing up the disc and the remaining three.

Birds of Rocky Shores

The only bird to be found regularly on the exposed rocky shores of Hong Kong is the sombre grey reef egret *(Egretta sacra)*; it makes a nest of sticks on the shore line and feeds extensively on crabs. The cormorant, *Phalacrocorax carbo* (Fig. 4.9) uses its large hooked bill to hunt underwater for bottom-dwelling fish. The bird's waterproofing is not as good as that of other diving birds and so in the winter months (when it is most commonly seen in Hong Kong) a cormorant will often be seen sitting on the rocks or on pilings with its wings outstretched to dry. The white-bellied sea eagle *(Haliaetus leucogaster)* (*see* Fig. 5.1) is a magnificent bird of prey, often to be seen soaring in search of the food. Several pairs breed in Hong Kong making their nests, or eyries, on some precipitous cliff; nevertheless they are often robbed of eggs or young. The peregrine falcon *(Falco peregrinus)* nests in similar situations but may be seen over the Mai Po marshes hunting for birds as large as duck, which it takes on the wing. A ubiquitous scavenger on the shore line, and especially in the harbour, is the black-eared kite *Milvus migrans* typified by its forked tail (Fig. 4.9). Occasionally a wading bird, on migation, may stop-over to rest on rocky shores, notably the common sandpiper *(Tringa hypoleucos)*. There is no resident colony of gulls (or terns) in Hong Kong but from October to April, Herring gulls and Black-headed gulls may be seen in migrating flocks offshore. Similarly the Caspian tern is a regular winter visitor roosting in small groups along the shores of Deep Bay.

Exposed Rocky Shores

FIG. 4.9. Common birds of rocky shores.
1. Black-eared kite *(Milvus migrans)*; 2. Cormorant *(Phalacrocorax carbo)*; 3. Caspian tern *(Hydroprogne caspia)*; 4. Herring gull *(Larus argentatus)*; 5. Black-headed gull *(Larus ridibundus)* winter plumage without black crown.

CHAPTER 5

Sheltered Rocky Shores

Rock Surfaces

From exposed outposts beaten by ocean swell we shall move to hard shores, more landlocked and enclosed within the shelter of bays. The south coast of Hong Kong is agreeably rich in this type of inlet. Crescentic sand beaches generally lie at the top of the bays. The outer heads, as at Cape D'Aguilar and Stanley Peninsula are tipped by granite, and the steeply zoned bed-rock at the sides of the bays is of rhyolite.

Deep Water Bay

The first example is taken from Deep Water Bay, lying behind the shelter of Middle Island (Fig. 5.1). At first view it seems to have little in common with Cape D'Aguilar. Strong waves seldom crash upon it, or sweep up the shore; and the whole zoning pattern is far more vertically compressed. The highest reaching periwinkles, still *Nodilittorina pyramidalis* and *N. millegrana,* in the same order, extend to only 2 or 3 metres above low water mark. The littoral fringe is a relatively narrow belt, without a *Porphyra* cover. The rock oyster *Saccostrea cucullata* has almost displaced the barnacles from the eulittoral zone. All that remain are occasional *Pollicipes* clusters, especially in crevices and wave runnels, and a scatter of *Tetraclita squamosa squamosa* below the dense oyster belt. This dense belt of oysters is bored by the date mussel *Lithophaga malaccana* and such bore holes are eventually colonized by a variety of nestlers such as the venerid bivalve *Irus mitus,* serpulid and polychaete worms, small crabs and the bottle-green anemone *Haliplanella luciae.*

Rock oysters are visually dominant on sheltered rocky shores in most warm and sub-tropical seas. At Deep Water Bay, they form a straight-topped belt; above them for a few centimetres runs a spatter of blackish-red *Hildenbrandtia,* a crustose red alga, that also covers the bare rock amongst the uppermost oysters. Small and thickly aggregated, the lowermost oysters carry a stubble of the red alga *Gelidium pusillum,* especially in shade or on north-facing vertical surfaces.

At some sheltered sites a true sub-littoral fringe hardly exists, for the rock

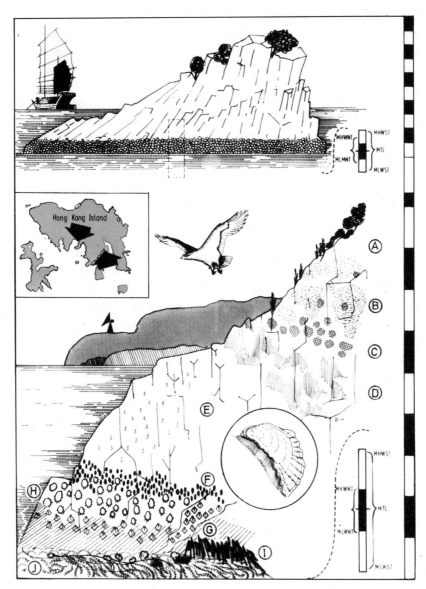

FIG. 5.1. Zonation of a sheltered shore at Deep Water Bay. *(Above)* Small islet at Deep Water Bay with *Saccostrea* zone and bare rock above. *(Below)* Zonation of rock face at Turtle Cove in December with Stanley in the background, with *(inset)* the white-bellied sea eagle *(Haliaetus leucogaster)*.

ZONES:
A. Terrestrial plants; B. Grey-green lichens; C. Yellow lichens; D. *Verrucaria maura*; E. *Nodilittorina* species; F. Splashes of *Hildenbrandtia*; G. *Saccostrea cucullata* with shell inset; H. *Tetraclita squamosa* merging into the basal coralline paint; I. *Petalonia fascia*; J. Lower zone of *Corallina*, *Sargassum* and *Colpomenia*. *Insets right* Tidal range for neap and spring tides and curves for maximum wave heights (compare with Fig. 2.3).

surface is sand-buried up to the eulittoral zone. The richest life is then to be found on loose sub-tidal boulders, lying just beyond low-water mark (*see* Fig. 7.6). At many places however the algal zonation from the middle of the oyster belt downwards is rich and complex; the mid-February picture is suggested in Fig. 5.2.

The living pattern can first be followed upwards beyond oysters and *Hildenbrandtia* into the littoral fringe and maritime zone. Above the oysters the rock is at first bare, with the grey of rhyolite or the faint pink of granite. Above this level, especially on vertical faces, the soot-black lichen *Verrucaria* appears. Higher again there are splashes of orange *Caloplaca*, and still further up a spread of grey or grey-green crustose lichens continuing upwards to mingle with the first green plants.

Tai Tam Bay

In the open reaches of Tai Tam Bay, facing seawards just south of the granite quarry on D'Aguilar Peninsula, there is an interesting shore easily accessible only by boat. The rock surface is smooth and unbroken, sloping at 30°–40°, and carries a clear-cut zonation, clearly visible from a distance, with a composition approaching that of open shores. Below the clean granite stretch with the *Nodilittorina* species in depressions, appears a 30 cm wide band of a blue-green alga, probably *Calothrix scopulorum*, slippery when wet. *Porphyra* is lacking from the rock face in winter months, but is densely loaded upon the shells of the limpet *Cellana grata*, giving it perhaps its only security from heavy grazing by these animals.

On the yellow-brown zone below the black *Calothrix*, are three other limpets, *Patelloida saccharina*, *Siphonaria sirius* and *Siphonaria japonica* all grazing among the minute algal sporelings. A little lower down are golden-brown *Pollicipes mitella* abutting on to a grey zone of large *Tetraclita squamosa*. Neither *Balanus tintinnabulum* nor the urchin *Anthocidaris* are present; *Tetraclita* is followed directly by a pink *Lithophyllum* zone, with a spread of brown algae (*Sargassum* and *Petalonia*), thickening rapidly in mid-January.

ALGAE

The algal zonation at Turtle Cove in early February is shown in Fig. 5.2. Strong algal cover begins with the lower oysters where *Gelidium pusillum* (B) becomes continuous. Below the oysters the prevailing hue is bright green (C), with at least three species of Ulvales. Tattered dark green sheets of *Ulva lactuca*, mingle with smaller bright green *Ulva conglobata*, with its frilled fronds clustered into a flower-like formation. Common also are the crinkled tufts of *Enteromorpha compressa*, with the cylindrical blades flattened to narrow ribbons.

Several more *Ulva* relatives should be mentioned in passing, and are shown in Fig. 5.5. One of the largest of its genus is *Ulva fasciata* with dark green blades, often reaching 100 cm long, it divides into several ribboned lobes with

FIG. 5.2. Algal zonation at Turtle Cove, Hong Kong Island in February.

ZONING SPECIES:
A. *Saccostrea cucullata*; B. *Gelidium pusillum*; C. Ulvoid zone; D. *Gelidium-Pterocladia-Gigartina-Ralfsia*; E. *Petalonia* with *Scytosiphon*; F. *Colpomenia* with coralline turf; G. *Sargassum hemiphyllum*; H. *Sargassum patens*.

ALGAE:
1. *Endarachne binghami*; 2. *Petalonia fascia*; 3. *Sargassum hemiphyllum*; 4. *Sargassum horneri*; 5. *Sargassum patens*; 6. *Ralfsia verrucosa*; 7. *Petrospongium rugosum*; 8. *Gelidium pusillum*; 9. *Gloiopeltis furcata*; 10. *Colpomenia sinuosa*; 11. *Galaxaura fastigiata*.

the edges ruffled. *Ulva reticulata* is to be distinguished by its marginal teeth, and pore-like perforations at the edges. The blades often form an open network of holes. In distinction from *Ulva* and *Enteromorpha,* the transparent blades of *Monostroma* are composed of only a single cell layer. *M. nitidum* is yellowish-green, softly gelatinous, and found in quiet pools near the high-water mark.

Of the *Enteromorpha* species, the largest is usually *E. prolifera,* often sold as food in the markets. The compressed tubular thallus can grow up to 40 cm long and attaches to virtually any solid object in inner bays. Smaller species include *E. compressa* and *E. linza,* flat and foliaceous like an *Ulva,* and living on rocks or in pools.

To return to Fig. 5.2, the space below the green algae (D) is thickly occupied by *Gelidium,* and wine-red immature fronds of a minute *Pterocladia*-like species. Below this band, especially on shaded faces, is small springy *Gigartina intermedia* sometimes forming a total cover, or sharing the ground with yellowish-brown scabs of *Ralfsia verrucosa.*

At the next level (E), the winter brown algae develop in full. The highest occurring are *Petalonia fascia* and *Scytosiphon lomentaria.* The first produces translucent golden-brown ribbons in early January, and by February has reached up to 60 cm long, and 2 or 3 cm across, becomes fertile and starting to disintegrate. Reaching its maximum at about the same time, *Scytosiphon lomentaria* has pale brown, tubular thalli, constricted into pods, and reaching 50 cm long. Often used as food, it is especially associated with fresh-water seepage. The brown alga *Petrospongium rugosum* forms small dark brown pads at *Petalonia* level, 2 or 3 cm across and heavily wrinkled.

From January onwards the brown alga *Colpomenia sinuosa* becomes common at this level and below (F). Beginning from small, thick-walled vesicles, it grows to 5 to 6 cm across by March, and become fleshy and convoluted before disintegrating.

The *Sargassum* species, always dominating the sub-littoral fringe, are three in number. *Sargassum hemiphyllum* is by far the most abundant (G), and shows rapid growth after January; the thongs reach their maximum in April, with the development of bladders and reproductive organs. By May most of the plants have broken away, often to form floating masses off-shore. By July or August nothing is left but a stubble of basal stalks. Further out grows the serrate-leafed *Sargassum patens* (H), generally at the edge of large beds of *S. hemiphyllum.* The third species, only occasionally seen, is *S. horneri,* distinguished by its twisted and furrowed stems with small spinules, its linear, pinnate leaves and its cylindrical bladders.

In drift from the sub-tidal, sometimes collecting between tides, other phaeophyceaens will be found, being typical of warm tropical seas and in Hong Kong usually sharply seasonal. *Padina australis* forms broad fans, generally in clusters

Sheltered Rocky Shores

and frayed or split into separate segments. Also sometimes found in pools or on open rock as well as sub-tidally is the golden-brown *Hydroclathrus clathratus*, related to *Colpomenia*; it arises in the same way as a gelatinous bag, but afterwards develops into a fleshy network full of large holes.

Often common on sub-littoral rocks are the fan- or kidney-shaped lobes of *Pocockiella variegata*, up to 5 cm long, and the stiffer *Zonaria coriacea*, also with fan-shaped thalli, erect and lightly hairy. *Dictyota dichotoma* forms clusters of thin, twin-branching ribbons, lying prostrate over low tidal rocks.

On sheltered shores the green algae, *Codium* and *Caulerpa* may be prominent off-shore as well as inter-tidally. *Codium intricatum* has short dichotomous branches, densely attaching to each other by rhizoids and forming dark green cushions, up to 5 cm across. In *Codium cylindricum* the dichotomous branches are elongate and cylindrical with pointed tips, up to 10 cm tall and 5 mm wide at the base.

The succulent *Caulerpa* species have an essentially creeping habit and a wide variety of leaf forms. *C. peltata* puts up clusters of small, mushroom-shaped leaves from the creeping rhizome. Other species, commonly found sub-tidally or washed ashore, are named from the resemblances of their leaves: the pointed *C. taxifolia*, and *C. cupressoides* var. *lycopodium*; and *C. sertularioides* with almost cylindrical pinnules. Two of these, *C. scalpelliformis* with thick erect leaves oppositely denticulate and *C. peltata* are seen in Fig. 4.6. These algae often colonize, in great profusion, the sandy spaces between two rocky headlands at sub-tidal depths, thereby intermingling with patches of coral.

There are several inconspicuous green algae made up of meshes of microscopic branching filaments. In *Boodleya composita* the branches are cross-linked and built into a spongy tangle, epiphytic on *Jania* or attached directly to the rock. *Struvea delicatula* forms tufts of pinnate branchlets, cross-anastomosing and needing a lens to see clearly. The whole thallus attaches by rhizoids to *Sargassum* holdfasts or *Corallina*.

A very different thallus, stiff and cartilaginous, is produced by *Dictyosphaeria cavernosa*, forming pale green pads from aggregations of big thick-walled cells.

A number of filamentary green algae should also be mentioned. In sheltered rock pools, *Cladophora delicatula* grows in dull green tufts, up to 5 cm long. *Rhizoclonium hookeri* forms pale, entangling masses in turbid littoral pools. Sometimes intermingled with it are thicker filaments of *Chaetomorpha brachygona*, from 2–7 cm long.

MOLLUSCA

Shores under moderate shelter are far richer in mobile gastropods than exposed headlands, and some important families of herbivorous grazers here are very noticeable, as shown in Fig. 5.3 for Deep Water Bay.

First, the littorines, *Nodilittorina pyramidalis* and—cutting off slightly below it—*N. millegrana,* reach to the top of the shore, aggregating in pits and recesses, well above high-water level. Moderately sheltered shores possess a third littorine, *Peasiella* sp., often going un-noticed because of its small size and habit of lying flush within little irregularities of the rock mass. Yellowish-brown, and only 2 mm across, it is convex above and concave below, with the edge sharply keeled. The typical habitat is the bare zone just above the rock oysters.

At the tops of sheltered bays, especially on wharf-piles and in low salinity, a fourth periwinkle, *Littorina scabra,* will appear. Living above the oyster level, and ranging into the branches and foliage of mangroves (*see* Fig. 11.8) it is relatively large, and smoothly conical, with fine spiral sculpture and brown zigzags on a pale ground. Also at bay-heads and in semi-brackish places, but avoiding mangroves and saltings, occurs another littorine, *L. brevicula,* small and squat, with strong spiral keels around the dark brown body whorl. The final periwinkle, *Littorina melanostoma,* is estuarine and almost totally restricted to salt-meadows and swamps. Pale and thin-shelled, with a dark brown area on the inner lip, it is more commonly mentioned when considering soft shores.

Even more important grazers, on sheltered shores, than the littorines are the limpets, both patelloid and pulmonate. Fig. 5.3 shows their vertical distribution at Deep Water Bay. The highest ranging, living in small high-tidal crevices and, further down, in the interior of empty oyster shells, is the pulmonate *Siphonaria japonica,* here no more than 7 mm long. In small pockets of shelter it deposits crescentic spawn masses that would desiccate in direct sunlight.

The remaining limpets are all true patellaceans (Acmeidae and Patellidae) shedding eggs and sperm, for external fertilization, directly into the plankton. Of the acmeid limpets, the high-built *Collisella dorsuosa* is absent, being restricted to exposed shores. It is replaced by some flatter-pitched species, at and above oyster level, *Collisella luchuana* and *Notoacmea concinna.*

The smaller, high-conical *Patelloida pygmea* is one of the commonest of the sheltered shore limpets. Radially splashed and streaked with black, it lives alongside *Siphonaria japonica. S. atra* is to be found within and around warm pools of standing water. The patellid *Cellana grata,* found at Cape D'Aguilar, is progressively replaced in shelter by a thinner and flatter limpet, *Cellana toreuma,* found abundantly in a variety of sizes and radially streaked colour patterns.

All those limpets, like the littorines, graze upon algal films, taking the almost undetectable fodder in crevices and upon the seemingly bare rock surface. A small chip of rock, acetone-extracted, will yield surprisingly high chlorophyll values. The productivity of this territory, including the contribution of short-lived algal sporelings, must be far higher than its meagre standing crop would indicate at any one time.

Sheltered Rocky Shores

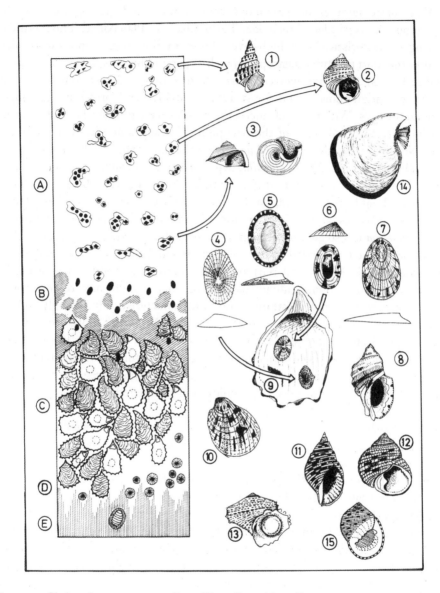

FIG. 5.3. Sheltered zone pattern at Deep Water Bay with molluscs.

ZONES:
A. *Littorine* zone; B. Upper eulittoral zone with *Hildenbrandtia*-like encrusting algae; C. *Saccostrea cucullata*; D. *Tetraclita squamosa* among the lower oysters; E. Film of basal coralline paint.

SPECIES:
1. *Nodilittorina pyramidalis*; 2. *Nodilittorina millegrana*; 3. *Peasiella sp.*; 4. *Siphonaria japonica*; 5. *Notoacmea concinna*; 6. *Patelloida pygmaea*; 7. *Cellana toreuma*; 8. *Thais clavigera*; 9. Empty *Saccostrea* valve with *Patelloida* and *Siphonaria*; 10. *Nerita albicilla*; 11. *Planaxis sulcatus*; 12. *Monodonta australis*; 13. *Lunella coronata*; 14. *Isognomon acutirostis*; 15. *Nerita undata*.

Four more families of gastropods contribute to the grazing force of the eulittoral: the Neritidae, Trochidae, Turbinidae and Planaxidae. The Neritidae belong characteristically to high-level shores in the tropics and sub-tropics. Unlike the most primitive gastropods (such as patellid limpets, haliotids, trochids and turbinids), they have evolved internal fertilization, and have complex genital ducts for sperm transfer and for secreting nutritive and protective envelopes around the eggs. With these advances, the neritids have produced both terrestrial offshoots (a foreshadowing of the true pulmonate snails and fresh-water lines). The Neritidae have also speciated greatly on the upper shore. On open rock, the commonest species is *Nerita albicilla,* depressed and hemispherical, and secured amongst the rock oysters almost like a limpet. Porcelain-white below, it is plain or streaked, and often eroded, on the upper surface. The eggs are deposited in small oval capsules, attached under stones or in moist crevices. A second nerite, *N. chamaeleon* is also common, ranging from moderate exposure to high shelter. This species is characterized by two spiral cords and distinct axial striae. Blotches of colour (black, brown and red) are spread across the body whorl against a cream background. Some individuals are more or less uniformly red or brown.

Two rare nerites, not illustrated, are both characteristically strongly ribbed. These are the black *N. costata* and the flesh pink *N. plicata.* Both are confined to the outer islands (e.g. Po Toi) and are uncommon in the Hong Kong region.

Planaxis sulcatus, the sole local representative of its family, is widely common in the tropical Indo-Pacific and in Hong Kong. Often highly gregarious on steep faces, related to the littorines, it is conical and chequered greyish-white, with strong spiral sculpture.

The top-shells, Trochidae, have conical or sub-spherical shells, with the operculum thin, circular and horny. *Monodonta australis,* the single species common on open surfaces, is found at mid-tidal level. The sculpture is of strong spiral ribs, greenish-grey flecked with darker splashes. The full range of trochids will be found later on boulder shores (chap. 7).

Distinguished from the top-shells by the convex shelly operculum are the turban shells, Turbinidae, generally grazers on the lower shore where *Gelidium* or *Corallina* begin to appear. The local common species, *Lunella coronata,* lives below the oyster zone; the shell is broad and low pitched, red-gold in colour, with beaded sculpture and smooth mamillae.

The common carnivorous thaid is *Morula musiva,* ranging through the oyster zone and boring holes in oyster and barnacle shells to insert the proboscis. Another smaller and more sharply ridged species is *M. margaritifera.*

A high-zoned bivalve of more sheltered rocky shores is the black-lipped *Isognomon acutirostris* occurring byssally attached, deeply embedded within crevices.

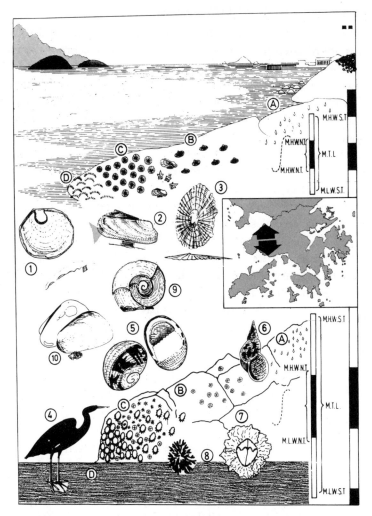

FIG. 5.4. *(Above)* Zoning pattern at Tai Lam Chung, near Castle Peak. *(Below)* Zonation at Tsim Bei Tsui, Deep Bay. Vertical scale in metres. Tidal range is shown for springs and neaps, with the curve for maximum wave height (broken line).

ZONES:
A. Littorines; B. Upper eulittoral with *Siphonaria*; C. Mid-eulittoral with *Tetraclita, Pollicipes* and *Barbatia*; D. *Corallina* turf.

MOLLUSCS:
1. *Siphonaria japonica*; 2. *Barbatia virescens*; 3. *Anomia achaeus*.

ZONES:
A. Littorines; B. *Euraphia withersi*; C. *Saccostrea* with *Balanus pallidus*; D. Mud reaching up to oyster level.

SPECIES:
4. Reef egret *Egretta sacra*; 5. *Nerita chamaeleon*; 6. *Littorina scabra*; 7. *Euraphia withersi*; 8. *Caloglossa leprieurii*; 9. *Serpulorbis imbricatus*; 10. *Modiolus elongatus*.

High Shelter with Low Salinity

Shores landlocked within harbours and bays have a diminished wave action. Zonation is abbreviated to lie between actual tidal limits. From the top of the shore the lichen zone disappears, these plants preferring clean, open rocks with strong spray. The oyster *Saccostrea cucullata*[1] is still dominant, and often densely crowded, with small individuals several layers deep, tightly attached to each other and to the rock. There is often no sub-littoral fringe of brown algae. This level is sometimes covered with silty sand, often strewn with boulders rather than intact bed-rock.

Several examples of harbour shores can be briefly mentioned. Near the Pavilion, at Picnic Bay, Lamma Island, rocky outcrops have three periwinkles in the littoral fringe: *Littorina scabra*, *L. brevicula* and *Nodilittorina millegrana*. On higher rock-stacks careful search may reveal a few *Nodilittorina pyramidalis*. Above the oyster zone, the upper eulittoral has a scatter of the smooth-shelled barnacle *Balanus amphitrite*, a species of low or intermediate exposure, often associated with brackish water. Below the oyster zone, scattered specimens of *Tetraclita squamosa* still occur; in the same situation there are usually reduced numbers of *Pollicipes mitella*. Both barnacles have a wide salinity range, reaching as far as the islands of the Pearl River estuary, but never existing in extreme shelter.

At the north-east coast of White Knights, in Tolo Harbour, a highly sheltered rocky shore showed algal growth well advanced in mid-January. Amongst the dense oysters are small mussels, *Septifer bilocularis* and *Brachidontes atratus*; the arcid *Barbatia virescens* and *Trapezium sublaevigatum*. Around their bases, nereid and phascolosomid worms form a miniature crevice faunule (*see* Fig. 7.16).

In January and February the algae form a well-marked series, with the Chlorophyceae prominent. Above the oysters is a dense layer of tubular *Enteromorpha compressa*; the *Saccostrea* zone itself is sprinkled with *Ulva conglobata*, and below this are thick tresses of an *Enteromorpha*. Near the low-water mark is a heavy fringe of yellowish-brown *Pylaiella*. *Ectocarpus* and *Pylaiella* are closely related primitive brown algae, forming dense tufts of silky filaments. They can be microscopically distinguished only by the position of the sporangia, stalked in *Ectocarpus*, intercalary in *Pylaiella* (Fig. 5.5). *Pylaiella* and *Enteromorpha* are so abundant in February they break off in large clumps and accumulate over the adjacent sand-flat. *Petalonia* and *Sargassum* are absent, but fleshy vesicles of *Colpomenia sinuosa* are abundant at low-water neap.

[1] With increasing shelter it is possible that a second species of rock oyster *Saccostrea echinata* becomes dominant. This differs from *S. cucullata* in the possession of long black species radiating from the uncemented upper valve. It has not yet been positively proven, however, that this is a true species. It may simply be that in higher exposure the spines are abraded away giving the superficial impression of two species. Throughout the Indo-Pacific, *S. cucullata* might be better regarded as a 'super-species', occuring in many habitats and in a variety of forms.

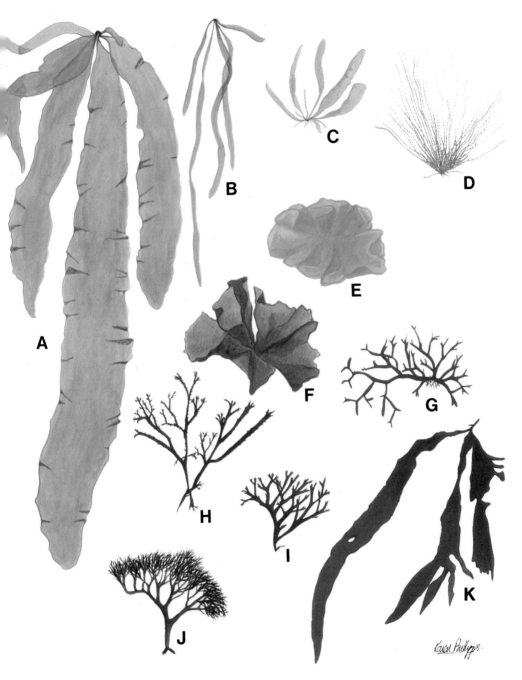

PLATE 5. Rocky shore algae: A. *Petalonia fascia*; B. *Scytosiphon lomentaria*; C. *Enteromorpha compressa*; D. *Chaetomorpha antennina*; E. *Ulva conglobata*; F. *Porphyra dentata*; G. *Gloiopeltis furcata*; H. *Gigartina intermedia*; I. *Gymnogongrus flabelliformis*; J. *Dermonema frappieri*; K. *Cyrtymenia sparsa*.

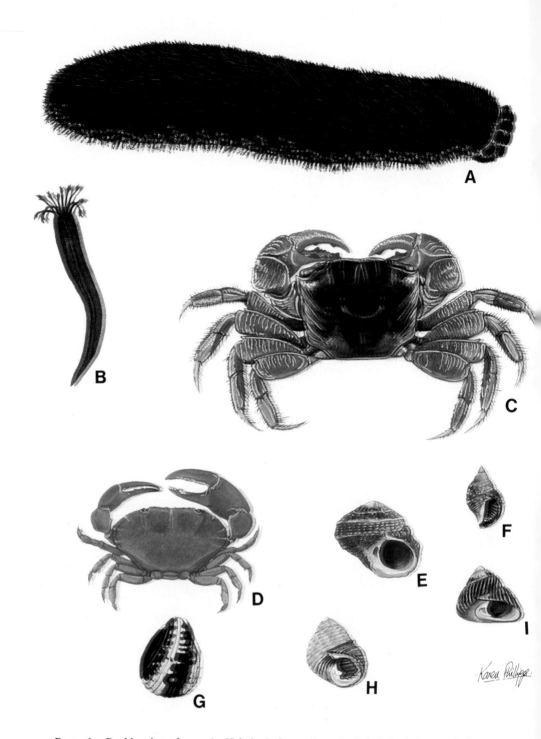

PLATE 6. Boulder shore fauna: A. *Holothuria leucospilota*; B. *Polycheira rufescens*; C. *Parasesarma pictum*; D. *Epixanthus frontalis*; E. *Lunella coronata*; F. *Planaxis sulcatus*; G. *Nerita albicilla*; H. *Monodonta australis*; I. *Chlorostoma nigerrima*.

Sheltered Rocky Shores

FIG. 5.5. Green and brown algal zonation on sheltered rocky shores. Above *(left)* at White Knights in Tolo Harbour and *(right)* Ma Wan.

ZONES:
A. *Balanus albicostatus*; B. *Enteromorpha compressa*; C. Oysters with *Ulva conglobata*; D. *Enteromorpha prolifera*; E. *Pylaiella*; F. *Colpomenia sinuosa*; G. *Porphyra suborbiculata*; H. *Tetraclita* and *Pollicepes*; I. *Ulva* sp.; J. *Scytosiphon lomentaria*.

ALGAE:
1. *Ectocarpus* (microscopic) with terminal sporangia; 2. *Pylaiella* with intercalary sporangia; 3. *Ulva lactua*; 4. *Scytosiphon lomentaria*; 5. *Ulva fasciata*; 6. *Enteromorpha compressa*; 7. *Rhizoclonium* sp.; 8. *Monostroma nitidum*; 9. *Enteromorpha prolifera*.

The sole littorine on this shore is *L. scabra,* but it accompanies patches of two species of high-zoned barnacles: *Euraphia withersi* and the minute, often crevice-dwelling, *Chamaesipho scutelliformis.* In the eulittoral zone a few *Nerita albicilla* persist. The thaid *Morula musiva* feeds on oysters at the same level. *Planaxis sulcatus* is abundant, but top-shells are absent. There is a single true limpet, *Patelloida pygmaea,* and small specimens of the pulmonate *Siphonaria japonica.*

At Ma Wan Island, between the mainland and the north-west tip of Lantau, the narrow channel (Kap Shui Mun) is swept by a strong tidal rip. Oysters and sheltered water balanoids are absent and in winter there is a strong *Porphyra* zone, with *Pollicipes* and *Tetraclita* in the eulittoral—the former in great abundance, the latter sparsely so. The algal diversity is greatly reduced and the winter dominants are two species generally regarded as pollution-markers: *Ulva lactuca* and the long tubular streamers of *Scytosiphon lomentaria.* This plant luxuriantly replaces *Sargassum* in those parts of sheltered, less saline harbours that are not so turbid as to inhibit algal growth.

Deep Bay

The shore illustrated at Tsim Bei Tsui, in full shelter and very low salinity (Fig. 5.4) has the shortest species list and the simplest zoning pattern of any place visited. Salinity falls to as low as $1^0/_{00}$–$2^0/_{00}$ in mid-summer. The upper shore crevices are colonized by small mangrove plants notably the young droppers of *Kandelia candel* and interspersed amongst them are tufts of the sea-grass *Zoysia sinica.* The rock surface is buried with mud up to the bottom of the oyster level, and the adjacent flats are probed by reef herons and white egrets. Oysters are still dominant, but are only thinly spread. Upon their shells around and between them grow the only two zoning algae, the small red species *Caloglossa leprieurii, Catenella nipae* and a species of *Polysiphonia.* All are also common on the trunks and branches of adjacent mangroves.

Above the oysters is an upper eulittoral belt of *Euraphia withersi,* the small barnacle typical of low salinity. This species, unlike the closely related *Chthamalus malayensis* on exposed shores, is confined to places of high shelter, such as Deep Bay, Castle Peak and the high reaches of Tolo Harbour. It is also found on the stems and leaves of the mangrove *Aegiceras corniculatum* at Tsim Bei Tsui. Still another barnacle, *Balanus pallidus,* is found at oyster level, with a very restricted distribution in high shelter and low salinity. Its only recorded habitat in Hong Kong is Deep Bay, in the brackish waters from the Pearl River. The few remaining gastropods of this shore include *Nerita undata, Littorina scabra* and *L. melanostoma.* The shell-less pulmonate *Onchidium verraculatum* can be found browsing on the upper levels of the shore. There are a number of species of *Onchidium,* as yet un-named, on Hong Kong shores. *Onchidium verraculatum* is the largest and possesses a brilliant blue foot and under-belly. It is also more typical of rocky substrates, the other species being more common on sand or mud.

MODERATE EXPOSURE WITH LOW SALINITY

The open coast line at Tai Lam Chung near Castle Peak is an unusual stretch, combining low salinity with moderately strong wave exposure. It helps to untangle the generally combined effects of brackish and sheltered conditions. Castle Peak has a number of characteristics of wave-exposed shores. Though far from oceanic waters, it has no oysters (though rock oysters may settle sporadically in spring when salinities are low), these being replaced by the typical exposed shore barnacle *Tetraclita squamosa* with *Pollicipes mitella*. The bivalves, *Septifer bilocularis, Barbatia virescens,* and the anomiid *Anomia achaeus* are also found in the middle and lower eulittoral. Below the barnacles is a dull pink zone of basal *Corallina* with patches of coralline turf *(Jania)*. In the winter months a turf-like assemblage of green algae including *Rhizoclonium* and *Chaetomorpha* spatters the middle shore. Two thaids, *Thais clavigera* and *Morula musiva* are plentiful. Other gastropods include *Nerita undata, Littorina scabra,* and, in addition, open-shore periwinkles *Nodilittorina millegrana* and *N. pyramidalis* here zoned high because of the presence of rock pools. The only limpet remaining at this low salinity is *Siphonaria japonica*.

FROM EXPOSURE TO SHELTER

The set of shores surveyed, with sloping, smooth rock faces, give a typical spectrum for the whole of Hong Kong. They show a simple situation, generally uncomplicated by pools or boulders; and illustrate the action of waves and tides in controlling and modifying a basic theme of zonation.

'Wave exposure' must cover a whole complex of physical effects that are still only partly understood. First, wave action will obviously increase the frequency and duration of wetting. At each wave break, swash flows up the shore to a level determined by the properties of the wave (height, length and velocity) and also by the shore profile and the wind direction. Increasing exposure involves increases in both wave force and wind, with predictable rises in the reach of swash, splash and spray.

Important, and potentially damaging, effects of waves include abrasion, pressure and drag. Abrasion is greatest where organisms are scoured by the sand and shingle held in suspension. Flexible algae will be damaged, too, when the waves lash them against the rock. Moving water exerts hydrostatic pressure that will be augmented by the air pressure built up in a hollow wave face. This, however, will act not directionally but uniformly, upon the bodies of plants and animals, and these—being water-filled and incompressible—will suffer little damage. The greatest wave hazard is likely to be drag, which is the frictional force of water over the organism and the dynamic pressure exerted in the direction of the flow. Drag effects will increase in magnitude as a plant or animal grows bigger and presents more resistance.

The upreach of surge, splash and spray is responsible for raising the levels of the successive zones, so increasing their vertical extent, passing through barnacles, littorines and lichens up an exposed shore. The species composition also changes further into shelter; not only do the zones become more compressed, but the pattern of organisms is markedly different. The diagram of barnacle distribution in Fig. 5.6, following the fortunes of the Hong Kong species from the most surf-beaten to the most sheltered parts of the coast, gives an example of such a trend.

A similar picture of the open coast bivalves (Fig. 5.7) reveals changes in the distribution of the dominant species. Thus the euryhaline *Saccostrea cucullata* dominates the shores to the west, with the stenohaline *Septifer bilocularis* occurring in the east. Between these two extremes, the byssally-attached *Anomia achaeus* and *Perna viridis* each occupy specific regions, with the ubiquitous *Barbatia virescens* able to colonize a wide range of shores.

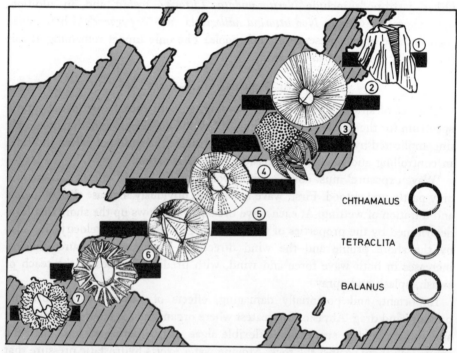

FIG. 5.6. The distribution of barnacles from shelter to exposure (from Tai Tam to Cape D'Aguilar).
1. *Balanus tintinnabulum volcano*; 2. *Tetraclita squamosa*; 3. *Pollicipes mitella*; 4. *Balanus variegatus variegatus*; 5. *Balanus amphitrite amphitrite*; 6. *Balanus albicostatus albicostatus*; 7. *Euraphia withersi*. A detail of the numbers and fusion of the valves of the principal genera are also given.

Sheltered Rocky Shores

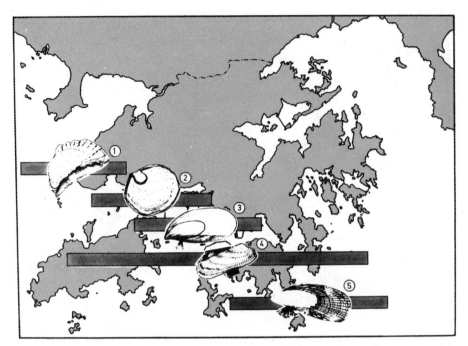

FIG. 5.7. The distribution of bivalves from exposure to shelter, schematically represented for the whole Hong Kong coast. Note that some sheltered species, such as the oyster, will be found also on coasts predominately exposed, in local situations of increased shelter.
1. *Saccostrea cucullata*; 2. *Anomia achaeus*; 3. *Perna viridis*; 4. *Barbatia virescens*; 5. *Septifer bilocularis*.

The exposure-shelter diagram in Fig. 5.8 is essentially a plot of vertical heights and composition of the zones, against the degree of exposure, measured along the horizontal axis. It would approximate also to a picture of an actual transition of shores. There is no comprehensive physical quantification of 'exposure', and the organisms and their zonal heights have been used as exposure indicators with an arbitrary horizontal scale. A biologically defined scale of 'exposure–shelter' is thus produced, by the arrangement of data from actual shores. On the same diagram, the ranges of the six littorine species have been superimposed.

The Hong Kong littorines are perhaps the best examples of how the species composition of a discrete group of animals can change with variations in shelter. Dr D. S. Hill,[2] has however produced a similar picture for Hong Kong's nerite snails (Fig. 5.9). Generally the genus *Nerita*, with eight species, occur on boulder shores at a range of vertical heights. *Nerita albicilla* is the only species regularly to be found on exposed shores. With decreasing salinity (and exposure), how-

[2] Hill, D. S. 1980. The Neritide (Mollusca: Prosobranchia) of Hong Kong. In *The Malacofauna of Hong Kong and Southern China,* ed. B. S. Morton, pp. 85–99. Hong Kong, Hong Kong University Press.

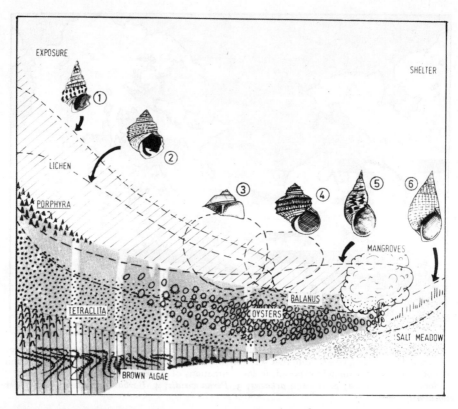

FIG. 5.8. Exposure-shelter diagram for Hong Kong shores. The range of the six littorine species are superimposed.
1. *Nodilittorina pyramidalis*; 2. *Nodilittorina millegrana*; 3. *Peasiella* sp.; 4. *Littorina brevicula*; 5. *Littorina scabra*; 6. *Littorina melanostoma*.

ever, other genera of the family appear on boulder beaches. In mangroves, as will be seen later, *Nerita lineata* can be found nestling in the crevices of the stems, with *Dostia violacea*, higher zoned on the mud. The genus *Clithon* dominates where streams drain on to sand beaches while *C.* cf. *retropictus* is a truly fresh water species. Thus the family Neritidae nicely demonstrates how one group of algal browsers have partitioned the aquatic environment under the dual influences of exposure and salinity.

Some of the physical components comprised within the useful but imprecise term 'exposure' could be measured. From a knowledge of wind directions, and with the use of a large-scale map 'exposure characteristics' can be established for a number of geographical sites. A provisional formula for wave action would be the number of days per hundred that the wind blows into the exposed aperture of the locality, this 'open-sea angle' being measured at a distance of 0.8 km.

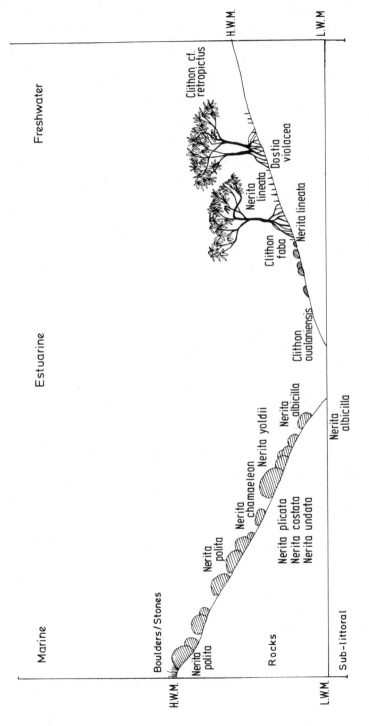

FIG. 5.9. Exposure-shelter (marine-freshwater) diagram, showing the ecological distribution of Hong Kong's Neritidae. (after, D. S. Hill, 1980).

Thus, on a straight shore—with an aperture, that is of 180°—and with wind direction uniformly distributed, there will be an exposure factor of 50. Not only should the direction and frequency of the wind be considered, but also its strength and the length of fetch over which it has generated waves. Full oceanic waves, it is generally estimated, can be developed over a distance of 320 km.

Measurement

Without too much expensive equipment, the height and force of incident waves can be ascertained on the shore. As a simple field exercise, a wave amplitude, i.e. half its distance from trough to peak, can be measured by holding up a graduated pole as the wave passes, and reading off values.

Wave-drag can be measured as the force exerted on a suitably mounted spring dynamometer suspended in the zone of wave impact. 50 lb spring balances have a circular disc of 2 sq. in. area attached by a line, to function as a drogue on which moving water acts to extend the spring. The maximum value is recorded by a pivoted arm remaining in the extreme position to which it had been pushed by the balance pointer (for details, see Jones and de Metropoulos[3]).

A second type of apparatus for measuring wave actions has been developed by J. G. Field.[4] The instrument is basically a Savonius-type current meter in which four rotating cups are held horizontally between two discs of perspex. Located at the end of a long pole, the apparatus can be placed at a variety of positions on the beach. The surging waves revolve the cups (in much the same way as the wind revolves an anemometer) and their number of revolutions per unit time can act as an indicator of wave strength.

By direct observation it can be determined how far a wave of ascertained height can wet or inundate a particular sloping shore. At a given wave height, the upreach of surge or swash could be compared upon sloping and vertical faces, in relation to the actual water level on a free-standing graduated pole. Special note should be taken of the surge up channels and narrow fissures, with its effect on the distribution of organisms.

For a given wave impact, the upward extent of splash and spray could also be measured. Absorbent pads could be exposed at regular distances up the supratidal shore, for determination of sodium chloride gain, or weight increase by water absorption.

Instructive comparisons can be made where two adjacent shores present large differences of exposure, as on the two sides of a breakwater or on exposed and more sheltered sides of Cape D'Aguilar.

[3] Jones, W. E. and de Metropoulos, A. 1968. Exposure to wave action: measurement of an important ecological parameter on rocky shores on Anglesey. *Journal of Experimental Marine Biology and Ecology* 2: 46–63.

[4] J. G. Field, 1968. The 'Turbulometer'—an apparatus for measuring relative exposure to wave action on shores. *Zoologica Africana* 3: 115–18.

CHAPTER 6

Wharf Piles

Wharves, embankments and groynes, mostly concrete but a few still of wood, make up a large fraction of Hong Kong's inter-tidal surface. The whole of Victoria Harbour, on both Hong Kong and Kowloon sides, is reclaimed and artificially embanked (Fig. 6.1). The natural communities follow the zonation of other sheltered shores, with rock oysters, periwinkles and the barnacles *Balanus amphitrite* and *B. reticulatus* and *B. variegatus,* as well as *Tetraclita squamosa.* Most reclaimed shores are, however, 'finished off' with a smooth veneer of cement which considerably reduces the number of niches available for settlement and thus sharply restricts the diversity of the flora and fauna that colonizes it.

Much of the Harbour is ecologically degraded and unrewarding for study. But in less polluted waters wharf piles offer many advantages for experiment. Their surfaces are smooth and uniform with a well-graduated exposure to insolation, illumination and water movement. Various combinations of these factors can thus be selected more simply than on the complex topography of a rocky shore. Wharf piles can thus form experimental islands in which a given species can be encouraged or excluded, or its predators held in check. Living space with any desired effects can be provided on long-and short-term settlement plates.

Though wharf piles can produce a zonation like a rocky shore, they have the important property of being vertical, with less exposure of a unit area to light and heat. Illumination can be reduced also by turbidity. Not only is algal growth restricted, but light-sensitive species that on natural shores are sub-littoral, can range up on wharves between tide-marks.

Fig. 6.1 shows typical members of the macrofauling community in Hong Kong. For a metre above low water there is a shaded 'hypofauna' that would elsewhere have to be looked for under stones. Such sites are rich in coelenterates (hydroids and anemones), ascidians, polyzoans and sponges; and these have luxuriant growth forms not achieved in the constricted space under boulders.

The upper part of the eulittoral zone is clad chiefly with barnacles and mussels. Large *Balanus amphitrite,* typical of sheltered waters, are especially common on wood. At some localities, *B. amphitrite, B. reticulatus, B. variegatus* and *B. trigonus*

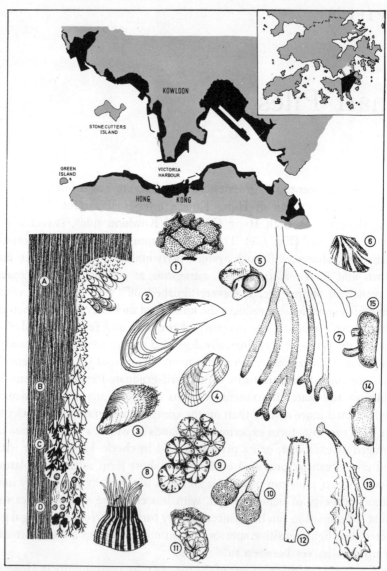

FIG. 6.1. Zonation on wharf piles. *(Above)* Victoria Harbour showing the extent of reclaimed or concrete embanked wharf piles. *(Below)* A wooden vertical surface in relatively clean conditions.

ZONES:

A. *Perna viridis* with *Balanus amphitrite amphitrite*; B. Ulvoid green algae; C. *Codium* with *Anomia achaeus*; D. Ascidians, *Bugula* and *Colpomenia*.

SPECIES:

1. *Iodictyum axillare*; 2. *Perna viridis*; 3. *Modiolus agripetus*; 4. *Ryenella cuprea*; 5. *Lunella coronata*; 6. *Electroma japonica*; 7. *Codium cylindricum*; 8. *Haliplanella luciae*; 9. *Botryllus schlosseri*; 10. *Eudistoma*; 11. *Styela plicata*; 12. *Ciona intestinalis*; 13. *Styela clava*; 14. *Ascidia sydneiensis*; 15. *Herdmania momus*.

are also found. The chief mussel is the long, smooth *Perna viridis,* somewhat parallel-sided and up to 8 cm long. Larger than *Septifer bilocularis,* its palest, jade-green forms are found sub-tidally and often appear in the markets. *Septifer* can be distinguished by its retention of an anterior adductor muscle lost in *Perna* located on a septum. In addition to *Perna viridis,* the common bivalves include *Barbatia virescens, Trapezium liratum* and the mussels *Musculista senhausia,* and *Modiolus agripetus,* all byssus-fixed. *Ryenella cuprea* is a small nut-shaped mussel with the special habit of embedding in ascidians. Another distinctive bivalve with byssal attachment is *Electroma japonica,* a thin-shelled member of the Pteriidae, related to the wing shells and the pearl oysters (p. 100). Rock oysters *Saccostrea cucullata* are frequent too, along with several species of *Chama.* In crevices, amongst the oysters, may nestle the small arcid *Striarca afra,* and the mis-shapen venerid *Claudiconcha japonica.*

The principal algae (early March) are large, well grown green *Ulva conglobata* and threads of *Enteromorpha prolifera.* The thread-like brown alga *Rhizoclonium riparium* is a common colonizer of wharf surfaces in Victoria Harbour. *Codium cylindricum* is recognized by the firm but velvety texture of its dark green, dichotomous branches. At low water and beyond grows *Colpomenia sinuosa,* also forming a seasonal band on nearby cobble beaches.

Wharf piles are the richest habitats for sessile and encrusting life forms, of almost every kind. But their diversity is greatest outside Victoria Harbour in places freest from pollution. The dark-green, red-striped anenome *Haliplanella luciae* is common in the locality figured. The species is highly polymorphic and the stripes may be red or yellow or even be absent, giving the impression of a range of species.

The Ascidiacea or sea-squirts come to their greatest abundance on shaded piles. The smallest individuals are massed in the bright-coloured crusts of the two colonial species, *Botrylloides perspicuum* and *Botryllus schlosseri.* Of the simple ascidians, five species can regularly be found on wharves and rafts. *Ascidia sydneiensis* has a smooth, spherical test, pale and translucent. *Ciona intestinalis* hangs vertically from under horizontal surfaces; the test is long and tubular, greyish-white and tinted with yellow spots around the siphon tips. *Styela plicata* is thick, tough-walled and rugose, attached by a narrowed base; while *Styela canopus* is squat, 1–2 cm high, carrying a leathery brown and heavily wrinkled test, with root-like extension posteriorly. *Herdmania momus* is russet-brown in colour with a thick test and very long siphons.

Polychaete worms should also be mentioned among the sessile organisms, and most typically of all, the serpulids with their calcareous tubes: *Pomatoceros triqueter, Hydroides elegans* and *Spirorbis foraminosus.*

Hydroids and Polyzoa

Encrusting colonial invertebrates play a large part in the communities of wharf piles and of shaded sub-littoral habitats. Important space occupants are sponges, hydroid coelenterates and Polyzoa. The knowledge of each is still very rudimentary for Hong Kong shores but fortunately many of their species are cosmopolitan members of the fouling communities, and they are known from studies in Japan. Field observations locally have already disclosed the genera present that will repay further study. The sponges are discussed when considering the fauna under boulders (p. 106).

Hydroids

The hydroid polyps, the smallest and simplest of the coelenterates, belong to the order Hydroida. Though they represent only the sedentary stage of a life history that includes a medusa (free-swimming or more rarely attached) these polyps are the only examples of their group that the student of shore biology is likely to find. All the hydroids are micro-carnivores, feeding by circlets of wide-ranging tentacles armed with stinging cells or nematocysts. They are by no means confined to the habitat beneath boulders, indeed many families are commoner epizoic on algae, and the most luxuriant colonial species frequent shaded wharf piles, at low tide or in the shallow sub-tidal.

The inter-tidal members of the Hydroida, as classified on the basis of the polyps, fall into two great series: the Thecata with a horny perisarc forming protective cups (hydrothecae) around the trophozooids (polyps) and the gonothecae round the modified reproductive polyps; and the Athecata, where the perisarc stops short at the base of the naked polyp. The characters of the principal genera likely to be encountered are identified in Table 6.1, and in the pictorial key (Fig. 6.2).

Polyzoa

The Polyzoa (Fig. 6.3) fall into two groups which are only superficially alike: the Entoprocta and Ectoprocta. The first is only briefly mentioned—note the small erect polyps of the genera *Pedicellina*, *Barentsia* and *Loxosomella*, the latter being found commensally on the crab *Pinnixa rathbuni* (p. 204). The second larger group, the Ectoprocta, is common and conspicuous, nowhere more so than on shaded piles and rafts and ships' bottoms, at or beyond low water. They form colonies of many shapes, ranging from calcareous or fleshy crusts and tufts, with erectly branched systems. All are built up of small chitinous boxes or 'zooecia', assembled in various ways, but each housing a single retractible polypide. Despite its polyp-like appearance, this is a minute but highly organized coelomate metazoan, feeding by means of a lophophore, a simple coil of ciliated tentacles surrounding the mouth.

TABLE 6.1 CLASSIFICATION OF HYDROIDA

I. Thecata

1. Campanulariidae: hydrothecae borne on pedicels, wide and bell-shaped. Gonothecae narrow-mouthed vases.
   Obelia: Hydrothecae on upright sympodially branched stems springing from a stolon.
   Gonothyrea: Monopodially branching, as in *Obelia*; margins of cups serrate.
   Clytia: Single hydrothecae rising from the stolon and rim-toothed.
   Orthopyxis: Hydrothecae direct from the stolon, with very thickened walls.
   Campanularia: Hydrothecae direct from the stolon with ridged walls and a toothed margin.
2. Campanulinidae: Creeping stolon and simple or branched stem. Cups slender and conical, with an operculum of four converging serrations.
3. Lafoeidae: Long tubular cups from a branched stolon; epizoic on *Synthecium* and *Sertularella*.
4. Haleciidae: Cups direct from stolon, or on sympodial branches, simple saucers with a flared margin, and sometimes in a pile of successive daughter hydranths.
5. Syntheciidae: Hydrotheca curved tubes, fused basally and in opposite pairs to the stem. Gonotheca arises from the interior of a cup.
6. Plumulariidae: Cups born along one side only of pinnate branches coming from a main stem; with small accessory cups (nematothecae).
7. Sertulariidae: Hydrothecae curved sessile tubes with a closing lid (operculum) of 1–4 flaps. In close series to give the stalk a serrated appearance.
   Sertularia: Opposite inserted hydrothecae.
   Sertularella; *Symplectoscyphus*: Short stems with zig-zagged alternately inserted cups.
   Amphisbetia: Long bear-like stems with oppositely inserted cups.

II. Athecata

1. Bougainvilliidae: Branched colonies with polyps like *Obelia* but with reduced cups.
2. Eudendriidae: Branched colonies, often epizoic on algae. Perisarc finely ringed and stops short at hydranth base. Gonophores produce fixed gonads on polyp body.
3. Hydractiniidae: Encrusting shells and rocks, with erect trophozooids and mouthless gastrozooids from a plate-like perisarc.
4. Clavidae: Filiform tentacles scattered over the hydranth rather than arranged in a ring.
5. Corynidae: Scattered capitate tentacles often in seaweed drift.
6. Tubulariidae: Hydranths large and pink with filiform tentacles arranged in proximal and distal rings.
7. Pennariidae: Both filiform and capitate (clubbed) tentacles, with colonies in feathery pinnate branches. Especially common on wharf piles.

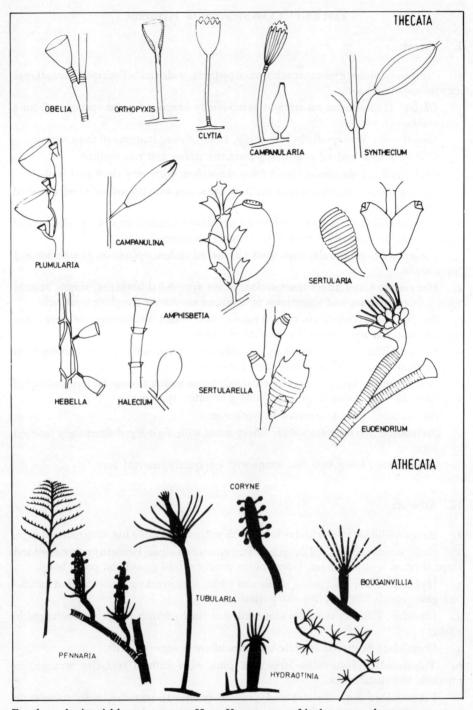

Fig. 6.2. A pictorial key to common Hong Kong genera of hydrozoan coelenterates.

Wharf Piles

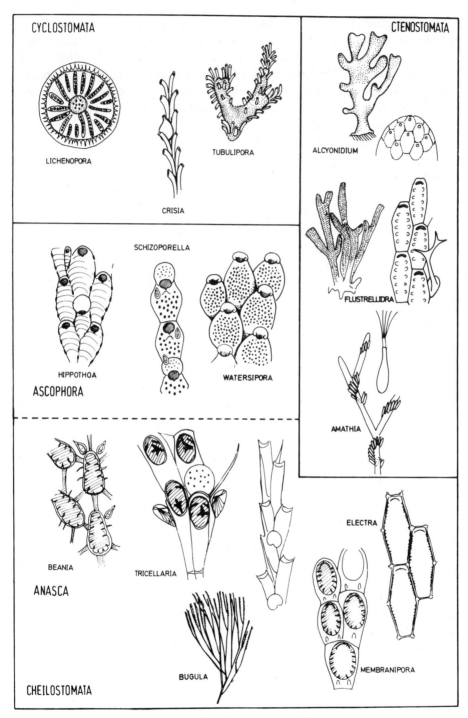

FIG. 6.3. A pictorial key to common Hong Kong genera of Polyzoa.

The ectoproct Polyzoa (Bryozoa) contain three groups. The largest are the Cheilostomata, found commonly on wharf piles, under boulders and epizoic upon algae. They are frequently polymorphic, with modified individuals called 'avicularia' and 'vibracula'. The latter are simple sweeping and guarding bristles, but the avicularia form complex bird's beaks, with snapping jaws, reduced to appendages of the main colony and serving a protective role. The unmodified polypides each have an operculum covering the zooecial mouth. The Cheilostomata are in turn divided into two sub-groups, the Ascophora and the Anasca. The first are the most fully calcified, appearing as pink, white, or occasionally brightly coloured crusts under stones. They often carry a beautiful microscopic sculpture or ornamentation of spinules. *Schizoporella unicornis* and *Hippothoa hyalina* are illustrated as common local examples. The aptly but falsely named lace-coral *Iodictyum axillare* (Reteporidae) forms brittle, pink, foliose masses resembling, at first glance a coral. Its polyzoan nature is, however, apparent, when a fragment is viewed under the microscope. The Anasca have the top of the zooecium uncalcified, and may grow in arborescent tufts. The family Membraniporidae have an encrusting habit spreading white lacework sheets over hard surfaces or the fronds of algae. *Membranipora* has zooecia forming elongate hexagons, free of ornament; but in *Electra angulata* the membranous frontal area is finely pitted and flanked with minute spines.

The principal branching family is the Bicellariidae, forming erect bushy colonies. The zooecia are long narrow boxes, placed side by side, two or more in a row. The commonest Hong Kong species on wharves and floats (but also found sub-tidally on boulder-strewn beaches) is *Bugula neritina*, with long slender dichotomous branches. The species is most abundant in the winter when it forms large, foliose colonies. Amongst its branches dwell a wealth of smaller animals and it is eaten by the polycerid nudibranch *Polycera fujitai* (Fig. 12.16) with brown marginal horns. As in all this family, the avicularia and vibracula are prominent. The related family Scrupocellariidae includes a fouling species common in Japan, *Tricellaria occidentalis*, forming smaller and more compact, greyish-white clusters amongst *Bugula*.

The *Beania* species, belonging to the same family, form a foliose reticulum, with the zooecia linked to their neighbours by cross-branches and the free surface covered with dense spinules.

The second class, the Cyclostomata are the most strongly calcified of the Ectoprocta. The zooecia are long slender tubes, opening terminally without the operculum found in cheilostomes. The *Crisia* species have a bushy habit with the zooecia long, erect and curving outwards at the apertures. Each branch is several zooecia thick and has at its base a swollen balloon-like zooecium forming an incubating chamber. *Tubulipora* colonies are by contrast prone and spreading, with the zooecia fused to each other and only their tips standing out somewhat

PLATE 7. Open sand beach at Tong Fuk, Lantau Island.

PLATE 8. *(Above)* The protected flats at Shiu Hau, Lantau Island and *(Below)* Tai Tam Bay, Hong Kong Island.

from the compact mass. *Lichenopora* grows in a plate-like colony similar to a minute inverted mushroom with the zooecia pointing radially outwards.

Unlike the other ectoprocts, the Ctenostomata are wholly uncalcified, and form either soft crusts and lobules, or slender stolons. The more massive and fleshy ctenostomes include *Alcyonidium*, forming thick branching and gelatinous crusts, with the hexagonal zooecia deeply sunken; and *Flustrella*, with the apertures bounded by double closing lips fringed with spines. The stolonate ctenostomes include *Amathia*, with fleshy stems around which the tubular zooecia are spirally disposed, and *Bowerbankia* where the zooecia arise from a creeping stolon.

Wood-borers

Of particular note upon timber piles and rafts are the small series of crustaceans and bivalves that are specialized for boring wood and ingesting its fragments as part of their food (Fig. 6.4). First a common amphipod of wharf piles, *Erichthonius pugnax*, does not bore, but forms vertical tubes of the fine sediment that may amass undisturbed on horizontal piles. With their prolonged antennae, fringed with setae, these crustaceans filter fine particles from the water current activated through the tube.

Distributed originally by wooden ships, several crustaceans that actually bore timber now range almost world-wide. The isopod *Limnoria tripunctata* causes most damage, making extensive burrows along the grain of the wood, perforating these with respiratory holes at intervals. Boring is undertaken by the mandibles, which form an asymmetrical pair working in concert, the left a rasping file and the right strongly pointed. *Limnoria* is able to digest cellulose, but may subsist also on microscopic fungi and bacteria from the rotting wood. The amphipod *Chelura terebrans* does less damage itself, living chiefly in the layers already bored by *Limnoria*, whose tunnellings it enlarges. It may also be found nestling in the empty shells of barnacles, along with another isopod, *Sphaeroma walkeri*, only recently introduced into Hong Kong.

The most specialized of all timber-borers and the most aberrant of the bivalve molluscs are the ship-worms of the Pholadacea. The tunnelling habit in bivalves is initiated by the pholads (Pholadidae) which will be found later (p. 124) in soft sedimentary rocks, but are represented in timber piles by the small *Martesia striata*. The true ship-worms (Teredinidae) are much further evolved than the pholads. Attenuated and vermiform, they consist almost wholly of a pallial siphonal tube, up to 10 cm long, with their bivalve character largely confined to the small anterior body and the diminished shell. The foot is a small attachment disc with which the animal adheres to the end of its burrow; boring is continued by the chisel-like margins of the shell, as the animal is rotated by asymmetric contractions of the pedal retractor muscles. The valves have no ligament, but

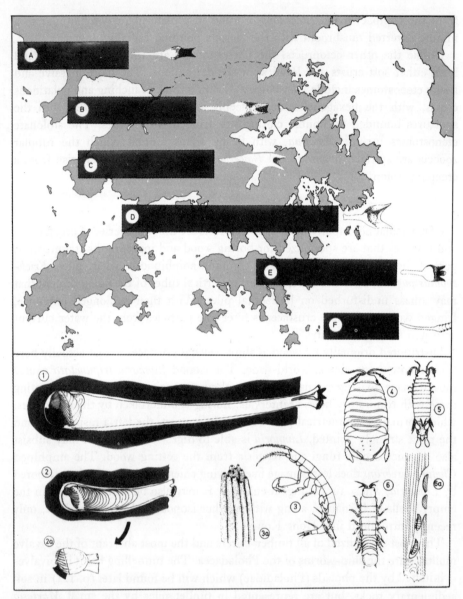

FIG. 6.4. *(Above)* Distribution of ship-worm species from west to east in Hong Kong with detail of their pallets.

SHIP-WORMS:
A. *Lyrodus pedicellatus*; B. *Lyrodus tristi*; C. *Teredo navalis*; D. *Teredo furcifera*; E. *Lyrodus medilobatus*; F. *Bankia carinata*.
(Below) 1. *Teredo furcifera in situ* in burrow showing extent of siphons and shell; 2. *Martesia striata in situ* showing adult condition with shell extension (callum) over the pedal gape; and 2a. young specimen with foot active; 3. *Erichthonius pugnax*, with 3a. mud tubes; 4. *Sphaeroma walkeri*; 5. *Chelura terebrans*; 6. *Limnoria tripunctata*, with 6a. detail of limnorid burrows with respiratory apertures.

rock upon each other from side to side by adductor contractions at a dorsal and ventral point. The naked tube is divided horizontally into inhalant and exhalant chambers separated by the gill (ctenidium). Food is obtained by filtering from the inhalant current and by the ingestion of wood fragments that are attacked by cellulose-digesting enzymes. The bases of the siphons are attached at the opening of the burrow, and the animal grows longer as the tube is enlarged. The withdrawn siphons are guarded by two shelly plates ('pallets') with distinctive sculpture which assist in separating the species. The distribution of these destructive borers in Hong Kong's waters has been studied by L. F. Fung and B. S. Morton[1] who have shown that each species has a limited distribution from exposed shores of high salinity in the east to shelter and brackish waters in the west.

---

[1] Fung, L. F. and Morton, B. S. 1976. Competition between limnoriids and shipworms in the coastal waters of Hong Kong. In *Proceedings of the IV International Congress on Marine Corrosion and Fouling, Juan-les-Pins, Antibes, France 1976.*

CHAPTER 7

Boulder Shores

Not all hard shores form steep intact surfaces. The inter-tidal can be gently sloped and diversified with pools and movable rocks. With its igneous geology and primary coastline Hong Kong is poor in inter-tidal platforms and the rock pools that form there. Loose rocky cover is, however, very common. Hard angular blocks are dislodged from their softer matrix by erosion of cliffs and backshore. These come under varying degrees of wave action. In the most exposed conditions, boulders become wave-rounded and smooth, being constantly mobile, or at least regularly wave shifted, according to their size.

Mobile shores can form ramps of several tiers on open coasts, their boulders grinding and rattling together at every high tide. Such boulder beaches are poor habitats for attached life. Wave swash is not transmitted far up shore, but seeps between boulders. The bare, low-pitched rock surface rapidly heats up in the sun. Smooth boulders neither retain water nor provide favourable surfaces for metamorphosing larvae to settle.

In Hong Kong most boulder shores are relatively permanent and immobile. The main physical variable of a boulder beach, that of stability, is obviously related directly to boulder size and inversely to strength of wave action. Around Stanley Peninsula, West Lamma and along the south coast of Lantau, the granite boulders are large and angular, bedding together with narrow inter-spaces, through which turbulent water moves constantly.

As well as size and stability, an important characteristic of a rocky shore is the substratum on which the boulders rest. This can vary from solid rock or other boulders, through pebbles, coarse or fine sand to mud. In the lighter sediments, boulders become firmly embedded; in the most sheltered inlets they can lie in anaerobic muds with a black ferrous sulphide content.

The faunas beneath boulders are undoubtedly the richest on the inter-tidal shore. But they have in the past usually been left out of tidy schemes of shore zonation. This must be in part due to the diversity and complexity that are revealed as soon as stones are overturned. All ideas of dominant species seem to break down. Sponges, ascidians, barnacles, polyzoans, tube-worms and molluscs appear to settle as opportunity affords. The patterns of life are far more like those rich communities forming a regular part of the sub-tidal, though normally seen only by divers. Such types of association can move up into the inter-tidal zone chiefly as sheltered enclaves beneath the boulders, taking advantage of the low illumination and temperature and freedom from the desiccation hazards elsewhere experienced between tides.

The top and sides of a single boulder, with their differences of aspect and insolation, often show a microcosm of the zoning over the entire bed-rock shore; and a regular succession of organisms can be detected, despite the complexity of boulder communities. Underneath the boulder, barnacles and tube-worms are the first species to arrive, generally followed by polyzoans, oysters and *Chama* species. Simple ascidians usually come next. The mature community beneath a long-stable boulder is finally dominated by sponges, bivalve molluscs and immobile gastropods, together with their predators. The climax of this succession is reached only by large and stable boulders at low water. Where boulders or cobbles are smaller, or frequently overturned by waves, or subject to regular emersion in the mid-littoral, the accumulation of species is held back at an earlier stage.

A Sheltered Boulder Beach: Tai Tam Harbour

Unlike its more open Bay, Tai Tam Harbour (properly so-called) is highly sheltered, with the bottom covered with sediment at low tide. On the boulder-strewn parts of its shore, the stones are small angular pieces of rhyolite, seldom 30 cm across until the lower shore. The water is turbid, but transmits sufficient light to allow algal growth, giving a subdued mauve or green colouring to the middle and lower shore (Fig. 7.1).

The upper shore corresponds broadly with a littoral fringe, forming a high bank of small rhyolite slabs. With reduced wave splash more terrestrial plants can invade the sea shore zone. In oblique fissures some two metres above high water mark can be found *Antirrhoea chinensis, Atalantia buxifolia, Asparagus lucidus, Morinda umbellata, Phyllanthus cochinchinensis* and *Sagaretia theezans*, with *Ficus superba* and *Hibiscus tiliaceus* overshadowing them all. The sea shore fringe community is wettened by high tide and includes the grasses *Cynodon dactylon* and *Zoysia sinica* with *Paederia scandens* (Rubiaceae) and *Vitex rotundifolia* (Verbenaceae). On the pitted surfaces of the rocks and amongst the fringe grasses in the crevices are found the periwinkles *Littorina brevicula,* and, less commonly, *Nodilittorina pyramidalis* and *N. millegrana. Ligia exotica* and the fast-running sesarmid crab *Parasesarma pictum* are abundant beneath.

The middle shore, approximately the eulittoral zone, forms a more gentle slope. The tops of the boulders have 'scars' of the barnacle *Balanus albicostatus;* many of the stones are green-tinged with an algal film of *Gomontia* sp. worked

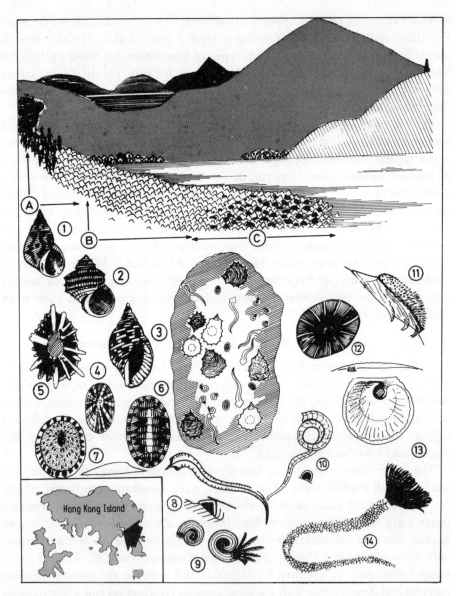

FIG. 7.1. Boulder and cobble beach in Tai Tam Harbour.
ZONES:
A. Upper shore with *Zoysia* grass backed by *Hibiscus*; B. Middle shore; C. Lower shore with *Colpomenia* and encrusting algae on boulder tops.
(Inset; middle) Mid-tidal boulder overturned to show oysters, chitons and tube-worms beneath.
SPECIES:
1. *Littorina scabra*; 2. *Littorina brevicula*; 3. *Planaxis sulcatus*; 4. *Patelloida pygmaea*; 5. *Siphonaria atra*; 6. *Lepidozona coreanicus*; 7. *Notoacmea concinna*; 8. *Pomatoleios kraussi*; 9. *Spirorbis foraminosus*; 10. *Hydroides elegans*; 11. *Chama reflexa*; 12. *Haliplanella luciae*; 13. *Anomia ephippium*; 14. *Sabellastarte indica*.

into their fine texture. The surface may also be mottled with two encrusting algae, the dull, red-black *Hildenbrandtia* and yellow-brown patches of *Ralfsia verrucosa*.

The lower shore (sub-littoral fringe) is composed on the whole of large boulders, 30 cm or more across. Much sediment collects beneath the boulders, but on the cleaner tops grow scattered *Saccostrea cucullata* (and *S. echinata?*) (see p. 66). Frequently the boulder tops are mottled with *Hildenbrandtia*, pink basal *Corallina*, or carry a stubble of *Gelidium pusillum*. In late winter the surface is transformed to golden hue by the maturing of the fleshy brown alga *Colpomenia sinuosa*.

Under-boulder Communities

Permanent sessile communities begin with the boulders of the eulittoral zone, and are composed chiefly of hard-shelled filter feeders. The commonest barnacle is a small, flat *Tetraclita* species up to a centimetre across. Serpulid tube-worms are also abundant. The largest is *Pomatoleios kraussii*, with a flattened tube, triangular in section, and with a median dorsal ridge that tapers to a spine over-hanging the aperture of the tube. The serpulid head carries a respiratory

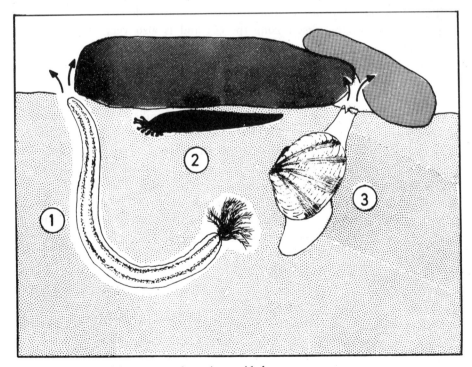

FIG. 7.2. A cobble lying on a sand matrix at mid-shore.
1. *Leptosynapta ooplax*; 2. *Polycheira rufescens*; 3. *Tapes philippinarum*.

fan of gill filaments or pinnae. These are highly sensitive when extended, and shadow or vibration will cause instant retraction. Each filament has fringes of ciliated secondary pinnules. Long lateral cilia draw a water current up between the pinnules to the centre of the crown. Food particles in this current are intercepted by shorter frontal cilia, to be carried down the pinnules and filaments to the mouth. One of the filaments on the left is modified into a stalked operculum which closes the tube when the crown is withdrawn.

A second boulder serpulid, *Hydroides elegans,* has a more slender tube than *Pomatoleios* and is cylindrical in section. The smallest and most numerous serpulids belong to *Spirorbis* species (e.g. *S. foraminosus*) with a tightly spiral coil, only 2 mm across. First to be found scattered under mid-littoral rocks, *Spirorbis* reaches right to sub-tidal levels. Its miniature crown functions like that of *Pomatoleios,* but has only five simple branchial plumes, with a sixth forming a hollow operculum within which the fertilized eggs are incubated.

The sabellid polychaete *Sabellastarte indica* similarly cements itself to the under-surface of boulders by building a thick membranous tube of mucus bound together with particles of mud and grains of sand. Its fan-shaped crown of brown tentacles filters the mud-laden waters but are quickly withdrawn when disturbed.

The anemone *Haliplanella luciae,* dark green with wide-spaced red lines, has been met with in the oyster zone (p. 56) and is also very characteristic of mid-littoral boulders.

These middle shore boulders have several characteristic bivalves. Small *Saccostrea cucullata* abound underneath, often bristling with channelled spines that exposed full-grown oysters have lost.

The Chamidae are one of the most important families of tropical bivalves. Looking superficially like oysters, they have evolved from a higher stock of bivalves, with the mantle margins prolonged into short siphons. Chamids are permanently cemented, the under valve being cup-shaped and the smaller upper valve forming a flat lid, both being spirally twisted. Three common Hong Kong species are *Chama reflexa,* bristling with thin lamellate spines; and the longer shelled *Chama dunkeri* (see p. 297) both attached by the left valve and *Pseudochama retroversa,* with a pronounced spiral, which is attached by the right side.

Ark-shells *(Barbatia virescens)* and saddle oysters *(Anomia achaeus* and, more commonly, *A. ephippium)* are also common, but can best be dealt with among the bivalves of the lower shore (p. 102).

The boulder gastropods of the middle shore will all by now be well known. The limpets are *Cellana toreuma, Notoacmea concinna, Patelloida pygmaea,* and, under clean conditions, *P. saccharina.* The common siphonariid limpet of such shores is *Siphonaria atra.* The top-shell *Monodonta australis, Planaxis sulcatus* and, at lower levels, *Lunella coronata* are each common. There are two abundant nerites, the depressed and black-streaked *Nerita albicilla* and *Nerita undata,* with strong spiral ribs, black and white speckling and yellow mouth. The common thaid is *Morula musiva.*

The small reddish brown chiton *Ischnochiton coreanicus* up to 2 cm long, is frequent under boulders at mid-tide level. It is separted from its smaller relative *I. lepidus* by longitudinal striations on the crests of the shell plates, which the later species lacks. R. A. van Belle[1] has produced a check list, with photographs, of some of Hong Kong chitons.

Crabs and their Habitats

Brachyura

Of all the animal forms adapted for the boulder habitat, crabs are the most distinctive. Just as molluscs, barnacles and algae are good indicators of different conditions on bed-rock, boulder and cobble shores can be classified by reference to their crabs. Great adaptive diversification is drawn from a basically uniform design.

Compared with a prawn or a lobster, the crab body has become obviously engineered for its special habitat. First, the long abdomen is almost lost, being reduced to a narrow flap, applied close beneath the thorax. The carapace is wide and dorso-ventrally compressed, generally wedge-shaped from front to back, for retreat backwards into narrow spaces. Crabs can move fast, backwards, or sideways but seldom forwards. The two pairs of antennae, forward-facing receptors in shrimps and lobsters, are hence reduced. The first thoracic legs have a pair of large claws (chelae) that can be brandished to cover a backward retreat. The other four pairs (pereiopods) are used for walking, though the nature and speed of gait vary greatly with different crabs.

Spending much time out of water, if seldom fully exposed to the sun's heat and drying power, crabs show respiratory adaptations graded to their level on the shore. Low-tidal species have a complete gill series, and full aquatic respiration. But in the high-tidal Grapsidae, and in the Ocypodidae, their counterparts on soft shores, where water circulation cannot proceed freely, part of the respiratory cavity comes to form a vascularized lung. In many species of the grapsid sub-family Sesarminae, the cheeks (the under-sides of the head outside the gill openings) are covered with setose ridges, through which small tides of water ebb and flow. This fluid is periodically withdrawn into the 'lung' after re-oxygenation outside.

The Grapsidae are pre-eminently the fast high-tidal crabs of sub-tropical and

---

[1] Van Belle, R. A. 1980. On a small collection of chitons from Hong Kong (Mollusca: Polyplacophora). In *The Malacofauna of Hong Kong and Southern China,* ed. B. S. Morton, pp. 33-35. Hong Kong, Hong Kong University Press.

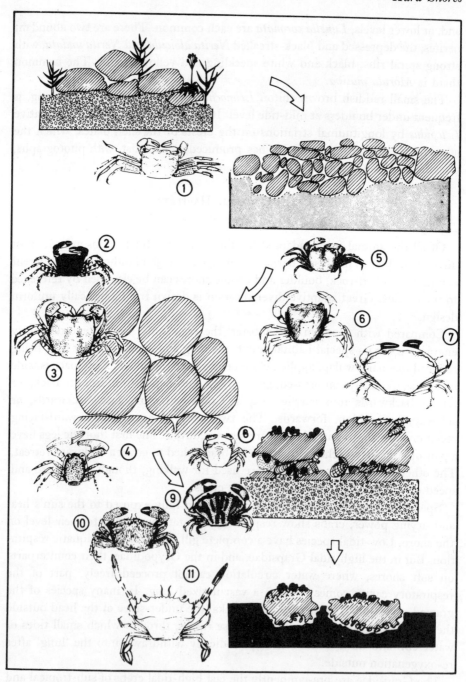

FIG. 7.3. The crabs of boulder and cobble shores.
1. *Parasesarma pictum*; 2. *Metopograpsus frontalis*; 3. *Hemigrapsus sanguineus*; 4. *Petrolisthes japonicus*; 5. *Cyclograpsus intermedius*; 6. *Gaetice depressa*; 7. *Epixanthus frontalis*; 8. *Nanosesarma minutum*; 9. *Sphaerozius nitidus*; 10. *Heteropilumnus ciliatus*; 11. *Thalamita prymna*.

Boulder Shores

tropical rocky shores. *Grapsus albolineatus,* seen already on exposed shores, has its typical habitat in crevices and recesses. There are comparable low-tidal grapsids, *Plagusia* and *Percnon* (p. 42).

On boulder beaches the high-level crabs are predominantly Grapsidae. The most numerous local group of crabs, unwieldy even to a specialist, is certainly the sub-family Sesarminae: swift grapsids with a sharp-cornered rectangular carapace and eyes set at the extreme corners. *Parasesarma pictum,* leaden grey and mottled with light brown, is commonest among high-tidal boulders, often along with the isopod *Ligia exotica.*

Living under clean, wave-rounded boulders at the same level (the typical home for its genus), is a moderately small crab, *Cyclograpsus intermedius.* Reddish-brown and smoothly polished, it is distinct from any other grapsid, by its rounded carapace edge and the lack of any lateral teeth.

Fast-running like sesarmids and *Cyclograpsus,* the smallish crab *Metopograpsus frontalis* is found at a slightly lower level, distinguished by its shiny black carapace, with straight sides, and the cross-barred, sparsely hairy legs.

At the next level down, in the middle eulittoral, live three very common Hong Kong grapsids. *Hemigrapsus sanguineus* belongs to a genus typical of middle-shore under-boulder habitats, in all warm temperate shores. Square-backed and with smaller legs than other grapsids, these crabs are still active runners. *H. sanguineus* has a greenish carapace, finely stippled with dark blood-coloured spots.

The grapsid *Gaetice depressa* is one of the commonest littoral crabs of Hong Kong. The carapace is less convex and almost flat across the back, and is camouflaged with a wide variety of colour patterns. Some individuals are blotched or marbled in dark brown, others are partly buff, rust-coloured or cream, on a ground colour of light grey. The tips of the chelae are white.

The smallest boulder crab, locally, is *Nanosesarma minutum,* hardly a centimetre in total span when adult. Leaden grey in colour, it is distinguished under a lens by the tuft of dense setae carried above the movable finger of each chela. Found under boulders in silty places, it is also prolific in oyster clusters and on wharf piles.

At the mid-tidal level are the family Xanthidae, especially important crabs in the tropics. The carapace is much wider anteriorly and bow-fronted, usually crenulate. The chelae typically have black fingers and in the male the right one is very unequally enlarged. Xanthids are characteristically short-legged and slow-moving. This is particularly so of the local common eulittoral species, *Epixanthus frontalis,* broad and elliptical, as smooth as a pebble and finely speckled in grey to reddish-brown. The small (left) chela of the male is exceptionally slender. The second eulittoral xanthid is *Sphaerozius nitidus,* also a slow crab, with a heavy right chela, and the carapace corrugated, and blotched in dark brown.

The crabs of the family Portunidae are generally round-fronted like the xanthids, but are lighter built and far more vigorous in their movements. The carapace front is sharp-toothed and the chelae, though slender, are strong and sharp-toothed. The fourth walking legs terminate in oval paddles used for active swimming. Some portunids are sub-tidal and live upon sedimented shores; they include some of the largest and best food species.

In Hong Kong markets, securely bound, can be found *Portunus sanguinolentus* with a grey-green carapace bearing three blood red spots; *P. trituberculata* which is a uniform dull green and *P. pelagicus,* perhaps the most visually striking with the male having a mottled carapace, brilliant blue walking limbs and purple chelae. Unusually, the female of this species is strikingly different with a dull brown carapace and chelae, but with the upper regions of the limbs still mottled, Juveniles of these species can be found inter-tidally. Members of the genus *Portunus* have the carapace drawn out into side spikes whereas in the second portunid genus *(Charybdis)* it is plain. *C. cruciata* is also found in the markets and the carapace is beautifully patterned in yellow and red with the paddles and chelae similarly colour-flecked. Other charybdids are common at low tide on boulder beaches and are aggressive, often unwelcome, finds. Five such species are illustrated in Fig. 7.4. A smaller portunid genus regularly represented by adults between tides is *Thalamita,* most frequently in the sub-littoral fringe. Straighter across the front than other portunids, *T. prymna* and *T. picta* are

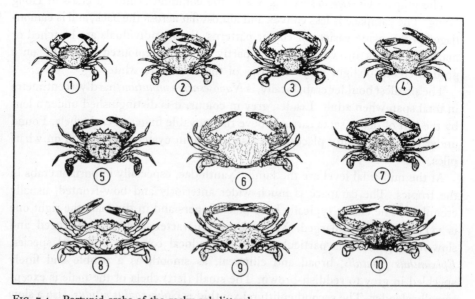

FIG. 7.4. Portunid crabs of the rocky sub-littoral.
1. *Thalamita picta*; 2. *Charybdis japonica*; 3. *Charybdis miles*; 4. *Charybdis truncata*; 5. *Charybdis bimaculata*; 6. *Charybdis cruciata*; 7. *Portunus gladiator*; 8. *Portunus trituberculata*; 9. *Portunus sanguinolentus*; 10. *Portunus pelagicus.*

greenish brown, with the chelae strongly spined and patterned, able to be fiercely brandished on alarm. They differ chiefly in the arrangement of the carapace spines. Commonest on muddy ground beneath stones *Thalamita* is, like all portunids, an active and aggressive crab.

The Dromiidae are slow-moving crabs effectively camouflaged with sponges or other organisms. One species, *Dromia dormia,* appears at low-water mark in late winter and early spring. The carapace is convex and sub-circular narrowing in front where it is covered like the legs with fine brownish hairs; but the principal feature of most dromiid species is a cap of living sponge or ascidian, cut out with the chelae and fashioned to fit the carapace closely. The cap is held in place with the reduced, upturned third and fourth legs, which are sub-chelate.

A second dromiid, *Conchocoetes artificiosus,* though living sub-tidally on shell gravel, must be mentioned here for the interest of its protective shield, an empty bivalve shell, held in place with the short fifth legs.

Anomura

Two common sorts of crabs on boulder shores belong not with the true crabs (Brachyura) but in another division of the decapod Crustacea, the Anomura, having a well-developed, though often asymmetrical abdomen.

The first family are the Porcellanidae (porcelain crabs or 'half-crabs'), ideally adapted crabs of the boulder habitat. The carapace is small and rounded, but a broad flat base given by the wide and prominently elbowed chelae. These are pressed close to the surface as the crab scuttles between boulders, or clings upside down to their under-sides. A common local species on cobble beaches is *Petrolisthes japonicus,* dull grey with the under-parts of the chelae pink. It displays well the key differences from a true crab.

The abdomen is longer and broad, and if the porcelain crab is placed on its back, can become mobile and used in swimming. A well-developed telson and tail fan, lost in the Brachyura, terminate the abdomen. The antennae are long and whip-like as in a prawn, and the walking legs are apparently reduced to three pairs. On close inspection a small fourth pair can be detected, folded so that their tips project into the branchial chamber, armed with tufted gill brushes. Finally, the third maxillipeds are large and finger-like, being rhythmically unfolded, either in unison or alternately. Their two terminal joints are fringed with long setae, intercepting food particles from the inhalant respiratory current. *Petrolisthes* is thus a microphagous filter feeder. When this crab is attacked, the chelae can be cast off (autotomized) and (still moving) attract the predator's attention while the crab makes its escape.

The squat lobsters (Galatheidae) are also widely distributed under stones in the eulittoral and the most commonly encountered species being *Galathea subsquamata.* In these anomurans the unreduced abdomen is carried under the body and the very long chelipeds are directed forwards.

The hermit crabs are far more specialized Anomura, with the well-known habit of living in disused gastropod shells (Fig. 7.5). The carapace is long, as in a crayfish, but strongly calcified only in its front half. The abdomen is altogether soft, being spirally curved, and attached to the axial pillar of the shell by the modified, asymmetrical telson and tail fan. The fourth and fifth walking legs are much reduced, the second and third remaining large, external and functional. The chelae are in some genera unequal, an enlarged and more bulbous one (right in *Pagurus*, left in *Calcinus*) serving as an operculum or stopper. In *Clibanarius*, both are equally developed. The antennae of hermit crabs are long as in porcelain crabs, and the third maxillipeds mobile and palpiform. In some species they collect particles as food, but most hermits feed actively on dead animal food, with the strong chelae.

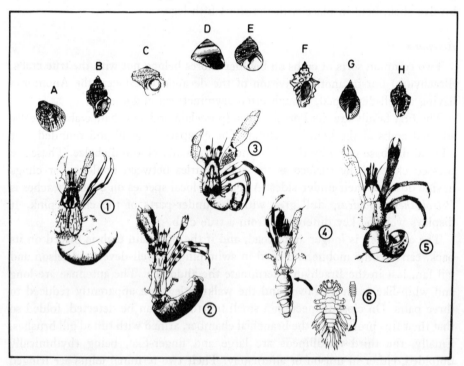

FIG. 7.5. The hermit crabs of rocky shores and their shells.

CRABS:
1. *Clibanarius striolatus*; 2. *Calcinus* sp.; 3. *Pagurus trigonocheirus*; 4. *Clibanarius bimaculatus*; 5. *Clibanarius virescens*.

ISOPOD:
6. *Pseudostegias setoensis*, larger female plus male.

SHELLS:
A. *Nerita albicilla*; B. *Bursa granularis*; C. *Lunella coronata*; D. *Chlorostoma rustica*; E. *Monodonta australis*; F. *Thais luteostoma*; G. *Planaxis sulcatus*; H. *Morula musiva*.

Boulder Shores

The genera of hermit crabs are distinguished by the relative sizes of the left and right chelipeds, and the species are commonly separated on the colour and ornamentation of the chelae and walking legs. The most abundant hermit crab of Hong Kong's rocky shores is *Pagurus trigonocheirus* with species of *Clibanarius*, especially *C. striolatus* also frequent.

A key to the hermit crabs most likely to be met with in Hong Kong has been constructed by Ms May Yipp and is given below:

TABLE 7.1 KEY TO THE COMMON HERMIT CRABS IN HONG KONG

(1) Chelipeds unequal . . . . (2)
    Chelipeds equal . . . . (5)
(2) Left cheliped larger . . . . (3)
    Right cheliped larger . . . . *Pagurus* sp. (4) [r]
(3) A movable process present between the eye-scales which are dorsal to the base of the eye-stalks . . . . *Diogenes edwardsii* [s]
    Process absent . . . . *Calcinus* sp. [r]
(4) Chelipeds bear blue spots . . . . . *Pagurus trigonocheirus* [r]
    Not as above . . . . Other *Pagurus* spp.
    (usually sub-tidal and rarely encountered on the shore)
(5) (a) 2nd and 3rd thoracic legs with longitudinal dull green and brown stripes . . . . *Clibanarius longitarsus* [s]
    (b) 2nd and 3rd thoracic legs of a homogeneous grey-black colour with 2 white rings in the last segment . . . *Clibanarius bimaculatus* [r]
    (c) as with (b) but without 2 white rings in the last segment
        Clibanarius virescens [r]
    (d) 2nd and 3rd thoracic legs with longitudinal blue and brown stripes
        Unidentified *Clibanarius* sp. [s]
    (e) 2nd and 3rd thoracic legs with longitudinal brown and orange stripes . . . . (6)
(6) Strong tooth present on the ventral surface of the cheliped merus [s]
    Clibanarius infraspinatus
    Tooth absent . . . . *Clibanarius striolatus* [r]

[r] rocky shores   [s] soft shores

Parasitic Isopoda and Cirripedia

The Isopoda belong to the sub-class Malacostraca of the Crustacea, that is they are associated with the higher crabs and shrimps. On rocky shores the omnivorous sea-slater *Ligia* has already been described and on wharf piles the wood-digesting *Limnoria*. Other isopods are, however, highly specialized parasites. Many of these parasites are out of the scope of this book, for example the blood-sucking ectoparasite *Nerocila phaeopleura* of sardines in the coastal waters of Hong Kong. Some of the most highly specialized isopod parasites, the bopyrids, are commonly encountered inter-tidally where they live upon crabs and shrimps. For examples of these animals, the student should look for unusual swellings under the carapace of shrimps. A recent study by Ms May Yipp of the bopyrid parasites of hermit crabs provides some nice examples of parasitism that the interested student could follow up. Attaching to the abdomen of *Clibanarius bimaculatus* is the distorted *Pseudostegias setoensis*. This blood-sucking parasite, lying unusually on its back, firmly grips the host's abdomen with overturned, hooked thoracic limbs and the mouth-parts. An extraordinary degree of sexual dimorphism is seen with the larger of a pair being female who is fertilized by a minute male hiding amongst her respiratory leaflets. On sandy shores *Diogenes edwardsii* is similarly parasitized by *Atheleges takanoshimensis*. *Diogenes* also possesses under the carapace the branchial bopyrid *Pseudionella pyriforma*. More reduced than the abdominal parasites this species too has a 'dwarf male' which must be sought for with care.

Not isopods, but highly specialized cirripedes related to the barnacles, are the rhizocephalan parasites of the genus *Sacculina*. These parasitize decapod crustaceans such as shore crabs *(Scylla; Thalamita)* and deeper water species e.g. *Eucrate alcocki,* and live partly within and partly outside the body. The portion within the host comprises nutritive roots, while outside, under the abdomen, lies a rounded chitinous sac that contains the gonads. So unlike a crustacean is this parasite that it can only be readily classified by means of the larval stage which is a nauplius.

Molluscs

Bivalvia

Communities under low-tide boulders are conspicuosly rich in bivalves, in various ways permanently attached to the surface. Oysters, thorny oysters and chamids are cemented by one valve of the shell; but most are flexibly fastened by a beard of byssus threads secreted by a gland in the foot. Free-moving bivalves have a transitory byssus as a post-larval organ; but in these rock-attached species it has persisted into the adult stage and the form of the animal has become profoundly modified.

The most primitive and the least specialized of attached bivalves are the ark-shells (Arcidae). The shell is roughly rectangular, with its anterior and posterior ends and their adductor muscles about equal. The byssus fibres are fused into a tough membrane that attaches the shell upright with its ventral edge against the rock. Three species of Arcidae are included in Fig. 7.6. *Barbatia virescens* is a large shell (4 cm long) with radial sculpture and rough black stubble. *Striarca afra* is small and plump without any hairy periostracum. *Arca avellana* erodes cavities in the rock, its coiled umbones and its flat posterior end being visible in the plane of the surface. The species will again be encountered as a nestler of coral heads (p. 281).

Boulder Shores

FIG. 7.6. Bivalves, gastropods and other attached animals of low-tidal boulders in Tai Tam Harbour.
1. *Ochetoclava sinensis*; 2. *Proclava kochi*; 3. *Clypeomorus humilis*; 4. *Striarca afra*; 5. *Barbatia virescens*; 6. *Bursa granularis*; 7. *Tectus pyramis*; 8. *Turbo argyrostoma*; 9. *Morula marginatra*; 10. *Septifer virgatus*; 11. *Serpulorbis imbricatus*; 12. *Cymatium echo*; 13. *Chama reflexa*; 14. *Corella japonica*; 15. *Balanus trigonus*; 16. *Chicoreus microphyllus*; 17. *Isognomon isognomun*; 18. *Pseudochama retroversa*; 19. *Hiatella flaccida*; 20. *Pinctada martensii* with shell in edge view; 21. *Pinctada furcata* with shell in edge view; 22. *Trapezium sublaevigatum*; 23. *Scutus unguis*; 24. *Diodora reevei*; 25. *Cypraea arabica*; 26. *Cardita leana*; 27. *Culicea japonica*; 28. *Arca avellana*.

The next line of adaptation, though not a direct evolution from the arks, is shown by the mussels (Mytilidae). The byssus has now moved towards the anterior end, pulling it to the rock and restricting shell growth until this end becomes narrow and pointed. In contrast, the posterior end continues to grow and becomes large and rounded. The anterior adductor muscle is reduced (in *Perna* lost), compensated by the large posterior adductor.

Three species of mussels will be found in harbour situations: the green-ribbed mussel *Septifer virgatus* (its ally the black *S. bilocularis* has been seen already on exposed shores and also widely met with in shelter); the smaller ribbed mussel *Brachidontes atratus* found among oysters and on mangroves; and a larger species, smooth and lightly built, *Perna viridis*, which is the characteristic harbour mussel. Dull brown colour between tides, it brightens in deeper water into jade or blue-green.

Further adaptive radiation of the mussels is shown by *Musculista senhausia*, enmeshed in a byssus nest in sand (p. 245), and species of *Botula* and *Lithophaga* boring into oyster shells and coral respectively (p. 245 and p. 269).

Two other wholly unrelated bivalve families, represented under boulders show this mussel-shaped trend ('mytilization'). The Trapeziidae *(Trapezium liratum* and *T. sublaevigatum)* are compressed and rectangular, with a small anterior byssus and rough growth sculpture. *Cardita leana* (family Carditidae) has strong ribs, radiating from the beaks (umbones) to the posterior end. White and brown speckled it is attached by a weak byssus, being descended from an unspecialized cockle-like ancestor *(Venericardia)* freely burrowing in sand and gravel and occurring sub-tidally in Hong Kong.

Byssus-attached bivalves reach a third grade of adaptation, with the complete loss of the anterior adductor muscle. With a straight edge dorsally, and the single (posterior) adductor now central, the rest of the shell becomes rounded, though there is much diversity of shape and habit. The foot is now solely concerned with byssal secretion and cleansing the mantle cavity. The byssus emerges from a notch in the right valve, and the shell lies on that side.

Good examples of this grade of adaptation are the wafer-thin Isognomonidae, often compressed and wedge-shaped. Fig. 7.6 shows *Isognomon isognomum* attached in clusters. The hammer shell *Malleus malleus* (p. 281) is a T-shaped production from this line. The family Pteriidae include the almost circular 'pearl oysters' *Pinctada martensii* attached beneath rocks at low tide, and the very flat *Pinctada fucata* living in rock fissures and between stable boulders. The former of these two species is used by the Japanese to culture artifical pearls. Blister pearls attached to the shell are formed by three other, rather deeper water, pearl oysters—*Pinctada maxima* is large with a whitish edge or lip to the shell, *Pinctada margaratifera* has a distinctive black lip while *Pteria penguin* has a large black shell, the hinge formed into 'wings' laterally. *P. penguin* used to be cultured in Tolo Harbour and specimens of this handsome animal are to be found byssally attached amongst coral heads.

In the Pinnidae or fan shells the hinge line has become long and straight, giving rise to the fan-shaped *Pinna* and *Atrina* (p. 195) taking flexible anchorage by the byssus in soft ground.

The scallops and their relatives, Pectinidae, have evolved in two directions from the byssus-fixed bivalves. Their most primitive members, such as *Chlamys* (p. 281), are still attached like the Pteriidae. From this early stage, the large *Pecten* scallops have become emancipated; the byssus is lost, and by clapping the light, circular valves with the adductor muscle they can take off from the bottom and swim jet-wise by expelling water 'backwards' at both ends of the hinge line. The final achievement of this evolutionary line is seen in the flat, fragile valves of *Amusium japonicum* and *A. pleuronectes,* the so-called 'sun and moon shells'

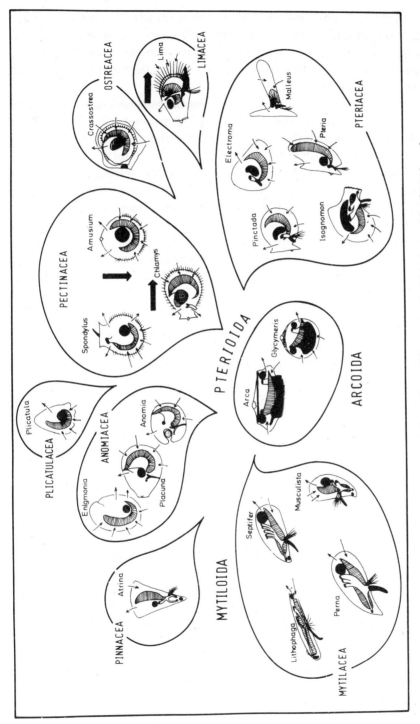

FIG. 7.7. Development of the different lines of 'monomyarian' bivalves, with adaptations for attachment or 'swimming'. The smaller entire arrows show the direction of the inhalant and exhalant currents. Direction of swimming in *Chlamys* and *Amusium* and crawling in *Lima* are shown in large black arrows. Each diagram shows the ctenidium, the foot (black) with byssus when present, the adductor muscle (hatched) and the pedal muscle insertions (black).

because of the red upper and white lower valves (p. 281). Found offshore in a few fathoms depth, these scallops are light, well-accomplished swimmers.[2]

Finally, there are scallop relatives that have become cemented to rock: *Spondylus barbatus* the thorny oyster (p. 281). Of similar habit but convergent appearance is *Plicatula plicata* (p. 115).

The Anomiidae or saddle oysters (e.g. *Anomia ephippium*) have become still more modified for close attachment. The semi-transparent left valve has taken limpet-like form, to match the irregularities of the rock. Beneath it is concealed the right valve, thin and almost functionless, and notched for the passage of the byssus. The fibres are massed into a short calcified cable, drawing the shell down when the byssus retractor muscle contracts. Just as with the scallops, some of the anomiids have broken free of the byssus. *Enigmonia aenigmatica* moves about like a limpet on mangrove leaves when juvenile but is more commonly found attached to the trunks of mangrove plank notably the pioneer *Aegiceras*. Another mobile anomiid is the widow-shell *Placuna placenta,* with two equal valves, flat and circular, living on Hong Kong mud-flats (p. 248).

Gastropoda

Of the mobile snails, far more species are found under boulders than ever appear on the upper surface. The families Trochidae and Turbinidae—unlike the dry-adapted Littorinidae and Neritidae—have their main habitat, with their largest species, under low-tidal boulders.

The largest local trochid is the flat-based and conical *Tectus pyramis,* sharply angled, with greenish-brown epidermis and fine growth lines. The columella is deeply spiral-threaded. Two smaller trochids are both common at low water: *Chlorostoma nigerrima* with black, obliquely set axial ribbing, and a green patch at the base, and *Chlorostoma rustica;* smooth and grey-brown with a flat base. In both these species the periphery is rounded. Still smaller low-tidal trochids include the species of *Clanculus* and *Euchelus,* having rounded shells, with beaded spiral ribs, and the columella strongly toothed.

Related to the trochids and found also beneath low-tidal stones are the Stomatiidae, with small wide-mouthed shells, like a *Haliotis* without perforations. The animal is broad-footed and too large to be withdrawn into the shell.

The Turbinidae are easily distinguished from trochids by their heavy calcareous operculum; they are represented at low water by the large *Turbo argyrostoma,* with rounded body whorl, yellowish-brown and corded with spiral ribs.

As well as these families, the Archaeogastropoda also include the keyhole limpets and their allies (Fissurellidae), that lose the spire in the adult and become

---

[2] Morton, B. S. (1980). Swimming in *Amusium pleuronectes* (Bivalvia: Pectinidae). *Journal of Zoology, London.* 190: 375–404.

bilaterally symmetrical. Unlike the true limpets, the mantle cavity retains two equal gills (ctenidia). Of the two common local species, the smaller is *Diadora reevei*, a high-conical and radially ribbed shell, with a small keyhole at the apex through which the respiratory currents and wastes are expelled. In the shield limpet, *Scutus unguis*, the animal far outgrows the shell. The soft parts are yellowish-brown, with grey mottling, and the shell is flat, like a duck's bill, covered in life by thin mantle lobes. Both these fissurellids appear to be sponge grazers, whereas trochids and turbinids take chiefly coralline and other fine algae. A third, strongly ribbed, fissurellid *Emarginula bicancellata* possesses a slit at the front of the shell.

Superficially limpet-like, the Calyptraeidae, or slipper limpets, are slow-moving or entirely immobile. The commensalistic *Syphopatella walshi* will be encountered later with the hermit crabs, but on boulder beaches a number of species are to be found, usually low down. They all collect their food by the ciliary currents of the gill. The small, brown, *Crepidula onyx* is associated with trochid gastropods, usually attaching to the smooth region of the shell around the aperture. The tall 'chinaman hat' shell *Calyptraea sakaguchi* is to be found attached to the under-surface of boulders or nestling within wide crevices.

The Cerithiidae are long, trailing gastropods, mainly found on soft shores (p. 222), where they feed on surface deposits or algae. Two species are illustrated from the rocky sub-littoral: *Ochetoclava sinensis*, with greyish green nodulose shell spirally lined in black, and the beaded, black and white *Clypeomorus humilis*.

Belonging to the Mesogastropoda, like the Cerithiidae and Calyptraeidae, are the cowries (Cypraeidae). Chiefly found sub-tidally on coral shores to the east, (p. 287), they have a common and relatively large species, *Cypraea arabica*, under low-tidal boulders in most parts of Hong Kong. Like other *Cypraea*, this cowry grazes on hydroids and other sessile animal food. The smaller *Cypraea erronea* is grey above, sometimes traversed by a brown band, and yellow beneath. The reflected mantle is grey flecked with bright orange, minute spots.

Three families of whelk-shaped carnivores, with pronounced anterior shell canals, are also exemplified in Fig. 7.6. The first are the tritons or trumpet-shells (Cymatiidae). Solidly constructed and conical, they are distinguished by the arrangement of their varices, strong ribs from former lips left behind by subsequent growth. Never more than two to a whorl, these do not, in cymatiids, connect longitudinally between adjacent whorls. Cymatiids often have a shaggy periostracum, especially strong in the species illustrated, *Cymatium echo*. The animal is generally handsomely coloured, and the diet is of ascidians, urchins or other molluscs.

The Bursidae or frog-shells, typified by *Bursa granularis*, are carnivores related to cymatiids. Their varices, as in trumpets, are only two to a whorl, but longitudinally connected from one to the next, to form prominent ridges down both sides of the shell. A larger species of *Bursa*—*B. rana*—is found in deeper waters on soft sediments.

The Muricidae or rock-shells are one of the largest carnivorous families, the opening of the shell generally having a long, tubular anterior canal. They have a variety of predatory habits, and—as with the thaids that belong to this family (p. 45)—bore through the shells of barnacles or of other molluscs. There are frequently three varices to a whorl, connected longitudinally to give usually the whole shell a triangular cross-section. The shell ornament is often of coupled channelled spines, as in *Chicoreus microphyllus* where the lip and varices are elaborately fringed.

The rocky shores of Hong Kong are extraordinarily rich in predatory muricids and Dr John Taylor[3] has investigated the distribution and feeding preferences of these species on a coral dominated beach in Tolo Harbour. In Fig. 7.8, 10 species are represented from a shore that is dominated at its lowest levels by a mixed community of *Saccostrea cucullata*, *Brachidontes variabilis* and *Balanus trigonus* a little lower down. Just below the low-water spring level a second rock oyster *Alectryonella plicatula* is found, with the under-surface of the rocks densely covered in *Spirorbis*. Below this level follows a thick band of coral to a depth of 10 m.

The muricids distribute themselves down the shore with the higher zoned species *Thais clavigera* and *Morula musiva* feeding mostly on *Saccostrea*. Both, however, also feed on *Brachidontes*, the former by pulling the two valves apart, the latter by drilling a hole in one valve. Lower down *Thais luteostoma* was found to feed exclusively on *Balanus*. *Cronia margariticola* was less discriminating in its choice of food and fed upon barnacles, *Spirorbis*, small bivalves and even the coral itself. *Cronia* thus possesses a wide depth distribution. *Mancinella echinata* and *Ergalatax contractus* fed on *Balanus* and *Alectryonella*, whilst *Drupella rugosa* was an exclusively coral feeder. The large *Chicoreus microphyllus* possess a wide range of prey, feeding on *Alectryonella* by drilling and also upon other bivalves and barnacles. This study by Dr Taylor is a good example of how any one area of beach is apportioned by various predators and goes some way in explaining how zonation patterns are not only established but also maintained. It would be an interesting experiment to selectively remove some of these predators from a strip of beach and to determine on the basis of stomach content analysis just how dietary preferences might change and also the extent to which the absence of predators alters the ecological structure of the beach communities.

The final family of low tidal whelks, the Pyrenidae, are of much smaller size. Their numerous species have sturdy, urn-shaped shells, with a short spire

[3] Taylor, J. D. 1980. Diets and habitats of shallow water predatory gastropods. around Tolo Channel, Hong Kong. In *The Malacofauna of Hong Kong and Southern China*, ed. B. S. Morton, pp. 163–80. Hong Kong, Hong Kong University Press.

Boulder Shores

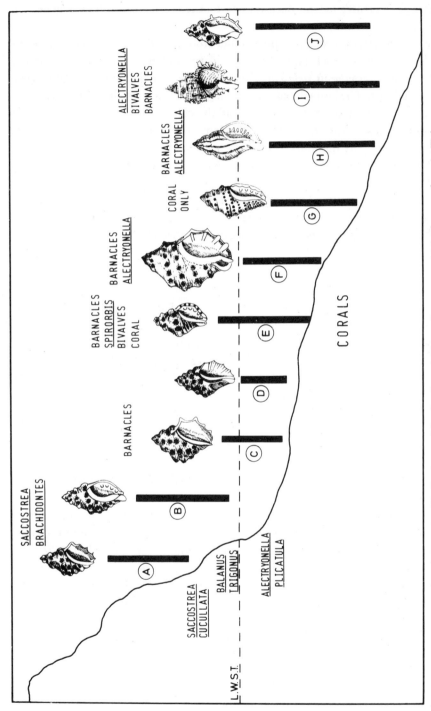

Fig. 7.8. The distribution and foods of muricid, predatory gastropods down a shore in Tolo Harbour.
A. *Thais clavigera*; B. *Morula musiva*; C. *Thais luteostoma*; D. *Thais tissoti*; E. *Cronia margariticola*; F. *Mancinella echinata*; G. *Drupella rugosa*; H. *Ergalatax contractus*; I. *Chicoreus microphyllus*; J. *Morula spinosa*.

and narrow aperture, having both its lips toothed. The operculum is generally reduced.

The most aberrant of all gastropods under stones are the Vermetidae, mesogastropods fused to the rock face like serpulid tube-worms. The common Hong Kong species. *Serpulorbis imbricatus* is coiled like a snake, and with granulose longitudinal ribs. The animal extended from the shell shows clearly its gastropod character. The foot terminates in a brightly coloured circular disc, and the operculum is lost. From the front of the foot arise two slender pedal tentacles to guide and support the mucous threads extruded from the pedal gland to trap planktonic food. Loaded with adhering particles, these are hauled back and pulled into the mouth by the radula. The gill filaments are long and narrow, permitting some measure of ciliary food collecting, after the mode of the bivalves. Lacking a protective operculum, *Serpulorbis* rapidly withdraws on disturbance into the lower reaches of its tube. The eggs are carried in drop-shaped capsules fastened to the interior.

Sponges

Little study has yet been made in Hong Kong of sponges although these are common under most sorts of boulder cover, except for the most mobile cobbles or rocks deeply embedded in sediment. The main life forms and genera likely to be encountered by a student on these shores are described.

The sponges (phylum and sub-kingdom Porifera) are the simplest multicellular animals, best regarded as colonial systems rather than unified organisms. They filter food from water currents taken in through vast numbers of pores (ostia) on the external wall. Water is passed through systems of narrow passages by flagellated cells of a special 'collared' type known as choanocytes. These not only strain off particles, but also ingest them. The converging exhalant currents from the spaces lined with choanocytes are expelled through large chimneylike openings called 'oscula'.

The simplest sponges are those small tubes and vases, with an osculum at the free end, and a supporting skeleton of calcareous spicules, belonging to the class Calcarea. Commonly found in clusters under boulders is the tubular *Scypha ciliata*. Others such as *Leuconia solida* comprise an aggregate of white, flattened, hollow lobes each about 1 cm high. The structure of the Calcarea is relatively straightforward and the textbook diagram of *Sycon* gives a general introduction to sponge morphology.

Most sponges, and all the largest, belong to the separate class Demospongea, with the spicules (when present) silicaceous, and supplemented by horny skeletal fibres of spongin. Expert identification must rely on microscopic spicule preparations; but the Demospongea have such diversity in colour, texture, build and 'feel', even in smell, as to offer many recognition marks to the field observer.

Boulder Shores

The Demospongea are divided into three sub-classes, the Tetractinellida, Monaxonida and Keratosa. Of these the Monaxonida are by far the commonest, especially intertidally. There are broadly four basic monaxonida demospongian patterns, the 'halichondrine', 'hadromerine', 'poeciloslerine' and 'haplosclerine'. The halichondrines—well typified by the 'crumb of bread' sponges—are light in build and crumbly, with the skeleton containing little spongin, and only a loose, irregular arrangement of spicules. The exhalant cavities open by funnel-shaped vertical oscula. Typical Hong Kong species are the yellow to buff *Halichondria japonica* and the black *Halichondria okadai*. Related to these are the 'haplosclerid' sponges, with the skeleton mainly of spongin and only small spicules. *Haliclona permollis* has tall conical oscula of a dull mauve tint. Whereas *H. chinensis* is

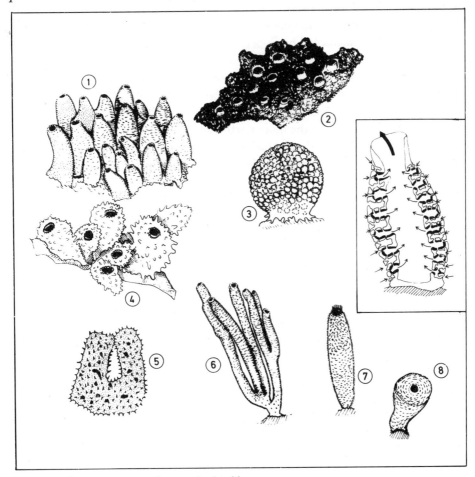

FIG. 7.9. Common sponges from under boulders.
1. *Haliclona permollis*; 2. *Halichondria okadai*; 3. *Tethya aurantia*; 4. *Callyspongia confoederata*; 5. *Ircinia* sp.; 6. *Siphonochalina truncata*; 7. *Scypha ciliata (Sycon okadai)*; 8. *Leucandra* sp. *Inset*: Diagrammatic section of a *Sycon*, showing water currents and flagellated chambers (black).

orange, the large branching *Sigmadocia symbiotica* is green but the colour is derived, strangely, from a red symbiotic alga *Ceratodictyon spongiosum* (a member of the Gracilariaceae) which invades the sponge tissues. A bivalve *Crenatula modiolaris* often lies buried in the sponge/alga mass, together with a small, commensal barnacle *Acasta sulcata*.[4] *Callyspongia confoederata* has a branched structure and *Siphonochalina* is produced into long tubular fingers. The 'poecilosclerid' sponges are often bright red, forming low compact crusts *(Mycale, Biemna, Neofibularia* and *Tedania)* permeated with spicules, with the spongin fibres reduced to a basal sheet.

The 'hadromerine' sponges have a firmer and more coherent structure than the halichondrines, relying on spicules rather than spongin. The family Tethyidae (common species *Tethya robusta*) forms small compact, orange to red balls, massively built, with a mammillose outer crust, and an inner core with long radiate clusters of spicules. The Suberitidae show the same compact structure, forming orange and brown crusts *(Polymastia)* in strong wave wash, or originating as shell-boring pustules afterwards fusing into a protuberant yellow mass. *Cliona celata* is very commonly to be found boring the shells of bivalves, notably of the Pacific oyster *(Crassostrea gigas)* cultured in Deep Bay, while another species bores coral heads in the more north-easterly waters of Hong Kong. Like *Suberites, Spirastrella vagabunda* forms a rounded mass that too possesses a commensal barnacle—*Pyrgopsella stellula*—embedded in the sponge tissue.

Further hadromerine sponges include the orange-brown branching, tree like *Acanthella* (Axinellidae) and the orange, encrusting *Agelas* sp. (Agelasidae).

Also grouped among the Demospongea, as a sub-class, are those rubbery sponges lacking spicules altogether, but with an internal skeleton of spongin, leading to the name Keratosa. Common species include the bath-sponge *Spongia reticulata* sub-tidally and elastic sheets of *Hircinia* and *Aplysilla* under boulders.

Special Habitats

Wave-exposed Boulders

On open or moderately exposed shores, granite boulders come under strong wave attack, as on the south coast of Lantau, the south-facing parts of Lamma Island and the outer reaches of Tai Tam Bay on Stanley Peninsula.

Under high exposure small boulders are continually kept in motion, and with storm waves they abrade each other to lose their sharp angles and become rounded. Up to half a metre in diameter, sometimes much larger, such boulders lie in steep ramps several deep.

---

[4] Reid, R. G. B. and Porteus, S. 1980. Aspects of the functional morphology and digestive physiology of *Vulsella vulsella* (Linné) and *Crenatula modiolaris* (Lamarck), bivalves associated with sponges. In *The Malacofauna of Hong Kong and Southern China*, ed. B. S. Morton, pp. 291–310. Hong Kong, Hong Kong University Press.

Boulder Shores

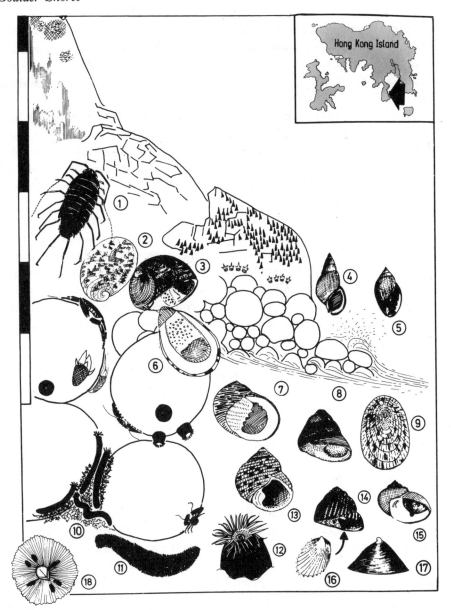

FIG. 7.10. The fauna of mobile boulders on a moderately exposed shore, Stanley Peninsula. The ramp of wave-washed boulders is shown from the upper mid-littoral downwards with the alga *Porphyra suborbiculata* and the stalked barnacle *Pollicipes mitella* growing immediately below. Detail of boulders and the animals in their interspaces, *Pollicipes mitella, Petrolisthes japonicus, Actinia equina* and *Polycheira rufescens* is shown at bottom left.

1. *Ligia exotica*; 2. *Nerita polita*; 3. *Nerita chamaeleon*; 4. *Zebina tridentata*; 5. *Laemodonta punctatostriata*; 6. *Nerita albicilla*; 7. *Nerita undata*; 8. *Chlorostoma xanthostigmata*; 9. *Cellana toreuma*; 10. *Polycheira rufescens*; 11. *Holothuria leucospilota*; 12. *Actinia equina*; 13. *Monodonta australis*; 14. *Chlorostoma argyrostoma*; 15. *Monodonta neritoides*; 16. *Crepidula onyx*; 17. *Calyptraea sakaguchi*; 18. *Tetraclitella purpurascens*.

Still larger boulders are heavy and angular enough to remain stable for long periods; and they develop a considerable fauna in their cool, dark inter-spaces. All but one or two species are mobile, unlike the communities of sheltered boulders; and the sessile acorn barnacles, tube-worms and bivalves are seldom found.

Crabs abound, especially the highly mobile grapsids. At the wave-break line of high-water neap, small *Grapsus albolineatus* and *Metopograpsus frontalis* run over and between the boulders, holding on to the convex surfaces with their wide, clawed limb-spread. Where high-tidal boulders rest on sand, *Cyclograpsus intermedius* is most typical; higher still *Parasesarma pictum* will be found, along with the isopod *Ligia exotica*.

Mid-tidal boulders, where coarse sand collects, are the haunt of *Hemigrapsus sanguineus* and *Gaetice depressa*. Shorter legged than other grapsids, these crabs are still active, but live beneath boulders rather than speeding between them. The most abundant of all boulder beach crabs is generally the half-crab, *Petrolisthes japonicus*. This species may owe its success to its microphagous feeding habit, rather than competing with carnivores for a share of the animal food. Slim, fast and aggressive, this crab is prettily grey flecked with the under-parts of the chelae salmon-pink.

The gastropods of clean boulders include both limpets and mobile snails. Of the former the commonest is *Cellana toreuma,* smooth and flat-pitched, clustering in great numbers over the tops and sides of large rocks. There is also the ubiquitous *Patelloida pygmaea. Planaxis sulcatus* is regularly found, and equally common is the rounded top-shell *Monodonta australis.* At a lower level the trochids *Chlorostoma xanthostigma* (smooth), *C. rustica* (strong-ribbed) and *C. nigerrima* (smaller ribbed) may be found. The small, low-pitched *Monodonta neritoides* is also typical of high-level boulders. With it, often in large local aggregations, is the conical hydrobiid snail *Zebina tridentata.*

The Neritidae are also common with three species. *Nerita polita,* usually on smooth boulders on clean sand, is smooth and glossy, variably streaked and mottled; some specimens have a vermilion spiral band. *Nerita undata* can be at once recognized by its strong spiral ribs, speckled with light and dark brown. *Nerita albicilla,* low-pitched and stream-lined, with a smooth shell splashed with black and white, is commoner on the under-surfaces of boulders.

In the narrow spaces between boulders, free from lodged sediment, a few sessile species are to be found. The most important and characteristic of these is the flat barnacle *Tetraclitella purpurascens.* This pale purple to dirty white, low, conical species is found deep within underwater caves but more typically on the undersurface of large boulders and around which waves surge. It is extremely susceptible to dessication and is never found exposed. The stalked barnacle *Pollicipes mitella* is scattered single or in small clusters, forming upright brackets against the boulder sides. Water flow between boulders lodges food on its outspread cirri. The purple-blue sponge, *Haliclona permollis* attaches itself to the protected under-surface of the boulders where it filters particles of detritus from the surging waters.

Two anemones are common on clean boulders. The deep red *Actinia equina* is found in many parts of the world attached to shaded surfaces and in dark recesses. This species is well adapted to live periodically out of water, in the cool, saturated atmosphere beneath boulders. The mode of feeding is still to be studied; the ciliated tentacles may—like the cirri of *Pollicipes*—collect fine particles from turbulent water. *Anthopleura nigrescens* is green with long thin, red/brown tentacles and a lighter oral area. *A. dixoniana* is beige with lighter striping and short, stubby tentacles. Both seek deep shelter in crevices of clean, boulder shores.

The commonest echinoderms of boulder shores are holothurians. At lower levels, where clean sediments accumulate, the small apodan, sea-cucumber *Polycheira rufescens* can lodge between rocks, and can sometimes be collected in handfulls. Eight to ten centimetres long, and jet-black or reddish-brown, *Polycheira,* like other holothurians, takes in water by the anus to fill its branched respiratory trees. When the tide is out it normally stays turgid but shoots out a strong water jet on disturbance. The circlet of ten feeding podia have terminal discs that can act as attachment organs, or pick up particles and convey them, each tentacle in turn, to the mouth.

The larger black cucumber, *Holothuria leucospilota,* with thick mammillated skin, feeds on deposits in the same way. Found lower down and also sub-tidally, upon disturbance it puts out intensely sticky, white collagenous filaments from the anus. A second large holothurian *H. arenicola,* is pale tan colour and is mottled with brown and white spots. It occurs under low-tide boulders but extends onto sub-littoral sandy deposits in much the same way as *H. leucospilota.*

Under smaller boulders upon silty sand, lives a burrowing worm-like holothurian, *Leptosynapta ooplax* (*see* Fig. 7.2). It is pale and translucent, with five longitudinal ambulacral lines, from which the tube-feet have been lost. These 'apodous' species burrow by peristaltic contractions of the body wall. They extend mouth downwards, with the finely branched podia collecting food and the anus taking in respiratory water from above.

A common ophiuroid echinoderm of boulder beaches is the red *Ophiomastix mixta*, its long arms snake-like and interwoven amongst the other encrusting organisms of the undersurface of the stones.

The fishes of boulder shores include two major ecological categories. The first comprises the blennies and gobies and their allies which spend much time in physical contact with the rock surfaces. In the gobies (Gobiidae) and the sucker fishes (Gobiescoidae) the pelvic fins have moved forward and fused, to form, with the lower rays of the pectorals, a shallow suction cup. Remaining close to the rock, they take an unbreakable grip at each rush of surge, at other times moving by rapid but spasmodic flicks of the tail. Two common Hong Kong sucker fish are *Amblygobius albimaculatus* and *Tridentiger trigonocephalus,* both abundant on the boulder shores of Tai Tam where they feed by biting small organisms from the sides of the stones.

The Blennidae, represented in Fig. 7.11 by *Dasson japonica* have no sucker, though the pelvic fins lie anteriorly. They nestle in interstices between rocks and retire deeply when disturbed. The Trachinidae or weaver fish are represented by *Paracentropogon indicus,* with sharp spines on the gill covers and anterior part of the dorsal fin, capable of inflicting a painful wound. The scorpion fish *Sebasticus marmoratus,* widely common in Hong Kong, and also much in contact with the bottom, has the same noxious capacity.

The second category of fishes of the rocky inter-tidal have a greater but varying independance of the substrate. The wrasses (Labridae) which are always brightly coloured *(Labroides dimidiatus, Halichoeres nigrescens)* elude potential predators by hiding between rocks or in crevices. Of the wide variety of inshore fishes found in Hong Kong's coastal waters mentioned here are the fast-swimming barracuda *(Sphyraena jello),* the bottom-feeding goat-fish *(Upeneus tragula), Therapon jarbua* (to be encountered again on sheltered sand-flats), the more sluggish snappers *(Lutjanus russelli* and *L. sanguineus),* and the groupers *(Epinephelus brunneus)* with the pelagic carangids *(Caranx kalla* and *C. malabaricus).* The rabbit-fish *Siganus oramin* is very common on all rocky shores and also possesses venomous dorsal fin spines. When alarmed it will lie on its side with fins raised, or if possible slip into a crevice. Mention must finally be made of the great schools of young fish to be seen feeding in the highly productive bays and river estuaries of Hong Kong. The grey mullet *(Mugil cephalus),* often captured by fishermen using beach seines and then to be grown to marketable size in the fish-ponds of the Yuen Long Plain, is of special importance.

Other small fish caught in this way are kept in cages suspended from rafts anchored in protected bays. These include: *Epinephelus akaara* (red grouper); *Epinephelus brunneus* (mud grouper); *Epinephelus awaora* (yellow grouper); *Lutjanus argentimaculatus* (mangrove snapper); *Lutjanus johnii* (John's snapper); *Chrysophrys major* (red pargo); *Mylio berda* (white sea-bream); *Mylio latius* (yellow finned sea-bream) and *Rhadosarga sarda* (gold line sea-bream). They are fed with trash food until of marketable size.

Cobble and Gravel Beaches

Intermediate between rocky and sandy shores, these beaches have some of the characteristics of each. Unlike boulder shores in close shelter, they have rather little silt because they are washed by waves of moderate strength. Their sands

Boulder Shores

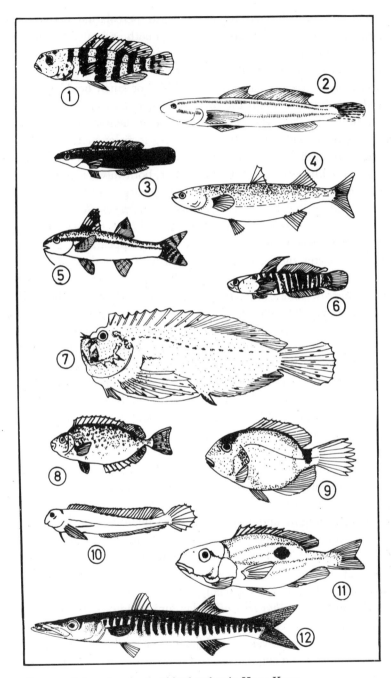

FIG. 7.11. Common fishes of rocky, boulder beaches in Hong Kong.
1. *Epinephelus brunneus*; 2. *Tridentiger trigonocephalus*; 3. *Labroides dimidiatus*; 4. *Mugil cephalus*; 5. *Upeneus tragula*; 6. *Amblyogobius albimaculatus*; 7. *Paracentropogon indicus*; 8. *Siganus fuscescens*; 9. *Lutjanus sanguineus*; 10. *Dasson japonica*; 11. *Lutjanus russelli*; 12. *Sphyraena jello*.

are coarse and well-drained by wave swash percolating between the small boulders lying on top. These are generally smooth and piled two or three deep; and must be lifted away to reveal the fauna of the sand or gravel beneath. Such beaches are characteristic of many reaches.

Few animals settle permanently on cobbles or small boulders. The beaches are hardly sheltered enough for rock oysters, but towards low-water mark, a different bivalve, *Plicatula plicata,* may be cemented to the rock in some numbers. Cream with reddish-brown rays, it belongs to a separate group unrelated both to the true oysters and to the scallops and thorn oysters. *Plicatula* attaches by the hinge region of the right valve, both valves being semi-triangular, and channelled by radial ribs, giving a zig-zag margin. In the same region of the beach a second bivalve, this time a true oyster, *Alectryonella plicatula* is cemented by the left valve in clusters on the rocks. As in *Plicatula* the shell margin, green in colour, is markedly incised giving a zig-zag appearance. *Alectryonella* is a typical warm water species often associated with a coral assemblage.

The coarse sand has some typical, shallow-lying bivalves. At the top of the beach is a narrow zone of *Atactodea striata,* a small, sturdy shell with a heavy growth sculpture. Belonging to the same family, Mesodesmatidae, is a second species, *Caecella chinensis,* thinner and ovate, with a golden periostracum, eroded near the beaks. Remarkable for its high tidal position, *Caecella* is found wedged in sand between small boulders, even on salt-meadow with *Zoysia* grass (p. 217).

Towards low-water neap, two larger bivalves become common. The venerid *Tapes philippinarum* is one of the best-known Hong Kong shells, equally at home in coarse sand, under cobbles, and on silty sand-flats (p. 167). The colour is highly variable, with hardly two individuals alike: buff or grey, radially ornamented with chevrons and finely traced zig-zags. A large bivalve, *Asaphis dichotoma,* lives near low tide, in pockets of silt mingled with coarse sand. Thin-shelled and purple-marked inside, it has coarse radial and growth sculpture. Like all the Tellinacea, to which it belongs, *Asaphis* has two long separate and highly mobile siphons and feeds on surface deposits drawn in by the much longer inhalant siphon. Jon Day[5] has plotted out the distribution of *Caecella, Asaphis* and *Tapes* at Ma Tse Chau nicely demonstrating how these three species are zoned in response to their degree of desiccation tolerance (Fig. 7.13).

Three sorts of crab are especially typical of cobble shores. *Cyclograpsus intermedius* and *Epixanthus frontalis* (high and middle shore) are both common at their distinctive levels; but the most abundant are generally large *Gaetice depressa,* showing a wide variety of blotched or marbled colour patterns.

At low-tide mark and slightly above, the holothurian *Polycheira rufescens* lives

[5] Day J. E. 1980. Correlation of gill physiology, emersion survival, and intertidal distribution of three bivalves from Hong Kong. In *The Malacofauna of Hong Kong and Southern China,* ed. B. S. Morton, pp. 211–17. Hong Kong, Hong Kong University Press.

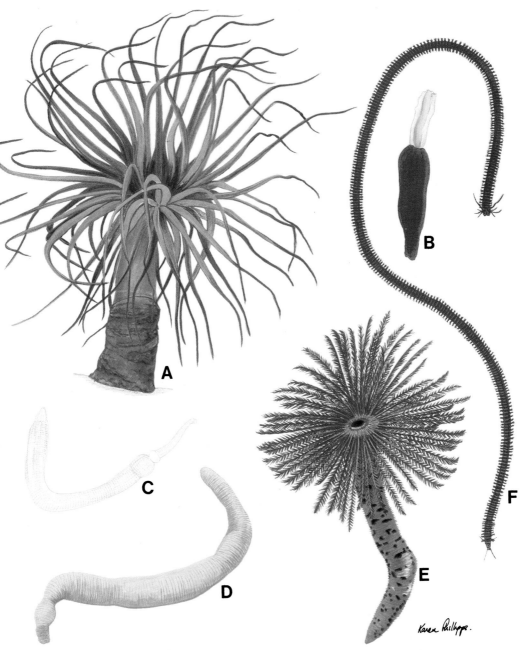

PLATE 9. Sandy shore fauna: A. *Cerianthus* cf. *filiformis*; B. *Ochetostoma erythrogrammon*; C. *Sipunculus nudus*; D. *Siphonosoma cumanense*; E. *Sabellastarte japonica*; F. *Ceratonereis* sp.

PLATE 10. Two aerial photographs of sheltered sand flats in Tolo Harbour. A. A number of small streams drain onto such a flat; note the luxuriant fringe of coastal vegetation including mangroves; B. A single stream discharges onto a beach forming a raised delta of inter-connected sand bars. Note wave defraction on the bars.

Boulder Shores

FIG. 7.12. Fauna of a cobble beach in Tolo Harbour. *(Above)* The islands, looking north-east from Shuen Wan Lei Uk. *(Middle)* A shore on Ma Tse Chau with metamorphosed shale and silt-stone of the Tolo Harbour formation giving rise to crevices.

ZONES:
A. With crevices between oblique shales; B. Cobble beach; C. Upper shore with *Zoysia* and *Scaevola*; D. Backshore with *Hibiscus tiliaceus*.
(Below) Cobbles with coarse sand showing burrowing bivalves *in situ* from left *Asaphis, Actactodea, Tapes* and *Caecella* (black).

SPECIES:
1. *Hibiscus tiliaceus* flowers and leaves; 2. *Alectryonella plicatula*; 3. *Plicatula plicata*; 4. *Asaphis dichotoma*; 5. *Tapes philippinarum* three variant shell patterns; 6. *Ceratonereis nuntia* var. *brevicirrus* with *(inset)* detail of head; 7. *Caecella chinensis*; 8. *Atactodea striata*; 9. *Epixanthus frontalis*; 10. *Gaetice depressa* three variant carapace patterns.

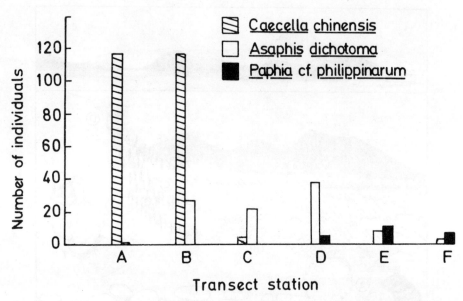

FIG. 7.13. The distribution of three species of bivalves down the beach at Ma Tse Chau.

beneath stones; and in the same places the pink, somewhat flattened *Ceratonereis nuntia* var. *brevicirrus* makes winding, superficial channels in the moist sand. Amongst the stones can be found the variably patterned nerite, *Nerita polita*.

Boulders Resting in Mud

In extremes of shelter, as in quiet pockets towards the top of harbours, boulders lie partly immersed in soft mud. The buried parts, and those areas underneath in contact with the top of the mud, are stained black with ferrous sulphide formed under anaerobic conditions. In places with heavy pollution, hydrogen sulphide may be generated in the surrounding mud. Generally however the substrate around the boulder is aerobic enough for tube-dwelling polychaetes and other animals to extend their tentacles for feeding and respiration.

The fauna on top of these boulders is sparse, generally with one of the barnacles *Balanus albicostatus*, *B. reticulatus*, *B. amphitrite* or *B. variegatus*. Sometimes a dull pink paint basal *Corallina* spreads from the sides of the boulder to beneath the dimly lit overhang. Small grey anemones (*Anthopleura* species) can flourish here.

Several kinds of sedentary polychaete immure themselves under these boulders, their tubes extending out to open at the edge. *Marphysa sanguinea* is a long-bodied euniciid worm, grey or rust-tinted. Dorsoventrally flattened, and with short parapodia, it forms winding galleries in the stiff sandy mud against the rock surface. The Euniciidae can be identified by the short red blood gills, along the

Boulder Shores

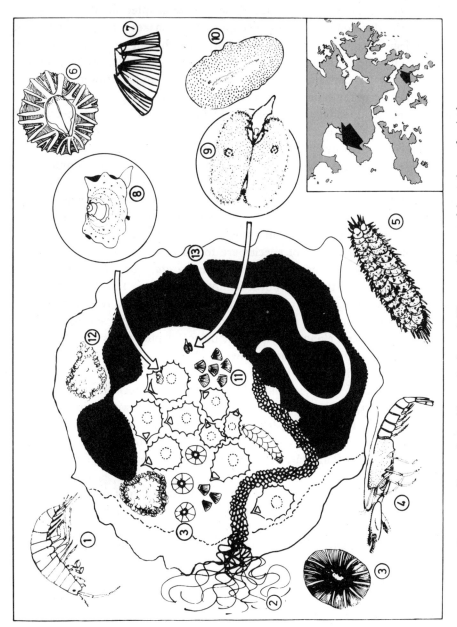

FIG. 7.14. Fauna of a boulder resting in mud. The anaerobic part of the under-surface is shown in black.
1. *Talorchestia brito*; 2. *Thelepus setosus*; 3. *Haliplanella luciae*; 4. *Alpheus bisincisus*; 5. *Harmothoe imbricata*; 6. *Balanus albicostatus*; 7. *Balanus trigonus*; 8. *Scintilla vitrea*; 9. *Galeomma (Paralepida) takii*; 10. *Stylochus ijimai*.

anterior part of the body, and the head appendages: one median and two pairs of lateral tentacles, and a proboscis with strong, black jaw pieces.

Worms of the family Terebellidae are still more adapted for tubicolous life. A common under-boulder species, *Thelepus setosus,* makes a tube of large sand grains and shell fragments stuck together with a parchment secretion, and firmly attached to the rock. The thorax is wide and highly contractile, and irrigates the tube with a water current. The abdomen is longer and narrower, extending through the rest of the tube. The head has neither eyes nor proboscis with jaws, but two sorts of appendage, branched red blood gills, and longer, pale-coloured feeding tentacles. These are deployed in a serpentine course over the mud, and their ciliated grooves carry food particles to the mouth.

The Cirratulidae are related to the terebellids, but lie free in the mud, with no secreted tube. Golden-brown in colour, they are delicate and fragile when taken from the ground. Thread-like tentacles spring out along the whole body, though denser in front, and these reach through the mud and rove about at the surface. The commonest species in Hong Kong is *Cirratulus cirratus.*

Under boulders a variety of species are found clear of the mud. In empty oyster shells may be specimens of *Harmothoe imbricata,* a broad, flat polychaete about 2.5 cm long, of the scale-worm family, Polynoidae. Two rows of fifteen detachable scales (elytra) alternate along the back, enclosing a respiratory space below. The head structure of *Harmothoe,* comprises two lateral and one median tentacle, four eyes and elongate palps. The muscular pharynx is cylindrical and extrusible, and equipped with horny jaws.

Two turbellarian flat-worms (Polycladida) may also be found beneath boulders on mud. *Stylochus* cf. *ijimai,* is thin and leaf-like, brown-speckled and just under 2.5 cm long. It employs two kinds of locomotion: crawling with a rippling wave over the substrate and adapting to its every irregularity, or breaking free and swimming with an undulatory wave. A second stylochid worm is thick and fleshy, the extended body rectangular, its organs seen by transparency from the lower side.

Near the mud, little concavities of the rock contain two of our smallest and most highly modified bivalves, *Galeomma takii* (Galeommatidae) and *Scintilla vitrea* (Leptonidae). The shell is translucent and fragile, the mantle extending over it in a thin fold covered with minute papillae. In the former, the valves are permanently held open at 180° and the animal creeps or attaches like a slug on a wide 'sole' formed by the fused mantle edges. The true foot is a vestigial tongue that can protrude through a small pedal gape. Though apparently free-living, these *Galeomma* species are related closely to other commensal bivalves of the super-family Leptonacea (p. 200) and are possibly living in loose association with the other members of the stone community.

The commonest crab under boulders on mud is the aggressive *Thalamita prymna*. In addition a snapping shrimp *Alpheus* sp.[6] may be found in galleries in the mud immediately beneath the rock. These shrimps are among the commonest Crustacea of warm seas, some have burrowing or commensal habits (p. 271), often living with gobies. Of the large pair of chelae, the left member is large and heavy, almost the size of the carapace, and can be snapped shut producing a loud crack. This chela is closed suddenly under strong muscular power, a peg on the movable finger fitting into a socket on the fixed one. The heavily calcified finger tips come sharply together, and a jet of water is emitted, displaced by the peg. The second legs also have chelae but are very slender, as are the other walking legs. The abdomen is well-developed, its pleopods drawing a strong current through the burrow.

Crevices in Sedimentary Rocks

The chain of islets on the north coast of Tolo Harbour is one of the few places in Hong Kong where sedimentary rocks play a formative part in the shore-line. The narrow fissures and crevices, produced by their laminar structure, confer upon the shore a special importance as a habitat.

Inclined rock slabs at eulittoral level can be sheared away with a hammer to display the community of animals taking refuge. Each crevice consists of an outer part, narrow but relatively open, and a deeper part packed with weathering products and accumulating sediment.

The most conspicuous crevice inhabitants press close together around the edge. The commonest of all is usually the small stalked barnacle *Ibla cumingi* (Fig. 7.16). Like a miniature *Pollicipes* the peduncle is covered with a dark hairy coat and may be oval and flat or elongate and compressed, variously mis-shaped according to the degree of crowding. The terminal shell plates are produced to a sharp point, the terga and scuta diverging to allow the feeding cirri to protrude. Like *Pollicipes*, *Ibla* holds the cirri immobile, taking advantage of the turbulence of the water through the narrow surroundings to bring food. Outside the crevice, at the same tidal level, *Pollicipes* itself lodges in irregularities of the open rock face.

Crowded into the same rank with *Ibla*, and impacted with sand and silt, is a brown sipunculid worm, a species of *Dendrostomum*, up to 2 cm long. With neither segments nor appendages, the body is thick and muscular, and can greatly alter its shape; the cylindrical proboscis can be everted from the mouth, fringed with clusters of branching tentacles that pick up edible material.

Of special interest on such shores in Tolo Harbour is the minute acorn barnacle *Chamaesipho scutelliformis* which cements itself in clusters around the periphery of the crevices. The somewhat larger *Euraphia withersi* is zoned higher up the shore.

Rock crevices harbour numerous polyclad flat-worms; a narrow ribbon-like

---

[6] The Hong Kong alpheids still need urgent taxonomic sorting.

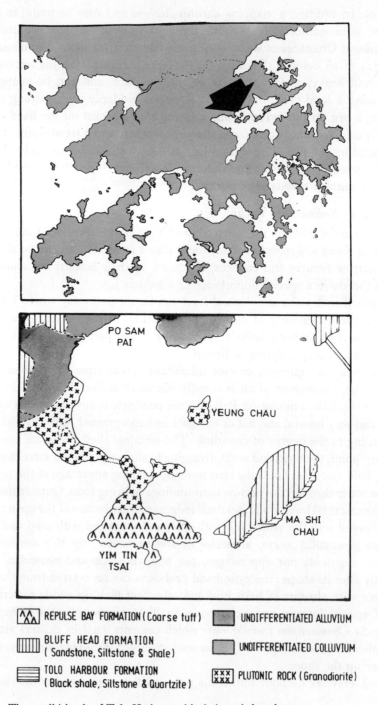

Fig. 7.15. The small islands of Tolo Harbour with their varied geology.

Boulder Shores

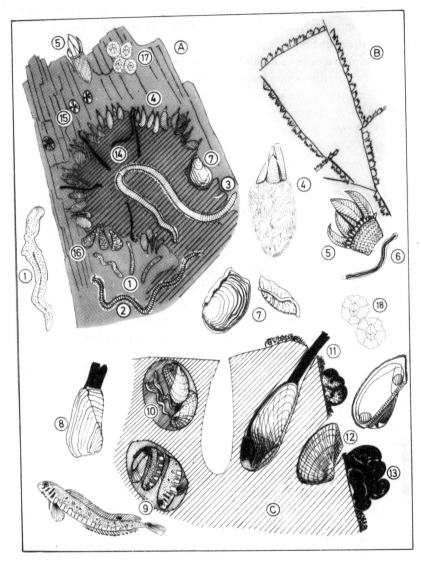

FIG. 7.16. Fauna of crevices at Tolo Harbour.
A. Crevice has been split open to show the arrangement of animals on the surface of a lamina. The zone of lodged sediments is cross-hatched and *Ibla* and *Dendrostoma* (black) are located around its edge; B. Open rock face with rows of *Pollicipes* following surface cracks; C. Low tidal silt-stone split open to show boring and nestling species.

ORGANISMS:
1. *Notoplana delicata*; 2. *Ceratonereis nuntia* var. *vallata*; 3. *Marphysa sanguinea* in burrow within sediments; 4. *Ibla cumingii*; 5. *Pollicipes mitella* with its internal commensal; 6. *Nemertopsis gracilis*; 7. *Claudiconcha japonica*; 8. *Hiatella orientalis*; 9. *Omobranchus loxozonus*; 10. Crevice sheltering *Trapezium liratum* and *Hydroides elegans*; 11. *Barnea* cf. *dilatata*; 12. *Arca avellana*; 13. *Colpomenia sinuosa*; 14. *Dendrostomum* sp.; 15. *Isognomon legumen*; 16. *Patelloida pygmaea*; 17. *Chamaesipho scutelliformis*; 18. *Euraphia withersii*.

species up to 2.5 cm long is commonly found, it is cream and buff-speckled, with rows of simple microscopic eyes round the margin of the head and extending back down the mid-line.

Inter-tidal crevices have in addition a heterogeneous and rather cosmopolitan fauna of small, air-breathing refugees from the land. Hong Kong crevices have not yet been fully explored, but should be examined for some of the typical life-forms already found overseas: small arachnids, including pseudoscorpions *(Obisium)*, maritime spiders, and marine mites *(Bdella* and *Hydrogamasus)*. Chilopods or centipedes also have crevice representatives as well as the primitive wingless insects, Apterygota, with a tail-less collembolan, *Lipura*.

The segmented annelid worms, belonging to the Polychaeta, are important occupants of rock crevices. The most familiar are probably the Nereidae, free-living or 'errant' worms, though they often retreat into narrow spaces, and burrow in sediment. The limb paddles are well developed and locomotion is aided by sideways undulations of the body. For the identification of nereids, like other worms with an armed proboscis, the head must be pressed to extrude this organ when alive, and ligated round the neck to keep it fully displayed when preserved. The special features of the nereid head are shown in Fig. 7.17. The small triangular prostomium carries four eyes and two short thick palps, and the peristomium has four pairs of slender tentacles. The proboscis possesses two sharp jaws and a pattern of minute denticles (paragnaths) that is diagnostic of the particular species. Like other polychaetes, the nereids of Hong Kong still await a proper study but *Ceratonereis nuntia* var. *vallata* is the commonest in the crevices of Ma Tse Chau.

Eulalia viridis is a dark green polychaete, thinner than a nereid, to be looked for in the open parts of crevices. Belonging to the family Phyllodocidae, it has broad, oar-like parapodia and a proboscis with papillae but no teeth. These worms scavenge on dead or moribund animal matter.

Deep in crevice sediments, some of the polychaetes normally seen under boulders on mud, are found. The euniciid worm *Marphysa sanguinea* not only burrows in sediment, but can erode galleries in soft rock, probably using its strong jaws and mandibles. The head structure is illustrated in Fig. 7.17. Like most of the Euniciidae, *Marphysa* is a scavenger, mobile enough to emerge from its tube and actively seek food before withdrawal. One of the tropical Pacific members of this family is the famed *Eunice viridis,* the palolo worm, annually emerging to spawn from borings in coral limestone.

The remaining families of crevice worms are permanently confined to tubes, and feed by gathering microscopic particles. The Terebellidae and Cirratulidae, already found under boulders on mud (p. 118) also occur in crevices. This is a good habitat also for serpulid tube-worms: *Hydroides* in the empty galleries bored by bivalves, and minute *Spirorbis,* scattered just inside the crevice entrance. Tubes

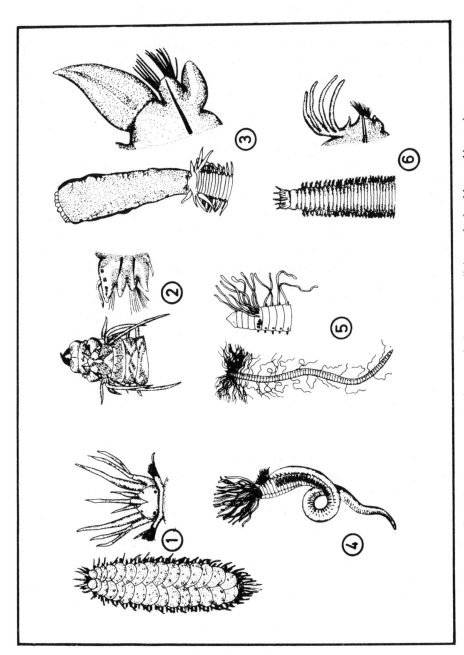

FIG. 7.17. Identification details of some polychaete worms living under boulders and in crevices. 1. *Harmothoe imbricata* with detail of head; 2. *Perinereis cultrifera* with detail of parapodia; 3. *Eulalia viridis* with detail of parapodia; 4. *Terebella ehrenbergii*; 5. *Cirratulus cirratus* with detail of head; 6. *Marphysa sanguina* with detail of parapodia.

of Sabellidae (p. 90) can also be found inserted in sediments and old bored shafts, with the branchial crown expanding beyond the crevice.

Towards the sub-littoral fringe, bivalve molluscs live in small cavities of their own, either eroding the rock themselves or nestling in the burrows left by other species. The piddock family (Pholadidae) are active rock-borers; they are found only in sedimentary strata soft enough to excavate, being wholly absent from igneous shores. The small species, *Martesia striata,* boring in timber piles was discussed earlier (p. 83). In soft rock, *Barnea* cf. *dilatata* can be detected externally by the tips of the long conjoined siphons, often squirting a water jet on disturbance. For further investigation, the bivalve must be excavated by opening the surrounding rock with a chisel or crowbar. The shell is widest at the deep (anterior) end, where it has a large gape. The plug-like foot which acts as a sucker emerges here from a tight pedal aperture, elsewhere the mantle is closed. The two valves come into contact only at a single dorsal point near the beaks. The ligament is reduced and their main attachment is by muscle. During boring, the valves can rock sideways upon their contact point. Contraction of the posterior adductor muscle brings about the divergence of the valves in front. The suction disc of the foot takes a firm grip to pull the shell forward until the file-like sculpture engages the wall of the shaft. At each such stroke the shell is rotated through a small angle, by asymmetric contraction of the pedal retractor muscles. A circular tunnel is thus drilled. Sediment dislodged is conveyed into the mantle cavity at the sides of the foot, and is carried by cilia to the base of the inhalant siphon where it is expelled by back-pressure. In *Barnea,* part of the anterior adductor muscle has spread out to be inserted on the external rim of the shell; here it serves in place of the elastic ligament to help open the valves in their abrading thrust. The exposed muscle is protected in this genus by a third shell-piece, forming a median shield.

A second bivalve commonly found in borings, and enlarging a shallow cavity by its own action, is the ark-shell *Arca avellana*. Wedge-shaped in front, its broad, truncate posterior end, with its coiled umbones, lies flush with the surface. *Trapezium liratum* is a familiar species nestling in abandoned shafts, but is attached by the byssus, and does not bore. Numerous young shells, 3 to 5 mm long, can be found in the open crevice. Small cornet-shaped *Isognomon legumen* also nestle in borings or lie in clusters alongside *Ibla*. Among the commonest nestling bivalves is generally the small and mis-shapen *Hiatella orientalis*.

The bivalve *Claudiconcha japonica* is a distorted venerid characteristic of crevices and shafts. For the first part of its life it is mobile and equivalved. With the selection of a suitable crevice for habitation, however, the right valve grows to closely embrace the other and its shape becomes very irregular. The same species, along with small *Trapezium,* is often found among clusters of oyster shells.

Finally, crevices and empty borings are the haunt of blennies. Fig. 7.16 shows a typical species, *Omobranchus loxozonus,* taking shelter by coiling up spirally within a disused piddock space.

PART THREE
Soft Shores

CHAPTER 8

Mobile Beaches

Origin and Moulding

The coast-lines discussed so far are produced where the initial slope is steep enough for wave erosion to carve back to bed-rock. But there are other shores so flat that the available wave energy cannot carry away the load of sediments that accumulate. These 'soft shores' include a whole spectrum of coasts from mobile sand beaches to mud-flats.

Soft shores can be classified into three sub-divisions according to the degree of wave action. The first group is OPEN SHORES: generally designated as 'beaches', and consisting of mobile sand or gravel, regularly subject to wave shifting. The second and third groups include those shores popularly known as 'flats'; they comprise PROTECTED SHORES, still composed largely of sand, but with increasing proportions of silt and clay, lying in estuaries and bays partly protected from wave action; and ENCLOSED SHORES, being those wide stretches of mud-flat, largely composed of silt and clay, landlocked in harbours or sheltered bays. Protected and enclosed shores are rich and diverse in burrowing organisms, but mobile shores are more 'difficult' habitats with a reduced species lists; and gravel shores (unless finer grades are also present) are almost biological deserts.

Beaches

The 'dynamic' coasts, that is to say, open beaches of fine sand, coarser sand and gravel, maintain a balance between erosion and deposition. These are the only shore habitats with a sub-stratum constantly mobile, where a single storm may alter the profile beyond recognition.

The structural terms used of beaches, 'backshore', 'berm' and 'foreshore' are explained by reference to the diagram in Fig. 8.2.

A beach has been described as 'a large wedge of sand detained between bed-rock, foreshore cliff and sea, the volume of which is adjusted to wave energy in such a way that input and output of sand are balanced'. Most of the beach sand is transported by waves from the sea-bed offshore; material eroded from

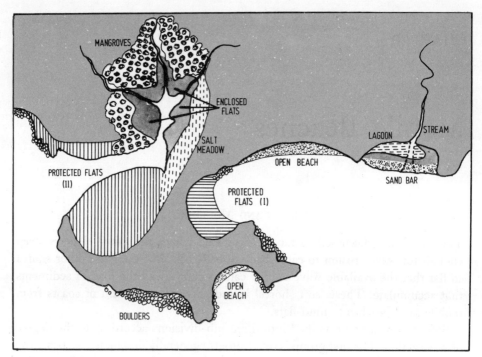

Fig. 8.1. A hypothetical map showing locations of open, protected and enclosed shores.

the headlands finally accumulates within the confines of the bays. On the advancing crests of the waves there is a net transport towards the shore, but little sediment movement under the troughs. Off headlands, in deeper water, wave transport is difficult; and the 'primary' coast is too indented and irregular for any sustained directional transport by offshore currents.

A great part of the beach sand of Hong Kong is derived by the weathering of disintegrating granite. The superficial zones around the large blocks of the granitic land-mass are loosened up, whether by cliff attack or by stream erosion, with subsequent transport. Much of the beach material so resulting consists of quartz grains, with some fragmented feldspar, and later addition of comminuted shell. There is a marked difference in sand grain size between beaches derived from granite weathering and those from rhyolite land-masses.

Most beaches are crossed by one or more streams, bringing down a highly unsorted material for deposit on the shore. Sorting is the subsequent work of the waves, with the largest particles tending to be retained progressively upshore, fine sand lower down and marine silts carried furthest offshore (*see* Fig. 8.2).

The plan of an open beach is ideally a smooth sweeping curve, fixed at either end by a promontory or headland (Fig. 8.1). The sectional profile from top to bottom differs from beach to beach and at any one beach over a period of time.

PLATE 11. *(Above)* Droppers of the mangrove *Kandelia candel*. *(Below)* Fringe of trees of the outer mangrove.

PLATE 12. An aerial view of the Mai Po marshes with the fringe of mangroves to the seaward.

Mobile Beaches

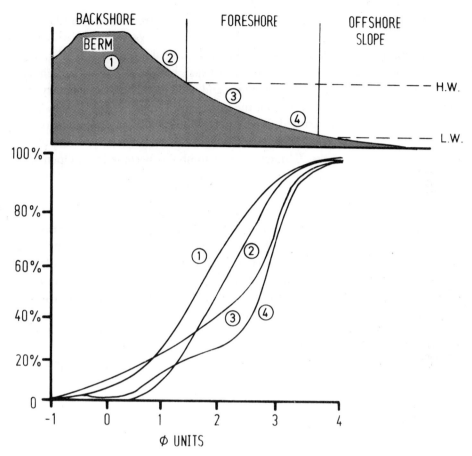

FIG. 8.2. *(Above)* Terminology of a sand beach profile. *(Below)* Cumulative curves for particle size at the levels 1-4 of the same profile. Grain size is in ø units.

The angle of slope is related primarily to grain size, becoming steeper with large size and increasing wave exposure.

Sharp Island and Pak Sip Lap, for example, have large cobble and boulder beaches where the slope exceeds 16°. Though already discussed—in biological terms—under the category of 'hard shores' (p. 86), such boulder beaches are indeed mobile aggregations of large particle size, obeying the same ultimate principles of wave sorting as sand, gravel and shingle.

Sand beaches built of coarser sediment particles are characterized by the reduced wave passage up their steep slopes, and the loss of the return flow (back-wash) by the percolation of water between the large particles. On a fine-particulate sand beach, by contrast, presenting a low angle, the waves run further up and percolate less; and there is thus almost as much water in the return flow as in the swash.

In Hong Kong, tombolos of sand are a typical feature linking two islands of hard granite. The island of Cheung Chau is a typical example. One of us[1] has shown how a tombolo has formed at Sham Wan on Lamma island (Fig. 8.3). Here a sand bar was created as a result of beach build-up. With a subsequent fall in sea level this bar has been stabilized and added to. This allowed entry into the bay so formed behind only from the northeast. Deposition in the bay from streams and the land resulted in greater stability allowing colonization by mangroves, to be succeeded by marsh *(Phragmites)* and ultimately, as at present by truly terrestrial grasses. Ultimately such tombolos become land. Open sand beaches have a definite form and a tombolo is no more than a stabilized sand beach.

As an environment of life, sand-beaches are characterized by their mobility and impermanence. Not only does the profile change markedly with the seasons; but also deposition and erosion of the surface layers are constantly happening with the rise and fall of each tide. Below the breakpoint, waves have been found to disturb sand to a depth of 40% of the breaker height—that is, 40 cm for a metre high wave. Disturbance is deepest at mid-tide where the action of waves is most efficient. On a steep, coarse beach, with waves breaking closer to the shore, energy is concentrated in a narrower band and the shifting of sand goes proportionately deeper.

In two excellent papers on the beach morphology of Hong Kong, Williams[2] studied erosion and deposition on the beaches of Repulse Bay, Stanley Bay and Shek O. Three clear zones can be recognized across the beach: shoaling waves, breakers and swash/back-wash. The whole sequence moves across the beach with each tide. At the water's edge the bottom is of fine sand, under strong back-wash action. Further out, near wave break, the bottom becomes very soft and mobile and the grain size larger. Where the waves are actually breaking, the bottom drops away in a step and a strong under-current can be felt.

As swash moves up the beach on a rising tide a thin veneer of sand is deposited (Fig. 8.4 *above* (1)) and behind this erosion takes place as back-wash moves down the beach by gravity (2). Where this meets upcoming swash, the load is dropped (3). Some of it goes back up the beach; the rest moves out diagonally to beyond the breaker zone, where deposition ceases, as a result of the high wave energy and scouring ability.

Such a sequence appears to be a general one for world shores; though Williams found that in Hong Kong the low waves associated with incoming tides may have

---

[1] Morton B. S. 1978. Molluscan evidence of environmental changes at Sham Wan. In *Sham Wan, Lamma Island. An Archaeological Site Study*. Hong Kong, Hong Kong Archaeological Society.

[2] Williams A. T. 1971a. An analysis of some factors involved in the depth of disturbance of beach sand by waves. *Marine Geology* 11: 145–58; and *idem* 1971b. Beach morphology and tidal cyclic fluctuations around Hong Kong Island. *The Journal of Tropical Geography* 32: 62–68.

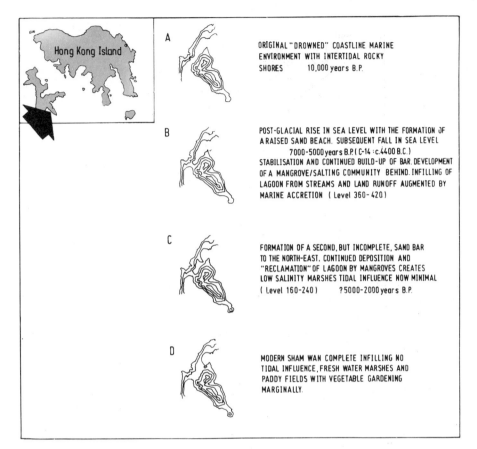

FIG. 8.3. The development of a Tombolo at Sham Wan, Lamma Island.

insufficient energy to move sediment shorewards, so that no deposition takes place on a rising tide. On falling tides, however, the strong back-wash, assisted by gravity, enables deposition to occur with the formation of a wave-break step.

Hong Kong beaches show marked seasonal changes, with variation of wind and waves. There is a cycle over the year, of summer accretion, except during typhoons, and winter erosion. But at Repulse Bay, seasonal accretion was never found by Williams to be greater than one metre, and a clear seasonal change is difficult to identify on account of wide fluctuation in conditions from year to year. Moreover, the embayed nature of Hong Kong beaches and the rather low wave energy generally keep shores relatively stable. Erosion predominates on a rising tide and during winter, deposition on an ebb tide and during summer.

The tropical summer cyclones and typhoons markedly affect the shores of Hong Kong. High seas generate short, steep waves that flatten the beaches,

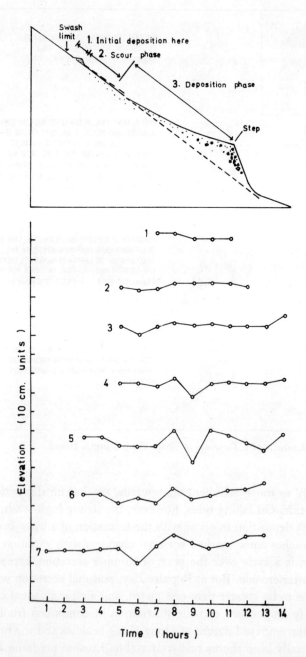

FIG. 8.4. *(Above)* Phases of beach changes in the tidal cycle (from Strahler, A. N., 1966. Tidal cycle of changes in an equilibrium beach, Sandy Hook, New Jersey. *Journal of Geology* 74: 246–68). *(Below)* Tidal cycle of erosion and deposition on a Hong Kong beach. Plots 1–7 are for successive positions from high to low tide.

coarsening the beach material by concentrating sand at the back. East-facing beaches gather the generated swell several days before the typhoon arrives. They recover rapidly by deposition, but full restoration may take several years.

There is further a pronounced raising of sea level due to the very low barometric pressure (980 m.b. or less at a typhoon centre). A typhoon passing during high tide may raise levels as much as 2 metres; long indentations amplify this further, with a tidal bore effect. Piling up of waters is further aided by strong winds. Tolo Harbour experiences a rise of several metres in sea level, 50% more than on the tidal gauge in Hong Kong Harbour, and paddy fields and market gardens are highly susceptible to flooding. Finally, heavy rainfall is associated with typhoons (over 200 mm has been recorded in 24 hours); rapid stream run-off causes slips and gulleying with heavy passage of sediment to the shores.

The marked seasonality in rainfall can have dramatic effects upon streams draining into the sea and across the mouth of which, long shore drift occurs. On long stretches of open beach, wave action holds in suspension large amounts of fine sediment and even sand. A strong longshore current, gradually transports such material until it is deposited as the current is blocked either by a headland or even the outflow from a stream or river. Thus at Pillar Point near Castle Peak a fairly strong ebbing tide moving eastwards at one knot, amply loaded with sediment from the Pearl river and mainland streams is steadily building a spit of sand across the mouth of a small stream, forming behind a sheltered lagoon in which stream sedimentation is steadily occurring.

A further stage in lagoon building is provided at Pui O on the southern coast of Lantau Island (Fig. 8.5) where again the ebbing tide transports material eastwards. Here the sand spit is over 6 m high and during the winter almost completely blocks off the stream mouth forcing it to be diverted to the east. In the summer months, rainfall swells the stream breaching the sand spit and depositing sand as a wide alluvial platform offshore. The seaward facing beach is sparsely colonized with ghost crabs and surf-clams *(Donax)*. Behind, however, an extensive area of lagoon has been formed, slowly filling in and because of its protected nature there has developed sand and mud flats with mangroves. The community of such beaches will be described later (Chapter 11).

Particles

The ultimate character of a soft shore, including its slope, and stability, and its capacities as a habitat, is dependent on the size and distribution of its particles. It is convenient to describe the successive levels of a beach, or to compare it with another, by presenting cumulative curves of particle size.

The scale of sizes generally used is that of the Wentworth series of sieves, logarithmically graded from pebbles to very fine sand. In sediment analysis, oven-dried samples are passed through meshes of descending size down to the

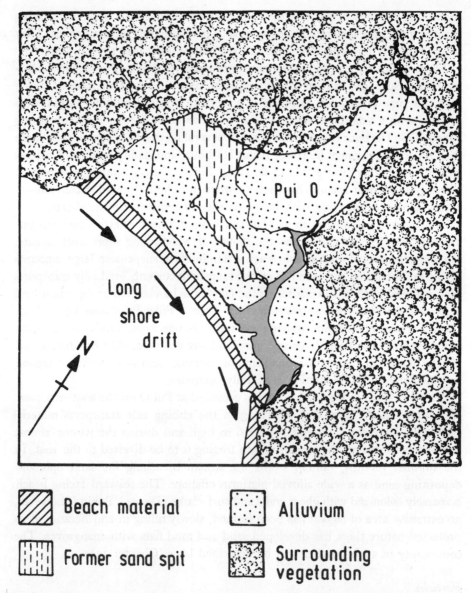

FIG. 8.5. The sand bar, with lagoon behind, formed at Pui O on the southern coast of Lantau Island by long-shore drift.

finest sand. The material passing this last sieve is then arbitrarily divided by fractional sedimentation from a column of water, over a given time, generally 30 minutes. The portion settling out is designated 'silt' and that remaining suspended and recoverable by filtering is 'clay'. Both silt and clay can then, if needed, be more finely sub-divided.

Particle sizes are today conveniently expressed as units on a ø (phi) scale, which is a logarithmic scale of particle diameters, defined as the negative logarithm to base 2 of the particle size in millimetres.

Sizes and slopes of beach material, together with their ø values, and the range of habitats are shown in Fig. 8.2.

Values for different particle sizes are plotted as percentages of the total dry weight of the sample, and may be set out either as cumulative histograms, or more conveniently as continuous curves. Four such curves for grain size distribution on the beach at Pui O, Lantau, are illustrated (Fig. 8.2), from results obtained by Henry Siu Chung Kit. Such a family of curves, from successive stations down the profile, or the *median grain sizes* extracted from them, give a compact quantitative description of a beach.

Where wave-sorting is happening, the largest particles will first be dropped out as the back-wash retreats down the beach. The finest grades thus tend to predominate lowest on the shore, and beyond low water. The *median grain size* (best expressed as the graphic mean $M_z = (ø\ 16 + ø\ 50 + ø\ 84)/3$) will thus be found to decrease passing downshore (Fig. 8.6).

Another important property obtainable from the sediment curves is the sorting coefficient of a sample. This is given by the formula *Inclusive Graphic Standard Deviation*:

$$O_I = \frac{(ø\ 84 - ø\ 16)}{4} + \frac{(ø\ 95 - ø\ 5)}{6.6}$$

which expresses the spread of the curve on either side of the most frequently occurring grain size. A small value indicates a considerable degree of sorting, and thus a high uniformity around the modal value.[3]

DONAX BEACHES

The coarsest of mobile shores, as represented by gravel and pebble beaches, lack both the relative stability of large boulders and the cohesion and water retention of sand. They are virtually biological deserts.

The most mobile of well-inhabited beaches can be typified by the clean, bay-head sand crescents, as at Shek O and Big Wave Bay, on Hong Kong Island,

---

[3] The manual by Folk, R. L. (1965), 'Petrology of sedimentary rocks' in *Syllabus for Courses at the University of Texas* (Geology 370k 383l, 383m) (Hemphills, Austin, Texas) is the best handbook available for the quantitative treatment of the sediments of particulate shores.

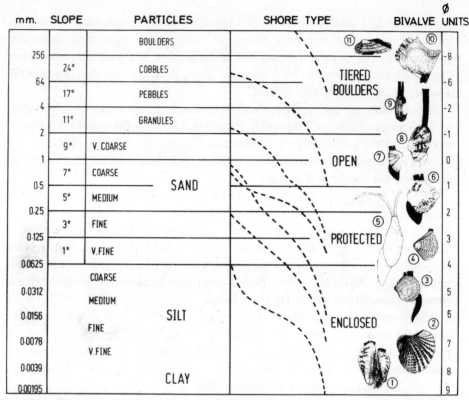

FIG. 8.6. The categories of particle size and their associated habitats and representative bivalve species on Hong Kong shores.
1. *Musculista senhausia*; 2. *Anadara granosa*; 3. *Fulvia aperta*; 4. *Anomalocardia squamosa*; 5. *Pharaonella perna*; 6. *Meretrix meretrix*; 7. *Donax cuneatus*; 8. *Tapes philippinarum*; 9. *Caecella chinensis*; 10. *Pinctada martensii*; 11. *Barbatia virescens*.

Tung O on south-east Lamma and the south-easterly facing sand beaches of Pui O and Cheung Sha on Lantau.

From their typical bivalves, these can be referred to as '*Donax* beaches'. As in the example of Big Wave Bay, shown in Fig. 8.7, such shore-lines present an attractive prospect at low tide. Swept by line upon line of white breakers, they are the most flawless in appearance of all the local coast forms. Clean and generally of white quartz sand, *Donax* beaches are firm and compact underfoot. They do not retain standing water, and are always perceptibly sloped; but the gradient is always gentle, seldom above 5°, the profile steepening only at the berm or where coarse sand gives place to gravel. The sediments of open beaches are free of silt, even-grained and well-sorted. Subjected to constant water movement, they are hence well-oxygenated, with no anaerobic black layer below the surface. After breaking waves have coursed up the beach, the sheets of back-wash seem largely to vanish into the sand, with only a lingering margin of white foam.

Mobile Beaches

FIG. 8.7. *(Above)* Crescentic sand beach at Sham Wan on Lamma Island. The beach crescent and the advancing waves are viewed from the granite ridge above. *(Below)* Profile of the sand beach at Big Wave Bay, showing A, the berm with backshore vegetation; B, the upper beach with *Ocypode*; C, the surf-swept lower beach with *Donax*.

SPECIES:
1. *Scaevola sericea*; 2. *Ipomoea pes-caprae*; 3. *Ocypode mortoni*; 4. *Donax cuneatus*; 5. *Donax semigranosus*; 6. *Hippa pacifica*.

The high-level berm is well exemplified at Big Wave Bay, raised as a flat-topped bank of sand. Its vegetation contains an assemblage of well-adapted sand-binding plants. In addition, the flora of such beaches is characterized by the dominance of herbaceous perennials, creepers, succulents and a large proportion of plants with stiff, spinose leaves including large caespitose grasses. The creepers, include most commonly, the ubiquitous beach convolvulus of the sub-tropical Pacific, *Ipomoea pes-caprae;* the shining green, 'goat's foot' leaves are as characteristic as the purple trumpet flowers, coming into flower around October to November. Common also on sand banks are several creepers of the Papilionaceae: the sea sword bean, *Canavalia maritina;* wild Kudzu vine, *Pueraria phaseoloidea;* and the pioneer creepers *Ixeris repens* and *Launaea sarmentosa* (Compositae).

The most familiar shrub of the sandy or rocky backshore is the so-called sea-lettuce, *Scaevola sericea,* standing as much as one metre tall, or on more exposed shores becoming prostrate and scrambling. The oval, semi-succulent leaves are frequently rolled to reduce transpiration loss. *Scaevola* belongs to the family Goodeniaceae, characterized by its small white flowers, like half stars with five petals in a semi-circle.

Common shrubs, often found on sandy shores include two members of the Verbenaceae: the sea-shore glorybower, *Clerodendron inerme;* and the beach vitex, *Vitex rotundifolia.*

A complex of coastal trees may develop behind the berm. The most familiar and distinctive is generally the horsetail tree, *Casuarina equisetifolia.* Accompanying species can include the two banyans *Ficus microcarpa* and *Ficus superba japonica;* the screw-pine *Pandanus tectorius* and the two succulent species: elephant's ear (*Macaranga tanarius,* Euphorbiaceae) and the introduced prickly-pear (*Opuntia dillenii,* Cactaceae). At Silvermine Bay on Lantau, *Spinifex littoreus littoreus* forms an impenetrable association with *Opuntia.*

Other sandy shore grasses include the large *Neyraudia reynaudiana, Reimarochloa oligostachya, Saccharum arundinaceum* and the ubiquitous *Zoysia sinica.*

Shiu-ying Hu[4] has reviewed the sea-shore plants of Hong Kong and shown that the sparest vegetation occurs on these unstable sandy beaches with greater variety on rocky shores and on mud-flats.

Towards the inter-tidal shore, the berm falls off to seaward on to a stretch of fine sand, only inundated at the highest spring tides. At this level, the ghost-crabs, *Ocypode,* are highly characteristic burrowers, drawing their name from their pallid colouring invisible against the white sand and their habit of running silently and swiftly ahead of the observer. Ghost-crab burrows descend to one metre or more deep, slanting landwards as they go down; and the holes can be detected by the seaward scatter of excavated sand. Burrow-digging is regularly

---

[4] Hu, S. Y. 1974. Sea shore plants of Hong Kong. *Journal of the Chinese University of Hong Kong* II: 315–44.

renewed as the crabs shift about. The adults rarely emerge by day, but the young run about more continuously.

The Ocypodidae are a family never encountered on hard shores; but from now on its representatives will increasingly be met with, burrowing in sand and on mud-flats. The eye-stalks are prolonged, almost half the length of the carapace, inserted close together upon the narrow 'front' that lies between them; they are freely erectile, or able to be folded back into long orbital grooves.

The genus *Ocypode* includes the largest members of the family, with three Hong Kong species, *O. mortoni, O cordimana* and *O. ceratophthalma*. The last-named (*see* Fig. 9.4) is easily distinguished by the long, horn-like prolongations of its eye-stalks beyond the eyes. *O. mortoni* (Fig. 8.7) has simple, rounded eye stalks while *O. cordimana* has rounded eye stalks with but a small horn appendage to the eye. Dr. R. George has shown how on Sai Wan beach on the eastern shores of the Sai Kung peninsula, these three ghost crabs have a differential distribution (Fig. 8.8) with *O. ceralophthalma* being the most seaward, *O. mortoni* occurring along the banks of streams entering the beach, and *O. cordimana* extending up and into the coastal vegetation. By vigorous digging, *Ocypode* can sometimes be recovered from its burrow. While in the sand it can often be heard to stridulate, by drawing the tubercles of the inner surface of the claw across a ridge on the arm: this may be a sexual communication or alternatively a warning that a burrow is occupied. Ghost crabs are easier to catch at night when they emerge from their burrows to feed.

On the sand surface, the ghost-crab moves like a leaf in the wind, supported on the tips of the long claws on only two of the legs. They run in curved salients with such agility that it is necessary to fall on them with a shirt or coat to make a successful 'tackle' or interception.

Like other ocypodids—such as *Scopimera*—the ghost-crab is a sand-shifter; but it is also adapt to catching insects, small crustaceans and even young turtles in the tropics. Like upper-shore grapsids, it only enters the sea as a last resort, and during the breeding season to liberate free-swimming zoea larvae.

Ghost-crabs apart, these open sand-beaches have disappointingly few other animals until the bivalve *Donax cuneatus,* or its close relative *D. semigranosus,* is encountered in permanently fluid sand at mid- to low-tide level. *Donax cuneatus,* a small bivalve with short siphons exists in huge numbers just beneath the sand surface. The shell is strongly built, smooth and triangular, sometimes radially marked with brown or mauve. Digging is immediate and active; and the stream-lined shells can be pulled almost instantly into an upright position, and drawn down rapidly into the sand. The properties of wet sand make it at once easily penetrable, when agitated to a sol phase, and also, as it reverts to a gel, capable of a thixotropic holding force on the burrowing animal. In many places the *Donax* species have been shown to migrate with the rise and fall of the tides.

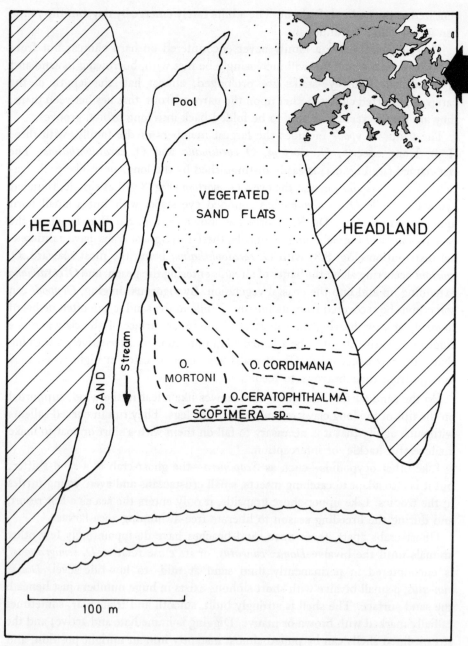

Fig. 8.8. The distribution of the three species of *Ocypode* and *Scopimera* at Sai Wan, N.T.

Donax semigranosus, on Japanese beaches, are said to react to the vibrations caused by the incoming waves of the flowing tide, by emerging from the sand and being carried up the beach. As the wave force lessens, they dig into the sand at the new level in a few seconds. On the receding tide they do not emerge until they detect the outgoing current of water, which then carries them down the beach to a lower level. In somewhat more sheltered situations *Donax* co-occurs with *Atactodea striata,* a mesodesmatid bivalve more common on cobble beaches with a coarse sand matrix.

On open sand beaches in the tropics and sub-tropics *Donax* is regularly accompanied by a sand-burrowing crustacean, the mole-crab, *Hippa pacifica,* belonging to the order Anomura. Ovoid and smooth as a pebble, this stream-lined crab burrows just beneath the surface, and filters particles from the water above. *Hippa* has been found like *Donax* to move with the tides up and down the beach. Mole-crabs have been recorded in Hong Kong only from Tai Long Wan, a superb stretch of open surf beach about 3.2 km long that deserves much more study. Its remoteness beyond road access at the end of the Sai Kung Peninsula is a positive advantage to its conservation.

The Strand Line

Open beaches normally possess a strand line of marine flotsam which is the best place to look for pelagic drifting animals cast up by the waves, often in large numbers.

The commonest of all drifters are probably the stalked barnacles *Lepas anatifera* attaching to floating timber, and *L. fascicularis* secreting its own raft of chitinous foam. Many jellyfish come ashore in the summer months, including the common 'lion's mane' jelly-fish *Cyanea nozakii* to which attaches the shell-less commensal barnacle *Alepas pacifica.* Other common medusae are the scyphozoans *Mastigias papua, Aurelia aurita* and *Pelagia panopyra,* with the trachyline (hydrozoan) *Liriope tetraphylla.* The portunid swimming crab *Charybdis feriatus* is often found clinging to the oral arms of the large jellyfish *Stomolophus meleagris.*

The colonial Disconanthae, closely related to the siphonophores, include the 'by-the-wind-sailor' *Velella velella* and the circular *Porpita porpita.* The ctenophorans or 'sea gooseberries' are egg-shaped predators, closely related to the Coelenterata. Two species, *Pleurobrachia globosa* and *Beroe cucumis* may be washed up on open beaches.

At sea, disconanthids are the prey of the violet shells Ianthinidae. *Ianthina globosa* is truly planktonic and constructs a buoyant raft of a tough, bubble-like secretion from the foot. After storms the beautiful, fragile and violet shells of this species are washed up onto the most exposed beaches.

The strand-line debris is the habitat of beach insects, notably the litter cockroach *Opisthoplatia orientalis,* the sea-shore tiger beetle *Cicindela anchoralis,* the darkling beetle *Gonocephalum pseudopubens* (Tenebrionidae), the staphylinid rove beetle *Bryothinusa* and sea-weed flies of the family Ephydridae. *Cicindela* is a rather large and fast-flying predator, hunting along the strand-line. The larvae occupy burrows on the high intertidal and they too are predacious. Female turtles, notably the green turtle *Chelone mydas,* used to come ashore to lay their eggs on such open beaches. Turtle Cove on Hong Kong Island is aptly named. Such events are rare today even though the turtles are regularly caught at sea, and when they do occur the eggs are quickly taken.

The Range of Soft and Mobile Habitats

Open beaches of mobile sand with *Donax* may be regarded as a simplified central point of Hong Kong's sea shore ecology. From this point the habitat spectrum extends in two directions. As particle size increases, the beaches of coarser sand and gravel (the most inhospitable of habits for burrowing animals), give way to cobbles (p. 88) and then to large stabilized boulders (p. 108) reaching progressively the distinctive biota of hard shores.

In the other direction, with diminishing particle size, a series of protected sand-flats, ranging from the clean shores with *Meretrix* and *Umbonium,* give way to shores of more silty sand, in the shelter of harbours and bays; finally to reach 'enclosed flats' where sand is virtually absent and the substrate varies from stiff clay to viscous or semi-fluid mud.

Fig. 8.6 shows the commonest species of bivalves found to be characteristic of each of these shore types, over the full range of particle sizes.

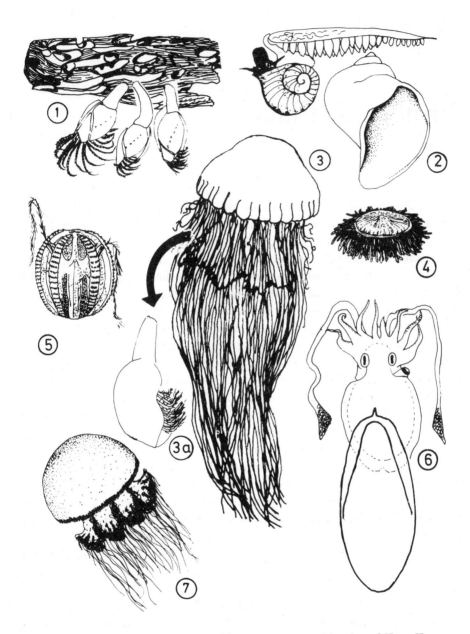

FIG. 8.9. Organisms found as flotsam and jetsam on open sand-beaches of Hong Kong.
1. Wood bored by ship-worms notably species of *Bankia*, *Teredo* and *Lyrodus*. Attached to it are the stalked barnacles *Lepas anatifera*; 2. *Ianthina globosa*; 3. *Cyanea nozakii*, with 3a. commensal barnacle *Alepas pacifica*; 4. *Porpita porpita*; 5. Sea-gooseberry *Pleurobrachia globosa*; 6. 'Bone' of cuttlefish *Sepia pharaonis*; 7. *Stomolophus meleagris*.

CHAPTER 9

Protected Flats

Of all soft inter-tidal shores these are the richest in life; and of the beaches visited in Hong Kong it is perhaps the sand-flats of Shiu Hau on south Lantau that are the most attractive and engrossing.

Protected flats are soft shores cut off from the access of long oceanic waves. They have well-recognizable differences from open shores. The waves that reach them are small and subdued, and the wave-line is far off at low water; foam-edged sheets of water seldom sweep the beach at wave-break, as with open shores.

The long inter-tidal stretch is almost level, sometimes with wide sheets of standing water. Far from being smooth, its surface can be corrugated with ripples, and is often strewn with the signs of burrowing life: the tips of projecting worm-tubes, heaps of spoil from diggings and piled up sand-castings.

Protected flats include a wide spectrum of shores. In most places, the sand continues to be shifted by waves, especially towards the lower shore; but more organic detritus settles as food than on open beaches. As shelter increases animals become freed from the constant shifting of the substrate and the continuous need to reconstruct their burrows. But they are still mostly designed, in contrast with species of estuarine shores, for continued mobility. Some permanent habitations, such as the flexible tubes of polychaete worms can be maintained. Compared too with the enclosed mud-flats found towards the heads of harbours, protected flats are more saline and better oxygenated; and there is less fine, clinging sediment to smother the gills of burrowing animals. Organic debris is however abundant enough on the surface to encourage an abundance of nassariid and potamidid snails.

The best examples of protected flats are found in the still remaining unspoiled outposts of the Hong Kong territory. The drowned 'primary' coast-line moreover provides relatively few situations where extensive soft flats can develop. Their distribution can be reasonably predicted from map study. First, there are shores with relatively slight protection from waves offered by the surrounding land-mass, often adjacent to fully open *Donax* beaches. The best-known example is that of Shiu Hau, referred to here as a *Meretrix-Umbonium* flat.

Protected Flats

Second, there are more protected flats lying within the wider reaches of bays and harbours. Examples will be described here from the cleaner stretches of both Tai Tam Harbour and Tolo Harbour.

MERETRIX-UMBONIUM FLATS: UNDER SLIGHT PROTECTION

The sand-flats at Shiu Hau are shown in Fig. 9.1, in relation to the open beaches of sand and boulders along the south Lantau coast. The protection by land from the full strength of swells from the south-east is obvious from the sketch map. Shiu Hau has a wider sand stretch and a richer fauna than the chain of open beaches, with *Donax* and *Ocypode*, stretching eastwards. Moreover the currents of the flowing and ebbing tides run in distinct and opposite directions, resulting

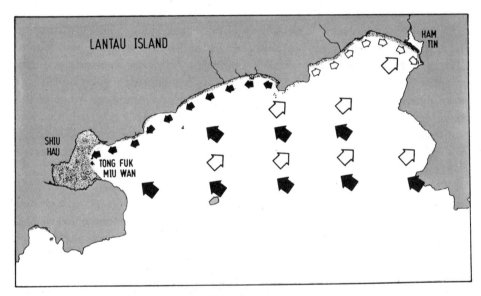

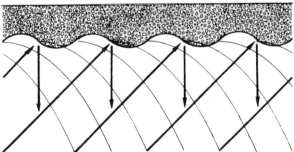

FIG. 9.1. Shiu Hau in relation to the open beaches of south Lantau showing *(above)* the opposing currents of the ebbing and flowing tides create protected flats at Ham Tin and Tong Fuk Miu Wan respectively. *(Below)* Longshore drift, caused by the swash approaching at an angle, carries material brought down from the shore by the backwash along the beach.

in the westward transport and build up of sediments within the landward hook that encloses the flats of Shiu Hau.

Looking down from the coastal road, there is a magnificent extent of the flats exposed at low water. Three natural divisions are visible; and these can be equated with distinct biological zones. First, there is the steep strip of white beach to the landward, corresponding to the *Ocypode* level on more open shores—the *'upper shore'*. The *'middle shore'* is a wide stretch of shelving flats, more or less completely laid bare even at neap tides. Beyond the middle shore, but curving with it in the broad contour of the inter tidal, is the *'lower shore'*, a further stretch only uncovered at low water springs. Gently sloping, so that an extra foot of tide-fall uncovers hundreds of metres of flats, the lower shore is protected from maximal wave attack by a scatter of rock-stacks just beneath the surface at high water.

The upper, middle and lower shores of protected flats can be approximately related to the littoral fringe, eulittoral zone and sub-littoral fringe of hard shores. The first zone is only occasionally submerged, with its occupants virtually adapted for atmospheric existence. The second is a zone of regular diurnal, or semi-diurnal, emersion and immersion; and the third is the mere fringe of the larger sub-littoral zone beyond, only occasionally and briefly emersed. The soft substrate and its biota are however so different from hard shores—with scarcely a single species held in common—that comparisons cannot be too far pressed without loss of precision.

The biological character of beaches depends primarily on the nature of their particles, and the degree to which these are allowed to stabilize, or are shifted by water movements. These properties are in turn a function of wave and current access, as determined by the surrounding land-mass. Preliminary geography, then, can give a reliable prediction of what the shore type will be, and of what animals will live in or upon its sediments. Further, the mollusc shells of the drift-line can be inspected even at high tide, and from these much about the ecology of the shore below can be reconstructed. It is, indeed, upon similar evidence from dead shells that the geologists have reconstructed their understanding of the palaeoecology of the shore and offshore habitats.

Key Species

From the large fauna of the *Umbonium-Meretrix* flats, a small cluster of species can be selected that abound on these shores but are virtually absent elsewhere. The most characteristic sand mollusc of the lower shore is also one of the smallest: the snail *Umbonium costatum*. A primitive archaeogastropod, *Umbonium* is classed alongside the top-shells, Trochidae, having a shell that is smooth and flattened, like a wheel or a biconvex lens. *U. costatum* is about 7 mm across, though a second and scarcer species, *U. moniliferum* reaches 20 mm. The colour pattern

FIG. 9.2. The protected flats of Shiu Hau, Lantau Island showing upper, middle and lower beach zones and areas of cobbles and boulders.
1. *Umbonium costatum* shell in top and side view with animal expanded in its burrow; 2. *Umbonium moniliferum*; 3. *Meretrix meretrix* with a sketch of a villager using the unusual clam rake used to catch *Meretrix*; 4. *Polinices melanostomus*; 5. *Tachypleus tridentatus*; 6. *Nassarius pullus* with 7. the commensal anemone *Paranthus sociatus*; 8. *Edwardsia japonica*; 9. *Paracondylactis hertwigi* 10. *Solen strictus*; 11. *Batillaria zonalis*; 12. *Batillaria multiformis*; 13. *Batillaria cumingii*.

varies widely, hardly any two individuals looking the same: a delicate network of lines and chevrons covers a pale ground, sometimes with high points of pink. Polished and stream-lined for burrowing in wet sand, *Umbonium* has a mobile, leaf-like foot, with which it penetrates the sand to lie just below the surface. The rippled sand is often pock-marked with little *Umbonium* holes. One sieveful may bring up thousands of specimens, each actively thrusting about with the long foot.

Uniquely among the Archaeogastropoda, the Umboniidae have evolved into ciliary feeders. The gill filaments are employed in a way similar to that of the bivalves. The mantle cavity takes in a water current by a short, wide inhalant siphon. Fast-beating cilia carry it between the filaments of the gill, where other cilia filter off particles of plankton and fine surface detritus. Material so collected is fashioned into a mucus cord and carried forward, out of the mantle cavity on the right side of the head, where it is pulled into the mouth by the teeth of the radula.

The second key species of clean flats is the handsome bivalve *Meretrix meretrix,* one of the largest of the venus-shells (Veneridae). The shell is smoothly polished with a lacquer-like periostracum, and measures up to 10 cm in length. The colour is highly variable, with a ground shade of rich cream to golden-brown, and a profusion of radial markings in grey and dark brown. The siphons are moderately long, fused almost to the tip, and active burrowing is effected with the broad, triangular foot. *Meretrix,* among numerous other bivalves, is regularly collected as food by the local people, scouring up the top five or six centimetres of sand by a specially constructed clam-rake, sketched in the small detail of Fig. 9.2.

The king-crab, *Tachypleus tridentatus,* is a third key species of clean sand-flats. A surviving member of the ancient lineage called the Xiphosura, early marine forbears of the Arachnida, living *Tachypleus* ('*Limulus*' as it is still widely called) is a zoologist's prize. Only in tropical south-east Asia, the Caribbean and along much of the Atlantic coast of North America can limulids be found surviving today. Of the two species in Hong Kong, *T. tridentatus* is dark green to almost black in colour, up to 10 cm in overall length, while a much larger grey species, *T. gigas* is found sub-littorally, being relished as food by the local people and often appearing in the markets.

The whole morphology of the king-crab is primitively complex. Like the related arachnids (spiders and scorpions) the body is divided into two parts: cephalothorax (prosoma) and abdomen (opisthosoma). The prosoma, the largest part of the body, is convex and rounded like a polished tin helmet, with two pairs of arachnid-type eyes set in its upper surface. The prosoma carries six pairs of appendages, of which the first is a small pair of chelicerae, with chelae, corresponding to the 'fangs' of a spider. Five pairs of short jointed legs follow these, corresponding to the pedipalps and four walking legs of a spider. The abdomen

Protected Flats

or opisthosoma is a nearly rectangular flat shield, movably hinged to the prosoma and strongly serrated at the sides. Jointed to it behind is the long telson spine. The abdomen carries six flattened appendages, concealing beneath them the thin plates of the 'gill books' through which respiratory water circulates.

The king-crab swims through the water by flapping the abdominal appendages. Most of its time is however spent bulldozing or burrowing in the sand, picking up the worms and other soft-bodied organisms on which it feeds. By turning the king-crab over, the thoracic limbs, normally concealed under the prosoma, can be seen in action. Each has a slender chela with styliform fingers, reaching into the substrate and picking up food like archaic sets of chopsticks. At the base of each limb pair are spiny jaws or gnathobases, passing the masticated food along the venetral mid-line to the mouth.

King-crabs deposit fertilized eggs that hatch into free-swimming larvae lacking the long tail of the adult. These too are found intertidally following metamorphosis.

The hingeless brachiopod *Lingula lingua* is common throughout the tropical Indo-Pacific on protected sand-flats. Though it has a wider local distribution than either *Umbonium* or *Meretrix*, it too is a constant species of protected flats. *Lingula* is the oldest surviving animal genus, being known from as far back as Ordovician times. Belonging to the more primitive of the two groups of brachiopods, it has two almost rectangular shell plates without a hinge, held together only by muscles. Thin and horny green in colour, the shell plates protect the main organs of the body, which gives rise behind to a long muscular peduncle. Up to 15 cm long this stalk serves to anchor the animal in the sand and, by contraction, to withdraw it deeper into its burrow. *Lingula* communicates with the surface by way of a long keyhole-shaped slit in the sand. Water currents enter at either extremity of this slit, the exhalant current leaving in the mid-line after filtering between the filaments of a large and complexly spiralled 'lophophore', the food-collecting organ corresponding to the bivalve gill, and occupying most of the space between the valves. Fig. 9.3 shows the burrowing behaviour of *Lingula lingua*.

Crustacea

Even before the molluscs, the Crustacea are universally the commonest animals of the soft shore. Increasing in numbers and diversity from the temperate zones to the tropics, they have evolved distinctive life-forms according to the tidal level, the nature of the sediments or the degree of shelter. The three levels of upper, middle and lower shores can generally be recognized, by reference to the crabs and related Crustacea found in them. The UPPER SHORE at Shiu Hau is a sloped beach, with a coarser grained sand, less well-sorted than lower down. The water table lies very deep and the sediments dry out between spasmodic wettings

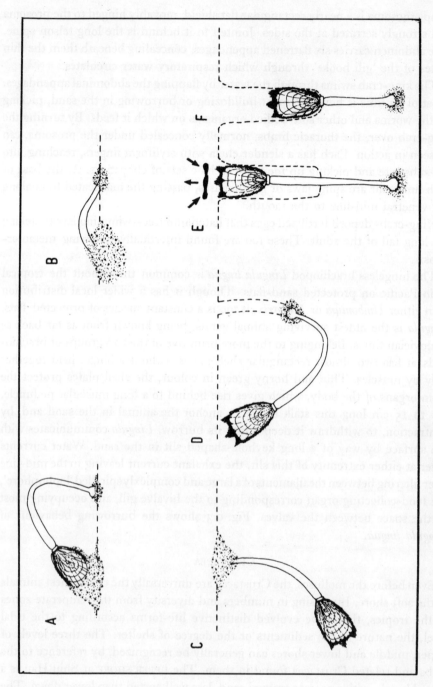

FIG. 9.3. *Lingula lingua*. The burrowing mechanism (A-E) after artificial removal from the sand and (F) the defensive withdrawal position.

by spring tides. In temperate lands, this stretch of the shore typically has talitrid amphipods or 'beach hoppers' burrowing near a moist surface, or aggregating under brown algal wrack. These favourable niches for small Crustacea are rarely presented in the tropics; here the crustaceans of the upper beach are large, and able to penetrate deeply to the moist layers. As on *Donax* beaches, there are large burrows of the ghost-crabs, *Ocypode,* reaching obliquely inshore, with a scatter of sand thrown out to seaward from recent diggings. The three species previously mentioned, *O. cordimana, O. mortoni* and *O. ceratophthalma* (p. 141), frequently turn up on protected flats. The distinctive eye-stalks of the last-named are not developed until the adult stage. Small *Ocypode* of one or more species may be found running about during the day; but the adults remain concealed, coming out to forage by night.

On some parts of the Shiu Hau upper shore, the narrow-leafed maritime grass *Zoysia sinica* can develop; and ghost-crabs are then replaced by the fiddler crab *Uca lactea,* associated with high-tidal cerithioid and ellobiid snails. These habitats develop locally in pockets of shelter, as behind sand-bars or berms. They form a link with the high-tidal saltings and salt-meadows, described in the next chapter; and, as will constantly happen, they break through any clear-cut tidy classification.

The MIDDLE BEACH, sometimes called the neap tide flat, is daily emersed at all but the most extreme of neap tides. Its most abundant animals are surface-trailing potamadid snails (*Batillaria* species), small scavenging whelks (Nassariidae), and small burrowing and sand-sifting crabs, *Scopimera* and *Mictyris*.

The smallest and commonest of all the ocypodid crabs, *Scopimera intermedia* can at once be detected by its surface traces. The burrow runs deep and vertical and the crab slips nimbly down it, leaving trails of sand balls, like lead shot radiating out along the paths of its recent feeding journeys. These are presumably formed when the sand is moist, just after the tide ebbs, to give the picture familiar at low tide.

Carefully dug out of its burrow, *Scopimera* is found to be about $1\frac{1}{2}$ cm across, with the carapace bluish-green, wider behind, and long eye-stalks either erect or folded into orbital grooves. The four pairs of walking legs have distinct oval patches on the femora. Like other ocypodids, *Scopimera* works over the surface sand for its food content, using the maxillipeds, or outer mouth-parts. The sand balls are the remains of this activity, being carried further out from the burrow as the crab moves along its feeding trail.

Scopimera has a convex back, and slender, sickle-shaped chelae, features that become more pronounced in the soldier-crab, *Mictyris longicarpus,* the most highly adapted of the sand-flat Ocypodidae. A widespread crab in the tropics and sub-tropics, *Mictyris* moves swiftly in close and crowded formations over the middle shore. The surface may become strewn with little convex mounds

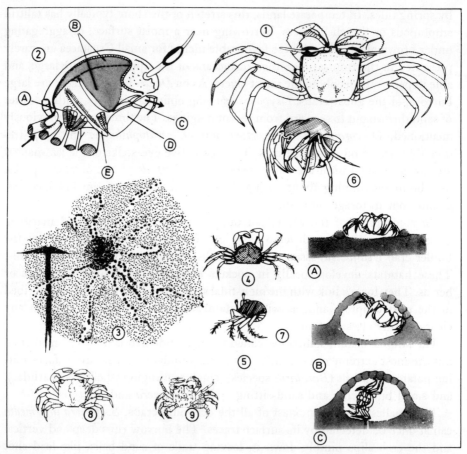

FIG. 9.4. Crabs of sandy and silty flats.
1. *Ocypode ceratophthalma*; 2. Sagittal section of body of *Ocypode* showing how the scaphognathite of the second maxilla creates a ventilating current from posterior to anterior through the branchial chamber. The walls of the branchial chamber are vascularized to supplement gaseous exchange on the gills.

    A. gill;   B. vascularized wall of the branchial chamber;
    C. scaphognathite;   D. second maxilla;   E. gill cleaner;

3. Detail of burrow shaft of *Scopimera intermedia* and opening in section and with radial trails of sand boluses; 4. *Scopimera intermedia*, dorsal view; 5. *Scopimera intermedia*, side view; 6. *Mictyris longicarpus*; 7. Diagrams showing the construction of the burrow of the soldier-crab (after Tweedie, 1952); 8. *Nanosesarma minutum*; 9. *Sesarmops sinensis*.

built like igloos over the shallow burrows. By rapid spiral movements *Mictyris* corkscrews its way into the sand, digging down with the legs of one side while walking backwards with those of the other. As the tide recedes, the soldier-crabs emerge from their burrows to forage. *Mictyris* is one of the few crabs adapted to walk forwards, and the slender, down-curved chelae pick up sand fragments as they progress towards the water. Pear-shaped pellets are left behind in their

Protected Flats

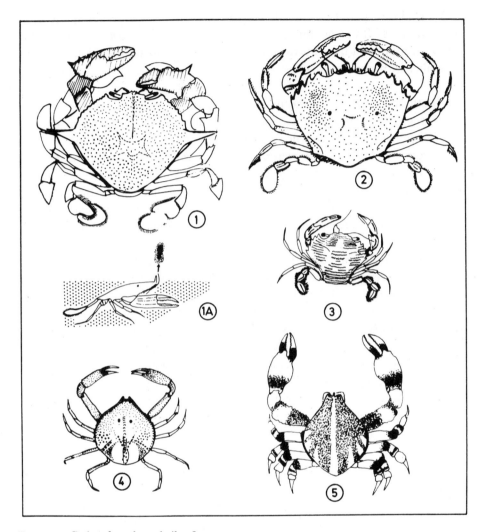

FIG. 9.5. Crabs of sandy and silty flats.
1. *Matuta lunata*, 1A. Shows in side view the position taken up by *Matuta* and *Ovalipes* in the sand; 2. *Ovalipes punctatus*; 3. *Macropipus corrugatus*; 4. *Philyra pisum*; 5. *Leucosia craniolaris*

paths. From scattered feeding groups, the soldier-crabs may form into large armies thousands strong, migrating several hundred metres before digging in anew at the return of the tide.

Taken up in the hand, a soldier-crab appears about the size of a cherry, and almost spherical, powder blue on the black, with the rest white, the leg-joints marked with pink. The chelae are held vertically, and with the body itself high-built, the eye-stalks are no longer elevated, but short and set close to the head.

A delightful account of the habits of *Scopimera*, and of the soldier-crab, is by M. W. F. Tweedie, published in *Malayan Nature Journal* 7 (1952). With a full description of their burrowing and feeding and social behaviour, he draws a parallel with human societies:

> Both species seem to me to exemplify, in an exaggerated degree, quite recognizable types of human social behaviour. The soldier-crabs are cheerful bohemians, living crowded together and out-doing in unrepressed and irresponsible behaviour even those human communities which aspire most strenuously towards this ideal. The timidity and reserve of the *Scopimera* presents a caricature of middle-class life; but here again the picture is overdrawn, for the refined inhabitants of Acacia now frequently venture beyond their garden gates and sometimes even speak to each other.

On the middle and lower shore, the hermit-crabs at once come to notice, forming a scavenging force as important as the nassariid whelks (p. 159). Like the rocky shores, sand-flats have their distinctive hermit species, commonest among them being generally *Clibanarius longitarsus*, typically occupying the long shells of *Batillaria* and *Cerithidea*. Generically determined by their equal chelipeds, the *Clibanarius* species are separated from each other by the colour pattern of their second and third thoracic legs: dull green-striped in *C. longitarsus*, homogeneous grey-black in *C. bimaculatus*, and orange-striped in the large *C. infraspinatus*. The two latter hermits occur on the lower reaches of the flats and occupy shells of sub-littoral gastropods, typically *Strombus isabella, Hemifusus ternatana, Tonna tessellata* and *Babylonia areolata*.

A second genus of hermit crabs, *Diogenes*, is commonly represented in Hong Kong's shores by the pale green *D. edwardsii*. This species and *C. infraspinatus* commonly possess a calyptreid snail, *Syphopatella walshi*, as a commensal. Shiny white, up to about 2.5 cm in length this flattened, limpet-like snail colonizes the inner surface of the lip of the hermit's shell and filter-feeds on the suspended material stirred up by the crab. Inside the hermit's snail shell is also rarely found the tiny commensal leptonid bivalve, *Curvemysella paula*.

On the LOWER BEACH at Shiu Hau, the sands are clean and fine-grained, fluid and mobile when the tide is in, and with the water table up to the surface, even during low tides. The crabs here are different from those of the middle shore. On the lower beach, living in semi-permanent burrows with their eyes just projecting from long stalks, above the surface of the sand can be found four species of *Macrophthalmus* (a further genus of the family Ocypodidae). Lindsay Bones has made a study of the macrophthalmines (sentinel crabs) from Shiu Hau and shown that the highest zoned is *M. convexus* with a bright red-purple inner face to the chelae and a dull grey body. Sharing this zone with *M. convexus* is *M.*

boteltobagoe with bright orange chelae and 2–4 teeth on the carapace edge. Lower zoned *M. latreillei* has a hairy inner surface to the claw and a brown-grey carapace. A fourth species *M. definitus* has a brown carapace with white markings and hairy palms to the chelae. It can be distinguished from the other species by the key-hole shaped gap between the chelae and its 'finger' when the two are opposed. These crabs are territorial, performing like all ocypodids, defensive and courtship displays using the chelae especially. The two upper zoned species build shallow, Z or J shaped burrows with a more complex structure in the female; the lower shore species have more simple, straight, deep burrows. All can wiggle into the soft sands when alarmed and away from their burrow: the two eyes alone projecting above the sand. Such crabs really like soft mud and they will also be found on sheltered flats (Chapter 10) and in front of mangroves. The most important crabs at Shiu Hau belong to the peculiar tribe Oxystomata, smooth-bodied and sinking obliquely into moist sand. The exhalant respiratory openings are two small holes lying close together in the mid-line, between the eyes and antennae, giving exit to outgoing water currents that pass along channels excavated beneath the third maxillipeds. The commonest and most active oxystomatid between tides is the sand-crab, *Matuta lunata*. The carapace, over 3.75 cm across, is smooth and circular, strongly serrated in front, and produced into a strong 'marlin-spike' on each side. The back is greyish-green, speckled in black, and under-parts are pure white. The yellow leg tips form pointed digging trowels, and on the fourth pair rounded swimming paddles. *Matuta* not only digs vigorously but, like the portunids, makes active use of its back paddles. It feeds by tearing soldier-crabs and sentinel crabs to pieces using the sharp-toothed chelae, and migrates actively into *Mictyris* and *Macrophthalmus* territory on the flowing tide.

To a second oxystomatid family, the Leucosiidae, belong the small pebble crabs, with smooth convex carapace and rather frail, slender legs. *Leucosia craniolaris* is grey on the back, with a median and two lateral white stripes. The legs are brown and cream striped. Common at Shiu Hau, it is often accompanied by the polished grey *Philyra pisum*. The closely related *Ebalia scabriuscula*, found in more silty sand beaches, is distinguished by its sculpture of strong granules.

The Calappidae or box-crabs are a third group of the Oxystomata, occurring from low-water mark to the fine sands of the sub-littoral. All these crabs carry a wing-like expansion at either side of the carapace, completely concealing the thin walking legs. The large chelae are serrated like cock's combs, and flattened against the front of the body. The movable finger of the right hand has a large tooth, fitting between two prominences on the fixed finger, and used to break open gastropod shells piece by piece searching for hermit-crabs. *Calappa* rests just beneath the sand, with the exhalant apertures flush with the surface, forcibly emitting a water jet when the crab is taken up in the hand. The inhalant current is drawn through the narrow space between the chelipeds and the cheeks, entering the gill chamber by an aperture at the base of the chela.

Calappa calappa is yellowish-grey and finely corrugated, while *C. hepatica* has strong, wide-spaced tubercles. *Calappa philargius* is cream with red blotches, the carapace being drawn out into short posterior spines.

The swimming crabs (Portunidae) also play an important role at low-water marks. *Ovalipes punctatus* commonly lives in clean shores, relatively free of silt. It is a vigorous crab, swimming fast by its back paddles and quick to take the offensive. The carapace is camouflaged by its sandy colour, relieved by two black spots, and the chelae are slender and needle-sharp, tearing animal food to shreds. *Ovalipes* digs in backwards by sustained strokes of the paddles, until it is buried obliquely with nothing apparent but the dumb-bell shaped respiratory hole.

At Tai Tam Harbour and other protected shores with some silt admixture, the common portunid at low water is *Macropipus corrugatus*. With further addition of silt *Thalamita prymna* is found (p. 92). Two minute grapsid crabs shown in Fig. 9.4 (*Nanosesarma minutum* and *Sesarmops sinensis*) are typical of the upper beach where some admixture of coarse gravel appears.

GASTROPODA

Gastropods creep freely over the middle shore, scavenging or gleaning organic food. Two families: Potamididae and Nassariidae are most typical of protected shores. Two species of long-spired potamidid are commonly found in about equal numbers: *Batillaria multiformis* and *B. zonalis* (*see* Fig. 9.2). The latter is at its maximum lower down the shore; and a third smaller species, *B. cumingii*, is occasionally encountered in areas with greater fresh water input. All species trail over the surface and ingest sand particles to remove the algal films that can give a greenish tinge to a whole protected middle shore. In more silty conditions they feed direct upon organic deposits.

Often found attached to the shells of especially *B. zonalis* and *B. multiformis* is the little, highly conical limpet *Patelloida lampanicola*, the ecological equivalent of *P. pygmaea* which colonizes stones on these sand-flats[1]. Most Cerithiidae (*Terebralia* and *Cerithidea* species) live upon enclosed and far more muddy shores, or amongst mangroves, but one species, the short, stout *Clypeomorus humilis* is often found with *Batillaria* on protected flats.

The typical scavengers of the middle and lower flats, in the tropics and sub-tropics, are the Nassariidae, small and active whelks sometimes outnumbering all the other gastropods. Lightly submerged in sediment, or creeping along

---

[1] Morton, B. S. 1980. Selective site segregation in *Patelloida (Chiazacmea) pygmaea* (Dunker) and *P. (C.) lampanicola* Habe (Gastropoda: Patellacea) on a Hong Kong shore. *Journal of Experimental Marine Biology and Ecology* 47: 149–71.

the surface, a nassariid can be recognized by the dark inhalant siphon carried aloft and moved rhythmically from side to side. In every sieveful of sand these snails will be discovered, thrusting about with the long narrow foot, and making frequent overturns and somersaults. The foot carries a thin operculum behind and is tipped by two filamentary tentacles. *Nassarius teretiusculus* is generally the commonest species at Hong Kong. Others are *N. pullus* with a heavy, smooth callus around the aperture, and *N. festivus* and *N. nodiferus;* all may feed on detritus, or congregate around dead molluscs, crabs, fish debris or other animal remains.

Where fresh-water spreads over the sand from a stream outfall, a small neritid snail *Clithon oualaniensis* becomes teemingly abundant. The shell is thin and smoothly rounded, olive-green or yellow, and variably marked in black, with chevrons, bars and intricate linework and tracery; with a high degree of polymorphism, hardly any two specimens look the same.

The cleaner sands of the lower shore are typically the haunt of the moon-shells, or Naticidae, though these occur also on the middle shore. Among the best adapted of all burrowing gastropods, naticids abound in great diversity on temperate and tropical beaches. Their presence can often be detected by the nidus or spawn coil lying at the surface, forming a coiled strap of mucus-bound sand, smooth or crenate, with the ovipositing snail still at its centre. The common Hong Kong moon-shells are depicted in Fig. 9.8. *Natica gautteriana* is attractively grey-coloured, *N. maculosa* cream and brown-speckled, both with white calcareous opercula. *N. alapapilionis* is cream and beige with delicate white and brown radial markings. *N. lineata* is cream transversely banded by fine brown lines. *N. vitellus* is uniformly dull brown with sharply irregular growth lines while *Eunatica papilla* is white and, uniquely, radially ribbed. Four common species of the second naticid genus *Polinices* occur on Hong Kong beaches. *P. melanostomus* is white with brown patches, *P. peselephanti* is smoothly round while *Polinices tumidus* is pointed and breast-shaped, milk-white with a horny brown operculum. The large grey shell of *Polynices didyma* is depressed and wide-mouthed. In *Sinum japonicum* the shell is extremely flattened, like a small *Haliotis*. The differentiation of these naticids can be achieved by a general inspection of the shell. Underneath the shell the columella forms an 'ear' that is characteristic in form, as for example that of the aptly named *P. peselephanti*.

The smooth naticoid shell is enveloped in life by the muscular side-lobes of the foot. This is extended in front into a large shield, the propodium, behind which the head tentacles and proboscis can narrowly emerge from the shell. By the taking in of water into the propodium and anterior part of the foot these structures are plunged into the sand like the foot of a bivalve. The broad foot is also used to grip and envelop bivalves taken as food, while their shells are bored by an acid secretion from a gland just beneath the tip of the proboscis,

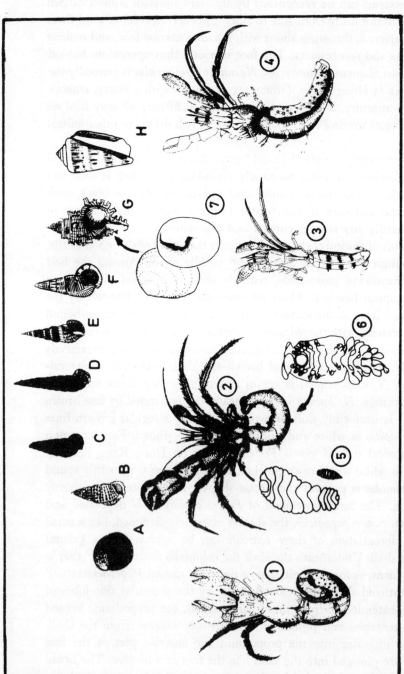

FIG. 9.6. Hermit-crabs of sandy shores and their shells.

CRABS:
1. *Clibanarius infraspinatus*; 2. *Diogenes edwardsii*; 3. *Clibanarius longitarsus*; 4. *Clibanarius* sp.

SHELLS:
A. *Nerita undata*; B. *Clypeomerus humilis*; C. *Cerithidea rhizophorarum*; D. *Batillaria multiformis*; E. *Batillaria zonalis*; F. *Terebralia sulcata*; G. *Murex torrefactus*; H. *Strombus luhuanus*.

COMMENSALS AND PARASITES:
5. *Pseudionella pyriforma*; 6. *Atheleges takanoshimensis*; 7. *Syphopatella walshi*.

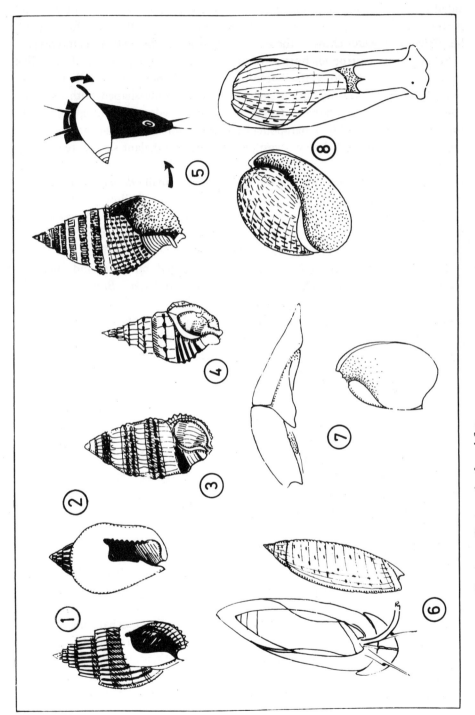

FIG. 9.7. Gastropods of sand flats.
1. *Nassarius nodiferus*; 2. *Nassarius pullus*; 3. *Clypeomorus humilis*; 4. *Nassarius teretiusculus*; 5. *Nassarius festivus*; 6. *Oliva annulata* shell and extended animal; 7. *Philine orientalis* shell and extended animal; 8. *Bulla ampulla* shell and extended animal.

aided by the scraping action of the radula teeth. Bivalve shells are often found perforated by *Natica,* especially near the umbones, where the main food tissues lie. *Polynices tumidus* chips at the edge of *Anomalocardia* shells to gain entry.

Several other gastropods burrow in inter-tidal flats, though some of them live in sediments far muddier than at Shiu Hau. The olive-shells (Olividae) are prosobranchs like *Natica;* but the shell is long and spindle-shaped, with a slit-like aperture. The foot is a large rectangle with its front edge crescentic, built up to form a sand-plough. Thick parapodia enclose the smooth shell at the sides. The tentacles are short and the eyes vestigial. A narrow inhalant siphon reaches up to the surface when the animal is buried. Olive shells are not commonly found between tides in Hong Kong, but the finding of the small *Oliva annulata* in shell drift at Shiu Hau indicates its presence in low tidal or shallow sub-tidal sand. Olives are a tropical element, with one or two large sub-littoral species around the islands to the east. They are certainly carnivores, but their diet, whether of bivalves, other molluscs or worms, varies from species to species.

The soft-bodied Opisthobranchia are represented on sand and mud by several sorts of bullomorphs or bubble shells. Of these, *Bulla ampulla,* is typically confined to mud-flats, but others occur in low-tidal sand. *Bulla ampulla* is the larger, with a more solid, smoothly rounded shell. Enveloped by parapodia at the sides, it is also fronted by a large convex head-shield, containing a gizzard with strongly calcified teeth, used to crush the shells of small bivalves and other molluscs swallowed alive. *Bulla* is found especially at the bases of small rocks partly immersed in low-tidal sand.

The second and more specialized sand-flat bullomorph, *Philine orientalis,* is much better adapted for permanent burrowing in search of small bivalve prey. The shell is thin and vestigial, completely internal, and the external parts are white and strongly muscular.

A third sand-flat opisthobranch is the beautifully purple, brown and red *Doridium cyaneum* (*see* Fig. 10.6).

BIVALVES

By their living biomass indicated by the number of empty shells, it is clearly the bivalve molluscs that dominate the local soft shores. Even a collection of drift-line shells can give a reliable indication of a beach's exposure, sediment type and ecology.

Bivalves can be divided, upon the basis of their feeding mechanisms, into two rather different groupings, though there may be some overlap in habits. Both kinds take fine particles from the environment above them, through the inhalant siphon. The first, and most numerous—the suspension-feeders—draw in a continuous water current and filter off the suspended organisms with the ctenidia (or gills).

PLATE 13. *(Above)* An aerial photograph of a small river discharging into the sea in Tolo Harbour. Note the development of deltaic cusps. *(Below)* The effect of artificial impoundment upon a river mouth in Crooked Harbour. Behind the barrier, water accumulates and marshy conditions dominated by grasses and sedges prevail.

PLATE 14. Contrasting coastal vegetation. *(Above)* An exposed coast of clean mobile sand with the back of the beach stabilized by *Pandanus* and *Scaevola*. *(Below)* An accreting shore. The almost natural stand of dwarf mangroves at Three Fathoms Cove in Tolo Harbour.

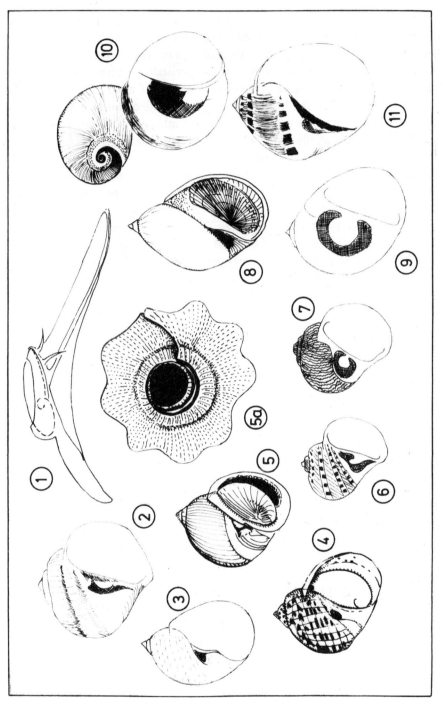

FIG. 9.8. Predatory naticid gastropods of sandy flats.
1. *Sinum japonicum*; 2. *Natica vitellus*; 3. *Eunatica papilla*; 4. *Natica maculosa*; 5. *Natica gauteriana* with 5a. *Natica* spawn coil; 6. *Natica alapapilionis*; 7. *Natica lineata*; 8. *Polinices tumidus*; 9. *Polinices peselephanti*; 10. *Polinices didyma*; 11. *Polinices melanostomus*.

The second group, the deposit-feeders (chiefly the Tellinacea: families Tellinidae, Semelidae and Asaphidae) take a far more concentrated diet, sucked in directly from the bottom layer of surface organic matter. In some species, this appears to be obtained chiefly when the tide is out. The detritus then lies as a concentrated soup, most easily gathered up by peristaltic contractions of the long inhalant siphon, without large accompanying volumes of water. The two siphons are wholly separate, and the longer inhalant is extended along the surface like a vacuum cleaner. In some deposit-feeders the inhalant siphon makes regular channels, diverging radially from where it emerges, showing the paths along which detritus has been sucked up.

Turning first to the suspension-feeders, the largest family is that of the venus-shells and their allies (Veneridae)[2]. There are numerous genera with varied shapes; but most venerids are strong-shelled and rounded, either smooth or with a generally concentric sculpture. The family is named, with a play of Freudian fancy, from the detail of the oval escutcheon (Fig. 9.9), the small area lying to the front of the beaks or umbones. The siphons are always separate, short or moderately long, according to depth of burrowing.

At the one extreme of shell form are the so-called cockles *Anomalocardia* which resemble the true cockles (Cardiidae) both in form and burrowing habit. They lie just beneath the surface, with short siphons and plumply rounded shell that moves only slightly when once embedded. The two Hong Kong species *(A. squamosa* and *A. flexuosa)* are prolific on silty flats near the mouths of harbours, but much less common at Shiu Hau. Their sculpture, both radial and circular, stabilizes the animal in the sand, contrasting with the smooth and mobile venerid *Meretrix meretrix,* common at Shiu Hau.

The low-tidal cockle *Placamen tiara* is another semi-immobile venerid, with a strong sculpture of concentric sharply lamellate ribs.

On the landward part of the middle shore, where coarse sand may be admixed with silt, the heavy and strongly sculptured *Gafrarium tumidum* will be found, ranging far on to muddy shores, and being also a typical bivalve of Pacific mangrove flats. The foot and siphons are short and the habit relatively immobile.

Both the middle and lower shores have several stream-lined venerids, smooth and laterally compressed. The circular *Cyclina orientalis* have relatively long siphons. The small *Circe scripta* is a shallow burrower, like a much compressed cockle, with strong concentric sculpture and very short siphons. The most difficult group to differentiate into species are the paphiid Venus-shells. This task for the Hong Kong species has been undertaken by Vivian Lam. Representatives of three genera are common. The commonest and most widespread is *Tapes*

---

[2] For a comprehensive guide to the Veneridae of China the reader should consult: Zhuang, Qi-qian. 1964. Studies on Chinese species of Veneridae (Class Lamellibranchia). *Studia Marina Sinica* 5: 104–24.

Protected Flats 165

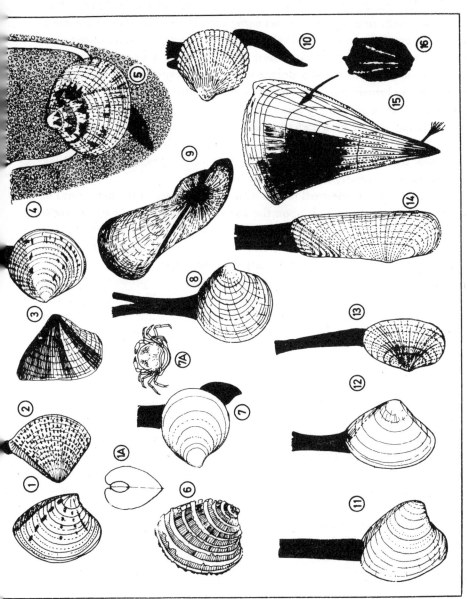

FIG. 9.9. Bivalves of sandy and silty flats.
1. *Gomphina meleagris*, 1a. showing lunule; 2. *Anomalocardia squamosa*; 3. *Anomalocardia flexuosa*; 4. *Circe scripta*; 5. *Codakia tigrina*; 6. *Placamen tiara*; 7. *Cyclina sinensis*, with 7a. *Pinnotheres cyclinus*; 8. *Pitar pellucidum*; 9. *Trisidos semitorta*; 10. *Fulvia mutica*; 11. *Mactra veneriformis*; 12. *Coelomactra antiquata*; 13. *Meropesta nicobarica*; 14. *Solecurtus divaricatus*; 15. *Atrina vexillum*, with attached to the shell, the gastropod; 16. *Amathina tricarinata*.

philippinarum, with a thin, lightly sculptured shell, and extreme variety of colour: grey chevrons and triangles on a cream, buff or reddish ground. Characteristic of coarse-grained sand, sometimes mixed with silt, it is very common at Tai Tam Harbour (p. 89) and at Tolo Harbour (p. 115) in sand between cobbles. Other paphiids likely to be found include *Tapes variegatus,* with radial ribs predominant. In deeper waters, but often washed ashore, are the larger more square species of *Tapes, T. dorsatus* and *T. literatus.* These are plain shells with weak concentric ribbing. *Paphia undulata* too is a deeper water species and is long and narrow with fine zig-zag brown markings on the rather smooth, cream shell.

The genus *Marcia* contains 3 species in Hong Kong. *M. hiantina* is orange-brown in colour with 3 or 4 radiating dark bands, *M. japonica* is uniformly yellow. The shell of *M. marmorata* is brown with trigonal figures of various sizes forming a pattern similar to that of *T. philippinarum.* Unlike *T. philippinarum,* however, it has no radial sculpture.

The small family Mesodesmatidae has two common species in coarse sand or amongst cobbles, generally high upon the shore: *Atactodea striata* and *Caecella chinensis,* already more commonly encountered from boulder shores (p. 115).

At a lower level on clean mobile sand, the two species of donacid, *Donax cuneatus* and *D. semigranosus,* have already been noted (p. 139).

The true cockles (Cardiidae) are found abundantly on the protected shores of the Indo-Pacific, the related and remarkably adapted Tridacnidae, which include the giant clams living in the full tropics. Cardiids are not abundant in Hong Kong, the single common inter-tidal species being the thin, radially ribbed *Fulvia mutica.* Cream and flushed with pink, *Fulvia* is, like most cockles, noted for the agility of its movements. The bright red foot is strongly compressed and can be flexed double so that the animal can make a forceful leap. The siphons are short and largely separate and the inhalant one has a marginal circlet of simple eyes. In deeper waters the thick-ribbed cockle *Vepricardium sinense* is common.

In the Mactridae or trough shells, the siphons are longer than in venerids or cockles and united down their whole length. Notably triangular, the shell is smooth and stream-lined for deep burrowing. Two common Hong Kong species illustrated are the thicker shelled *Mactra veneriformis,* and *Coelomactra antiquata* which is light-shelled and polished grey brown. Belonging also to the Mactridae is the thin-shelled, somewhat posteriorly gaping *Meropesta nicobarica.*

The order Myoida, typically comprising deep burrowing, immobile inhabitants of soft muds *(Mya)* is sparsely represented in the tropics. At around low tide on soft shores in Hong Kong can be found, however, the inequivalve *Corbula* in which the animal lies on the surface on its enlarged right valve with the reduced left valve uppermost.

The Superfamily Lucinacea has numerous members on tropical protected shores. *Codakia tigrina* is occasionally found at low water at Shiu Hau and other

Protected Flats 167

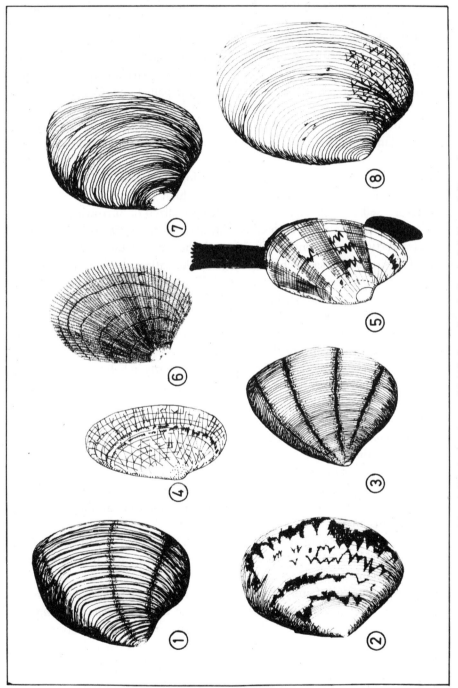

FIG. 9.10. The shallow burrowing paphiid bivalves of sandy flats and shallow sub-littoral waters. 1. *Marcia hiantina*; 2. *Marcia marmorata*; 3. *Marcia japonica*; 4. *Paphia undulata*; 5. *Tapes philippinarum*; 6. *Tapes variegatus*; 7. *Tapes dorsatus*; 8. *Tapes literatus*.

clean or moderately silty sand-flats, as at Silvermine Bay, Lantau. Superficially like a venus-shell, it can be distinguished by the divergent pattern of the sculpture and the absence of the escutcheon. Unlike any similar bivalves burrowing in sand, true siphons are lacking. The posterior mantle edge can instead be extended into a concertina-tube that does duty for an exhalant siphon, without any retractor muscles to form an embayment in the pallial line. *Codakia* embeds itself with the rounded ventral margin downwards, by the alternate elongation, anchoring and contraction of the piston-like foot. The same organ can be used to form inhalant and exhalant holes at opposite ends through the sand above the buried shell.

The muddy shore species of the Lucinacea belong to the family Diplodontidae and will be dealt with when considering this habitat (p. 249).

The razor-shells, Solenidae, are the most specialized of burrowing bivalves, long and pod-shaped, open at both ends. The shell of *Solen strictus* is beautifully adapted for vertical movement within the sand, the foot burrowing as fast as one can dig. The shape has been achieved by great elongation of the posterior half, the umbones lying at the extreme anterior end. The long, piston-like foot occupies half the mantle cavity, and can be pointed anteriorly to plunge into the sand. Once fixed, the tip dilates with blood to anchor the animal, which is then pulled downwards by muscular contractions, the process being repeated so fast that it is almost impossible to capture the animal.

Members of the deposit-feeding Tellinidae vary greatly in habit, sculpture and solidity, though they are all relatively compressed and streamlined. The thickest shell is that of *Arcopagia inflata,* roughly circular in shape; its close relative *A. diaphana* is still thick, but more elongate. *Tellina vestalis, Tellinella virgata* and *Pharaonella perna* are by contrast best adapted for fast burrowing in clean sand. The thin, wide foot thrusts into the semi-fluid substrate, taking fast hold like an anchor as the sand around it sets, and drawing the shell downwards by the quick shortening of the pedal retractor muscles. The foot can be inserted under the shell when lying on the surface, and the whole animal flipped over or thrown some distance.

The posterior end of many tellinids is truncated or squared off, and beaked or inflected to one side. The long, separate siphons are thus guided upwards as the buried shell lies on its side.

The second tellinacean family, Psammobiidae, has several species common on protected flats. The thinly compressed, wafer-like shells of *Soletellina diphos,* and *S. olivacea* are stream-lined, for active burrowing. The periostracum is varnished in shining, deep brown, interrupted in *S. diphos* by two pale rays. In all these species the interior is an attractive mauve. The larger and heavier *Asaphis dichotoma* belongs not to sand beaches, but to coarse sand, intermixed with silt, as on the cobble beaches of Tolo Harbour.

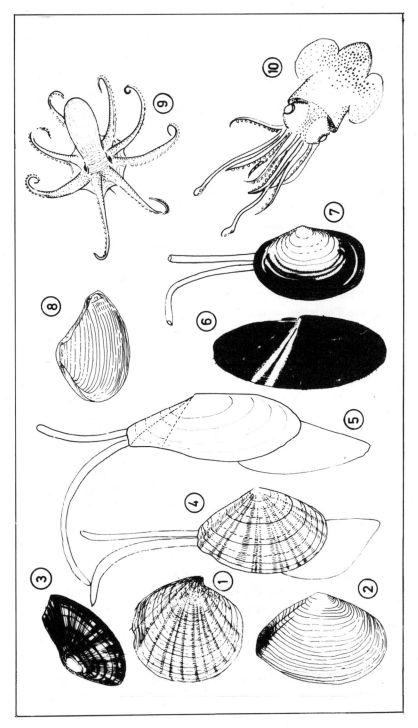

FIG. 9.11. Bivalves and cephalopods of sandy shores.
1. *Arcopagia inflata*; 2. *Arcopagia diaphana*; 3. *Tellina vestalis*; 4. *Tellina virgata*; 5. *Pharaonella perna*; 6. *Soletellina diphos*; 7. *Soletellina olivacea*; 8. *Corbula* sp.; 9. *Octopus membranaceous*; 10. *Euprymna berryi*.

The Solecurtidae, also tellinaceans, contain the narrow rectangular *Solecurtus divaricatus*, with the habit of a razor-shell, long fleshy foot and apparently basally fused siphons.

Final mention must be made of a member of the primitive Arcacea (ark-shells). In the elongate *Trisidos semitorta* it seems as though a pair of hands have twisted it about the umbones to produce a highly characteristic form. The animal lies inclined in the sand with the flat face of the left valve projecting on to the surface.

COELENTERATES

All anenomes are carnivores feeding on small dead or living animals which they capture by means of stinging cells, the nematocysts. The anenomes, though scarcely mobile, in places form an important part of the fauna of protected flats. A common species in sand but also found on boulder and cobble strewn beaches, *Haliplanella luciae,* retains the usual anenome shape, being attached by its broad base to shells and buried stones. *Nassarius pullus* shells sometimes carry the commensal anenome *Paranthus sociatus*.

The burrowing anenome *Paracondylactis hertwigi* differs in being long and pedunculate, with its terminal disc embedded deeply in the sand into which it can withdraw. The tentacles stretch out over the sand surface. Unlike *Cerianthus filiformis* (p. 191) of protected mud-flats *Paracondylactis* does not secrete a mucilaginous sheath within which it is housed. *Edwardsia japonica,* which is about 5 cm long, is another cylindrical anemone modified for peristaltic burrowing. Exposed at the surface, the oral disc carries a spread of 16 tentacles. Below this a narrow neck (scaphus) is followed by the wider body, rust-brown coloured from the adherence of particles. The free-living, bottle-green anemome *Bolerceroides mcmurrichi* possesses a very reduced scaphus and lives on the surface of protected flats, swimming by gentle undulations of its tentacles.

BURROWING WORMS

No naturalist has ever begun to make an adequate study of the worms of Hong Kong's shores. Yet these are, like molluscs and crustaceans, a constant element of sand- and mud-flats; the various families are rather easily recognized and their species are attractive to observe in life.

The most numerous and highly organized worms are the Polychaeta, segmented marine Annelida, with lateral parapodia bearing bristles or setae. Care should be taken to dig polychaetes out without injury. Identification details should first be looked for with the worm alive, observing head, gills and parapodia with the binocular microscope. The best world-wide reference works to the genera and families of polychaetes are those of Fauvel[3] and Fauchald[4]. Polychaetes are by tradition divided into two broad groups: ERRANT, moving about freely, and with head appendages complete, and eyes generally present; and SEDENTARY, living in tubes or burrows, with head organs specialized or reduced, and taking particulate food, either suspensions or deposits.

The first errant family, the Nereidae, have already been described in relation to boulder shores. Less common in sandy shores, several species burrow in estuarine muds (see head detail of *Nereis succinea,* Fig. 9.12). The abundant errant forms of sand-flats are the fast-burrowing Nephtyidae and Glyceridae, carnivores superficially like nereids. *Nephtys* species (e.g. *N. ciliata*) burrow with a rapid side-to-side undulation of the body. Movement is initiated by shooting forward the thick proboscis to take purchase in the sand; its retraction together with body undulation draws the worm downwards. The Glyceridae (e.g. *Glycera rouxii*) are rounder in cross-section, where nephtyids are flat. Their parapodia are smaller and the body is sometimes thrown into spiral coils. In *Nephtys* the proboscis is papillose and lacks teeth; in glycerids it is armed terminally with four recurved teeth, like dark rose thorns.

One of the commonest polychaetes on soft shores is the long (15 cm), extremely thin and deep red *Ceratonereis sp.* It builds long, thin brown tubes in the sand that extend deeply downwards. Digging easily breaks the tubes and the worms but with care one will recover more them 10 with each spadeful. The exact life style of *Ceratonereis* is unknown but is probably either a scavenger or, like most of its clan, a predator of other small, soft-bodied invertebrates.

The Orbiniidae *(Orbinia* and *Scoloplos)* are slender and slower moving, their reddish-brown bodies often tautly stretched when clods of sand are broken up. The segments are short and numerous, flattened on top and more rounded below. Their parapodia form undivided lobes, those over the first fourth of the body giving off red blood gills dorsally (see *Scoloplos armiger,* Fig. 9.12). The head is a pointed cone without eyes or tentacles, the proboscis a frilled rosette, lacking jaws and ingesting fine deposits. The tail carries two long cirri.

The Capitellidae are long, string-like worms, which are blood-coloured deposit-swallowers of silty and muddy shores, producing conspicuous castings. The proboscis is wide and eversible and the parapodia are much reduced. In *Capitella* the posterior part of the body carries tufted gills that can be withdrawn into pits.

Some worms pass their whole lives in tubes, sometimes parchment-like, others of mucoid cement binding sand grains or debris. The tips of the tubes generally rise clear of the ground for about 2.5 cm and with a little care tubes can often be dug

---

[3] Fauvel, P. 1927. *Faune de France :* (v) Polychetes errantes (1923); (xvi) Polychetes sedentaires.

[4] Fauchald, K. 1977. *The Polychaete Worms. Definitions and Keys to the Orders, Families and Genera* (Natural History Museum of Los Angeles County, Science Series 28), pp. 188.

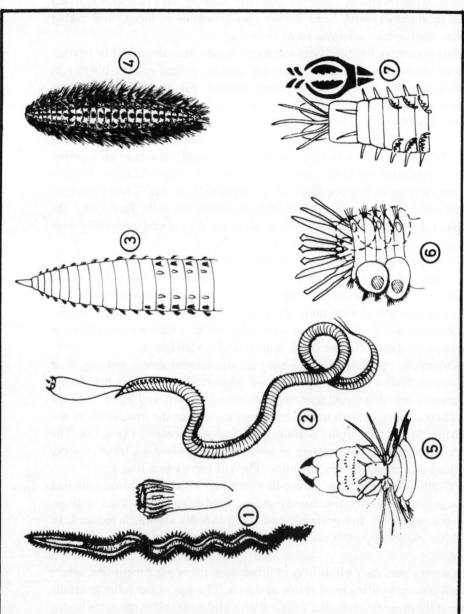

FIG. 9.12. Errant polychaete worms and their identifying features.
1. *Nephtys ciliata* with detail of proboscis; 2. *Glycera rouxii* with everted proboscis; 3. *Scoloplos armiger* with detail of anterior end; 4. *Chloeia flava*; 5. *Nereis succinea* with everted proboscis; 6. *Paralepidonotus ampullipherus* with detail of head structure; 7. *Eumice indica* head and anterior end with details of jaws and denticles.

Protected Flats

out with the occupant intact. The first family is the Euniciidae, already introduced by *Marphysa,* living in crevices on rocky shores (p. 122). Eunicids are technically classed as 'errant' from their well-developed parapodia and head appendages. They can partially emerge from their tubes to forage for animal food.

Diopatra neapolitana (Fig. 9.13), the local common sand-beach member of the Onuphidae, makes a characteristic tube, untidily studded with shell fragments, small pieces of plant debris and sand. There are two frontal and five occipital antennae, a number of pairs of blood-gills with spiral pinnae borne on the branchial stalks.

The Spionidae are small to nereid-sized worms, easily distinguished by their head structure, and living in projecting tubes of fine sand. The head bears two pairs of eyes, but neither palps nor prostomial tentacles. A good recognition feature is the pair of long tentacular cirri from the peristomium, serving to intercept food and convey it to the mouth in ciliary grooves. From the dorsal cirri of the parapodia red blood-gills are reflected over the back.

The Maldanidae construct thick and rather brittle tubes of sand, projecting slightly above the surface. Taken out of its tube, a maldanid is unmistakable, with segments like the long inter-nodes of bamboo, and the parapodia reduced to small enlargements at their junctions. The head is protected by a fleshy shield and the short, thick proboscis swallows surface deposits (see *Maldanella harai,* Fig. 9.13).

The family Chaetopteridae appears to have several species on Hong Kong flats. Related to the Spionidae but much more specialized, they live in tough, parchment U-tubes, sometimes tapering towards the entrance. The tubes of *Mesochaetopterus japonicus* with entrance and exit wide apart, extend down for about half a metre, and careful digging is needed to secure the occupant. This is fragile and elaborately constructed. As in spionids, the head has long peristomial cirri. Behind it lies a narrow region, with broad, forward-directed 'wings' intercepting current-borne food. Next behind are three membranous flaps, derived from parapodia and driving a water current through the tube. Some chaetopterids secrete a filtering net of mucus, from a glandular cup lying in front of these flaps. The habits of the Hong Kong species will repay investigation.

One of the most beautifully adapted tube-worms is *Pectinaria hyperborea,* belonging to the family Amphictenidae, with a curved, tapered tube built of a fragile mosaic of sand grains. The narrower end projects from the sand, alternately receiving and voiding the respiratory current, pumped by contractions of the dorsal body wall. Emerging from the broader end of the tube, the head bears two combs of stiff, golden setae, used both for digging and as a protective operculum. Thin tentacles radiate from the head, bringing food by ciliary grooves; there are also two pairs of blood-red gills.

Three sedentary families, the Terebellidae, Sabellidae and Cirratulidae have

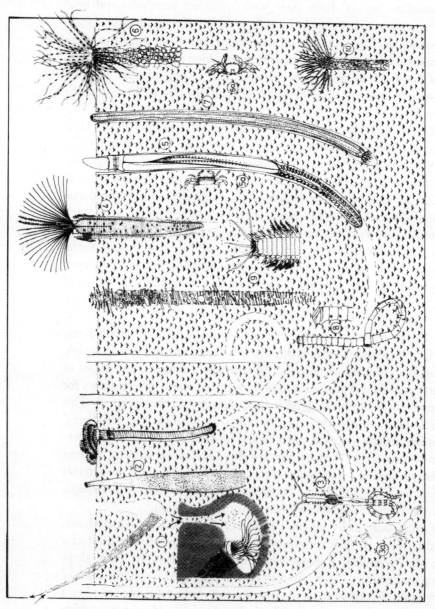

FIG. 9.13. Sedentary polychaetes, also *Balanoglossus* and *Protankyra*, shown *in situ* with tube details and some commensal crabs.

1. *Pectinaria hyperborea* tube in burrow with detail (*below*) of head; 2. *Mesochaetopterus japonicus* inhalant end of tube; 3. *Mesochaetopterus japonicus* animal removed from tube, with 3a. the commensal pea-crab *Tritodynamia rathbuni*; 4. *Sabellastarte japonica* with crown of tentacles expanded; 5. *Balanoglossus misakiensis*, with 5a. the commensal pea-crab; *Pinnixa balanoglossana*; 6. *Loimia medusa*, with 6a. the commensal pea-crab *Tritodynamia horvanthi*; 7. *Protankyra bidentata*; 8. *Maldanella harai* with detail of anterior end; 9. *Diopatra neopolitana* tube with detail of head; 10. *Lanice conchilega* fringe of tube.

already been encountered on muddy shores with boulders, and are also discussed later in relation to low-tidal mud-flats.

Sipunculids and Echiuroids

Both these groups have a thick muscular body wall, but they are not segmented and do not possess parapodia. Both move through the sand by peristaltic contractions, and are common in sandy shores; in their morphology and mode of life they show, however, many differences. The best guide to the identification of these little studied worms is that of Stephens and Edmonds[5].

The Sipuncula have two common species in protected flats, *Sipunculus nudus* and *Siphonosoma cumanense*. The first is shorter and as thick as a finger, with the pearly grey body wall netted by annular and longitudinal muscle bands. There are neither eyes nor appendages of any kind; the anus may be found on one side, slightly behind the head. The proboscis is short and invaginable, covered with minute tubercles and when fully extended has a crown of five flattened tentacles, fringed with side branches.

In *Siphonosoma cumanense* the body is smoother and more slender, dull brown and sometimes thrown into a spiral twist. The only ornament is of minute horny denticles, which are especially prominent in rows along the eversible proboscis. When the proboscis is extended a fringe of branched tentacles is revealed, serving to pick up surface deposits.

The echiuran worms have two common sand-flat representatives in Hong Kong, both thick and sausage-shaped but capable of great peristaltic change of form. Unlike the sipunculids, the anus is at the posterior end, and the head carries a soft, permanently external proboscis, a tongue-shaped appendage, with a groove running down it to the mouth. *Thalassema fuscum* is dull brown with a pale yellow proboscis, extended up to the surface from a short vertical burrow (Fig. 9.14). In life the proboscis is extended horizontally over the sand surface and, covered with attached particles of detritus, periodically withdraws. The sand around the burrow is thus etched with radiating feeding lines.

In *Ochetostoma erythrogrammon* (Fig. 9.14) the burrow is far more spacious, smoothly mud-lined and up to 2.5 cm wide, constricting at the oral aperture to a narrow neck. A deep vertical shaft turns into a horizontal gallery, up to half a metre deep, opening vertically by a vertical and tapered anal shaft. The two apertures lie some distance apart, and deep digging may be needed to capture the occupant. As the animal retreats down the vertical shaft, water can often be seen bubbling in the long stretch it has passed through. The opening of the anal shaft can sometimes be detected first by the pile of fine, black sand around it. The direction of the horizontal shaft and the identity of the oral shaft can then

---

[5] Stephens, A. C. and Edmonds S. J. 1972. *The Phyla Sipuncula and Echiura*. London, Trustees of the British Museum (Natural History).

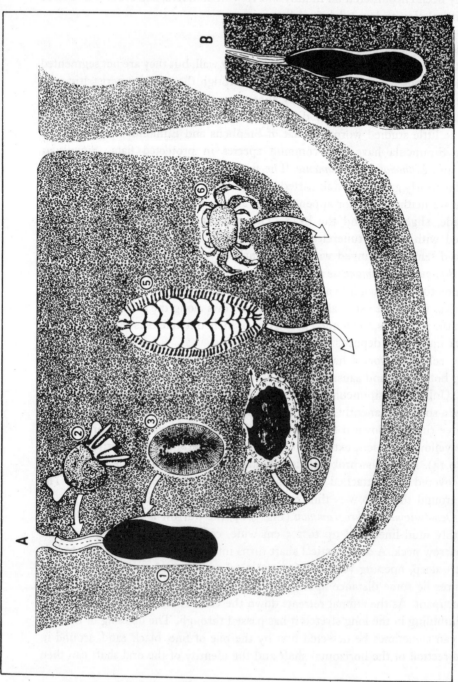

FIG. 9.14 A. The 'Innkeeper' and its associates. 1. *Ochetostoma erythrogrammon* in the burrow; 2. *Sigaretornus plana*; 3. Un-named red polyclad; 4. Un-named leptonid bivalve cf. *Achasmea* sp.; 5. Un-named scale-worm; 6. *Mortensella forceps*. The approximate location of each commensal in the burrow is indicated by an arrow. B. *Thalassema fuscum* in its burrow, showing the difference in burrow shape between the two echuiroids.

be ascertained, by injecting water into the anal shaft and seeing where it flows out.

The animal is firmly muscular, reddish-brown, with a bright yellow proboscis. The feeding mechanism of *Ochetostoma* is still uncertain. In the related American echiuran, *Urechis caupo,* a cone of mucus is secreted from the head end, to form a funnel for intercepting particles from the water current pumped peristaltically through the burrow. Loaded with food, the mucous cone is periodically ingested.

Like *Urechis, Ochetostoma* has an intriguing range of commensal or 'guest' species within the burrow, procuring for it the name of 'inn-keeper'. In the upper part of the oral shaft of *Ochetostoma,* a small scarlet flat-worm glides over the burrow lining. Clusters of specimens look like drops of blood. Also in the upper reaches may live a group of the curiously flattened snail, *Sigaretotornus plana.* The deeper parts of the burrow are inhabited by a small blind crab, *Mortensella forceps,* an un-named leptonid bivalve (probably an *Achasmea* sp.), and an unidentified polynoid scale-worm.

Nemertines

From rocky shores we have encountered the nemertine *Nemertopsis* living commensally within the mantle cavity of the barnacle *Pollicipes mitella.* Two other nemerteans may occasionally be dug up on sandy-muddy beaches. Both are probably species of *Lineus,* but their specific names require confirmation. The most common is a chocolate brown species 1 cm wide but up to 1 metre in length and characterized by a white stripe running medially along the length of the back. A second, somewhat shorter species has a white head gradually changing to a beautiful rose-red posteriorly and terminating in a delicate tentacle. Both are eyeless, each with a wide mouth and are probably predators of other worms. They are extremely delicate, difficult to unwind because of their length and secrete copious amounts of mucus.

CHAPTER 10

Sheltered Flats

Clean sand-flats with *Meretrix* and *Umbonium* are not very numerous in Hong Kong. The most extensive of the soft shores lie in further shelter, generally in the mouths of long harbours or inlets, lower down than the true mud-flats. The relationships of all these communities are suggested in Fig. 10.1, showing that 'protected shores' include two sub-divisions more sheltered than *Meretrix-Umbonium* flats. The most widespread are the *Uca-Anomalocardia* shores. Good examples are to be found in the flats of Three Fathoms Cove, Long Harbour, and Wu Kwai Sha, (both parts of Tolo Harbour); and in the small, much visited tidal flats of Tai Tam Harbour. The second sub-division is for softer strips of low tidal shore, characterized by *Atrina* and the anemone-like *Cerianthus*. The example described is at Hoi Sing Wan (Starfish Bay) close to Wu Kwai Sha.

These flats differ from *Meretrix-Umbonium* shores in that the sediments are less sorted, and contain much higher fractions of silt and clay. The water becomes turbid at low tide by the stirring up of suspended silt, and in some patches an intermixture of stiff clay can make sieving difficult.

As could be expected, there is some overlap with *Meretrix-Umbonium* flats; but the faunas on the whole are fairly distinct. The upper shore—in silted harbour flats—is pre-eminently the territory of the fiddler crabs *(Uca)*. The middle and lower shores are dominated by the venus-cockles *Anomalocardia squamosa* and *A. flexuosa*. At low water live the true cockle, *Fulvia mutica*, and the star-fish *Archaster typicus*, with the urchins *Salmacis* and *Temnopleurus* both common. *Meretrix* and *Umbonium* never occur; and the ghost-crabs *Ocypode*, with the fast-burrowing bivalves *Tellinides* and *Tellinella*, and *Soletellina diphos*, will generally be looked for with less guarentee of success. Of the crabs, *Scopimera* remains common in sand stiffened with clay. *Mictyris* is found only on the cleaner areas of *Anomalocardia* flats, where the sand fraction is predominant; its predator *Matuta lunaris* is rather scarce.

At the other extreme, these flats of silty sand must be distinguished from 'mud-flats', which belong to a category of shores wholly enclosed within the shelter of the surrounding land-mass. Silt and clay are brought down by streams

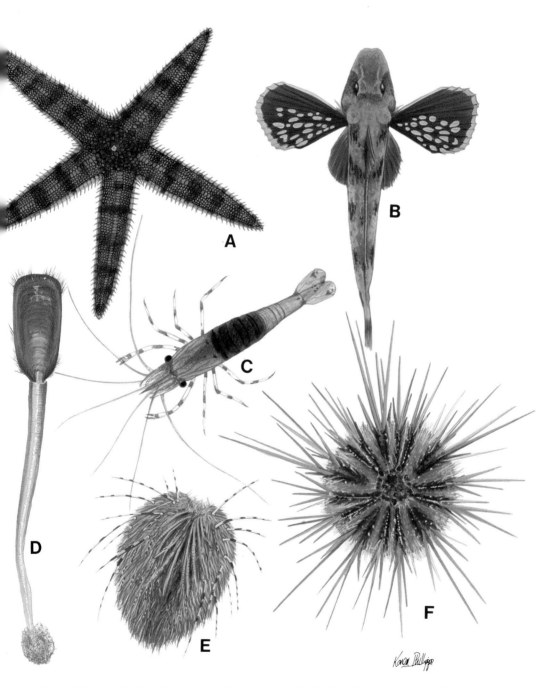

PLATE 15. Sandy shore fauna: A. *Archaster typicus*; B. *Chelidonichthys kama* (male, in courting colours); C. *Palaemon serrifer*; D. *Lingula lingua*; E. *Lovenia elongata*; F. *Temnopleura reevesi*.

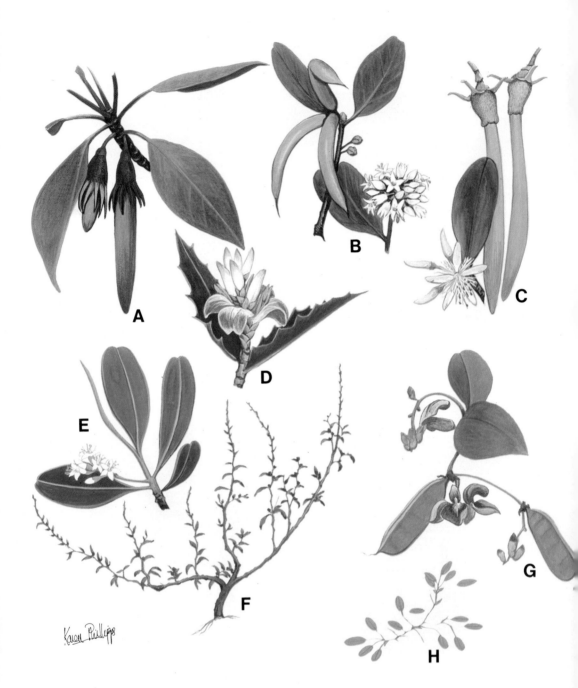

PLATE 16. Mangrove plants: A. Droppers of *Bruguiera conjugata*; B. Droppers and flowers of *Aegiceras corniculatum*; C. Droppers and flowers of *Kandelia candel*; D. Flower of *Acanthus ilicifolius*; E. Flower of *Lumnitzera racemosa*; F. *Suaeda australis*; G. Flowers and seed pods of *Canavalia maritima*; H. *Halophila beccarii*.

Sheltered Flats

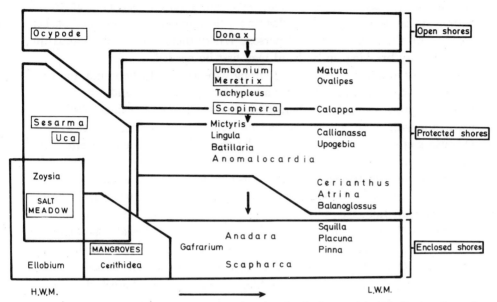

Fig. 10.1. Sequence of open, protected and enclosed soft shores with their key species and divisions from high water *(left)* to low water *(right)*.

and deposited on mud flats with little subsequent re-arrangement by waves and tidal currents. The surface layers are sticky and adherent, and progress on foot is uncomfortable or hazardous. The key organisms are very distinct from those of *Anomalocardia* flats.

The Study of Soft Flats

As compared with rocky shores, beaches and flats are apt to be neglected or passed over unknown by the uninstructed. The staple food species, which are the bivalve molluscs, are traditionally taken by a clam-rake, constructed after the pattern in Fig. 9.2, drawn at Shiu Hau. But for the rest of the burrowing life, rich as it is, soft flats do not yield their information without some persistence, and even discomfort.

At an average low tide, and through daylight hours during the whole winter, much of the shore will be covered by 30 cm or more of water. The sediments must thus be sampled under water, by wading in, equipped with a shovel and sieve. Plimsolls and shorts are the most serviceable garb, but the upper part of the body should ideally be kept warm during a wading exercise that may last for an hour or more.

Where the sediments have natural cohesion from their clay content, an ordinary garden fork may be used, a spade being too resistant to stiff clay and the handle likely to break. For less compact substrates, including clean sand, a suitable shovel can be constructed, with edges raised like a sugar trowel. Sediments can be dropped from this into a sieve under water and well shaken, ideally with the help of a second person. Where there is some clay admixture, a preliminary breaking up by hand may first be needed.

A 1 cm or even a 0.5 cm mesh of strong wire will hardly be found too large, serving to pass the sand rapidly and retain the molluscs and other macrofauna. For worms and the smaller, more elusive Crustacea, a finer mesh can be used; but with these species the violent technique of sieving is best avoided. Quantities of sediment should be carefully scanned for small animals after being lifted clear of water.

Tropical and sub-tropical protected flats tend to have large numbers of species, mostly with few individuals. Repetitive and patient sampling will thus be needed to discover the full species list. If the number of new finds after each sieving is plotted cumulatively against the sample number, a long series of digs is required before the curve begins to flatten out at the top. By contrast, with the simple fauna of a *Donax* beach, one or two digs will establish the whole list. This procedure can be a convenient working measure of the faunal diversity of different soft shores.

Glass-bottomed viewing cylinders can be used for observing surface species, such as starfish, dragonets and gastropods, along with *Natica* spawn and the openings of burrows, with piles of spoil or faecal castings. All these should be looked for before the sediment is stirred up by digging and walking about.

On favourable days when a stretch of the lower shore is left dry by a good spring tide more information is available. Burrows can now be followed downwards into the sand, and carefully opened up to reveal the occupants. Their course and shape should be carefully noted, with exits as well as entrances if separately present. Above all it is important to note an animal's position in its burrow, and the presence and site of any commensal species. All these details of mode of life and ecology will be missed if a heap of specimens is simply shaken in a sieve.

To study soft flats, then, a student must be prepared to get wet and dirty; or, on dry ground (when low spring tides allow) to keep an alert watch for details, with head and eyes close to the ground. The tide tables should be scanned for days with good low tides; these will be infrequent enough, so should not be wasted. Finally, for an extensive sieving programme, a small place like Tai Tam Harbour should if possible be avoided. Any specimens beyond modest reference needs should be returned alive to the sand.

Tai Tam Harbour: *UCA-ANOMALOCARDIA* Flats

Tai Tam Bay and Harbour form a deeply incised, drowned inlet on the southern coast of Hong Kong Island, some 3.2 km long and 500 metres across at the mouth (Fig. 10.2). Originally the Bay would have taken the waters from mountain streams that have now been dammed to create Tai Tam Reservoir. Today relatively little fresh-water run-off is received; the ecological balance is dependent on the interaction of marine conditions at high tide with the influence of the rather small streams that cross the Bay as the tide recedes. The deposit of terrigenous material by these streams has created the present soft shore, including the sand-bars of the Bay that now form ecologically stable protected flats.

Though only of small extent, Tai Tam has an important scientific potential, being the only part of the coast of Hong Kong Island (predominantly a drowned rocky shore) to develop protected inter-tidal flats. Its environment is today under high pressure from the village farmers, gardeners and boat people, and from recreation and industry, including the extension of an unsightly quarry on the inner shore of Cape D'Aguilar.

Undergraduates, school classes and professional biologists are also making heavy use of Tai Tam. The area is too restricted to survive these demands, even those of the biologists, without careful conservation and management. Two protective measures immediately needed are:

1 Immediate halt to all proposed (authorized or unauthorized) building or agricultural plans for the Bay, with suitable compensation for people on the adjoining land already reclaimed by tipping.

2 Inclusion of the Bay and Harbour as the nucleus of a multi-purpose Nature Conservation and Recreational Area.

FLORA AND FAUNA OF TAI TAM

The protected shore at Tai Tam Harbour is shown in a general view in Fig. 10.3. The same three levels, upper, middle and lower, can be distinguished as on the sand-flat at Shiu Hau. But the UPPER SHORE is consolidated with clay, and fiddler-crabs *(Uca)* dominate the burrowing community, entirely replacing the ghost-crabs *(Ocypode)*.

At Tai Tam the upper shore, as far as the reach of high spring tides, is clad with the halophytic grasses *Zoysia sinica* and *Sporobolus virginicus*; it forms the first example of the 'saltings' or salt-meadows that fringe the harbour coasts (*see* Chapter 11).

The trees and shrubs of the backshore form a characteristic back-drop gradually giving place in front to halophytes. The yellow *Hibiscus tiliaceus* is almost everywhere common, generally with *Pandanus tectorius, Clerodendrum inerme, Derris trifoliata, Litsea glutinosa,* and, furthest shoreward, the succulent shrub *Scaevola frutescens*.

Further out than other trees, or actually growing on the beach, are the mangrove-associate *Excoecaria agallocha* (one of the Euphorbiaceae) and, until recent destruction, small stands of the mangrove *Avicennia marina*.

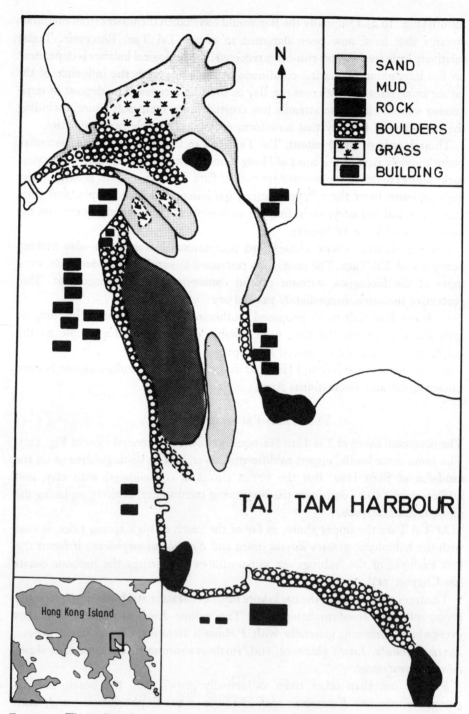

Fig. 10.2. The various shore types of Tai Tam Bay and Harbour.

Sheltered Flats

FIG. 10.3. The protected flats of Tai Tam Harbour. The zones from shoreward are A. *Zoysia* salt-meadow with cobbles and backshore of *Hibiscus*, *Lantana* and *Excoecaria*; B. middle shore with *Scopimera*; C. lower shore.

SPECIES:
1. *Parasesarma plicatum*; 2. *Tmethypocoelis ceratophora*; 3. *Uca chlorophthalmus* male; 4. *Uca lactea* male with 4a. large chela and (*right*) smaller feeding chela; 5. Detail of ucid burrow; 6. *Scopimera globosa* with (*arrowed*) a section through the burrow; note the radial arrangement of the sand boluses; 7. *Alpheus brevicristatus*; 8. *Upogebia major*; 9. *Callianassa japonica*.

Crustacea

In front of, or amongst the *Zoysia* grass, there is a rich community of salt-meadow gastropods described in Chapter 11. Here the characteristic crabs of silty flats are described, beginning with the upper shore.

The fiddler-crabs, forming the large genus *Uca,* are a colourful part of the fauna of every tropical muddy or silty shore. Of small to middle size (2–3 cm across the carapace), they run freely over the surface or sideways into shallow but complicated systems of branched burrows. By squatting motionless with a pair of binoculars they can be clearly observed, and much can be discovered about their ecology and entrancing behaviour.

The *Uca* species are characterized by the one grotesquely enlarged claw of the male (generally but not always the left). Bright and distinctively patterned, this claw is often as large as the rest of the crab, and has an intra-specific signalling function. Not only are its colours 'chic' and attractive, but its movements are highly ritualized. The crabs emerge from their burrows to engage in the arm-waving distinctive of their species. The large chela is used both for beckoning to females, and also in ritual inter-male contests. Dr Jocelyn Crane[1] has given a detailed account of the different display patterns among the *Uca* species. Generally, however, *U. vocans,* lower-zoned, has a less sophisticated repetoire of behavioural characters than the higher-zoned species.

The remaining chela (both of them in the female) is minute, with the hand hardly larger than the other segments. But the finger tips are beautifully specialized, being converted into narrow spoons, fringed with setae for picking up sediment and passing it to the mouth. The fringed maxillipeds are also specialized for sand-sifting.

On protected flats, *Uca vocans* dwells in silty sand, generally lower down the shore. The large claw is pale mauve, with dull yellow. The colour and shape easily distinguish it from *Uca lactea,* where the adult claw is slender and less inflated than in *U. vocans,* being pale yellow and white.

The highest zoned uciid at Tai Tam, often living amongst the beach vegetation of *Zoysia,* is *Uca chlorophthalmus.* The carapace varies from deep red or orange with black outline, to cross-marbled with bluish-green and black. The enlarged chela is bright scarlet and white. The largest of the Hong Kong ucids, *U. arcuata,* lives chiefly in mangrove muds (p. 228) and will be dealt with later.

As at Shiu Hau, the middle shore has both potamidid snails and scopimerid crabs. At Tai Tam this stretch looks dull green from a distance, its coarse grains of quartz sand being filmed with microscopic algae. Shells of *Batillaria multiformis, B. zonalis, B. cumingii* and *Clypeomorus humilis* are all common either

---

[1] Crane, J. 1975. *Fiddler Crabs of the World (Ocypodidae:* genus *Uca).* Princeton, Princeton University Press.

alive or hermit-occupied. The neritid *Clithon oualaniensis* is found in places of brackish influence.

TABLE 10.1 CHARACTERISTICS OF HONG KONG UCIIDS

Species	Large chela	Adult male carapace	Legs
U. lactea	Smooth; yellow, white-tipped.	Mottled, grey-brown.	Speckled brown.
U. chlorophthalmus	Smooth; coral-red, white-tipped.	Jet-black, splashed with blue.	Black.
U. vocans	Granulose; pale mauve with dull yellow.	Brown in front, yellow behind.	Dull yellow.
U. arcuata	Granulose; deep, rust-red, with orange; white-tipped.	Black, streaked with light blue.	Dull brown.

Scopimera globosa is as common at Tai Tam as *S. intermedia* is at Shiu Hau; but the soldier-crab *Mictyris* occurs, if at all, only on the clean, north-eastern part of the Bay. Two other small ocypodids are found along with *Scopimera*, though these are much less common. The first species, *Tmethypocoelis ceratophora* is a little larger than *Scopimera*, being recognized by its buff-coloured carapace with eye-stalks prolonged into slender horns, and the long, wide-elbowed chelipeds with expanded palm-like hands. The second and smaller species, *Ilyoplax tenella,* is common in patches, being bluish-grey, and distinguished from *Scopimera* and *Tmethypocoelis* by the more normal shape of its chelae. The males use the claws for signalling as in *Uca,* but they are of equal size and moved rhythmically up and down.

Much more remains to be learned about the burrowing ucids and other ocypodids. Prolonged observation, with careful digging and study of habits during a whole ebb tide, could form an engrossing project.

The middle shore and the lower shore at Tai Tam are the habitat of some well-adapted burrowing shrimps. On the middle shore, the snapping shrimp *Alpheus brevicristatus* (Fig. 10.3) makes long vertical burrows. It differs by its more slender chelae from the heavier alpheid *(A. bisincisus)* living in sediments under boulders. Another common burrowing alpheid of soft shores is *A. bellulus.*

In the lower flat at Tai Tam, completely uncovered only at low spring tides, live the two filter-feeding crustaceans, *Callianassa* and *Upogebia,* belonging, with the hermit and porcelain crabs, to the Anomura. Both are so specialized they never leave their burrows.

The ghost-shrimp *Callianassa japonica,* pallid milk-white and sometimes flushed with pink, is much commoner. Numerous small specimens about 2.5 cm long may be left in the sieve; large ones of 7.5 to 10 cm are occasionally secured at or beyond low water. A second species *C. petalura* has a cheliped that is long and delicate compared with the broad and robust cheliped of *C. japonica*. *C. petalura* has been recorded from Hoi Sing Wan in Tolo Harbour.

The vertical callianassid burrow tapers up to a narrow entrance and sometimes has side-shafts and turning bays. Its occupant is frail-looking and loosely built, being helpless if taken out of the sand. The enlarged left cheliped of the male is broad and lamellate; its function is uncertain, though it has been suggested that, held across the water current, it can deflect or modify the flow. The anterior thoracic limbs are weakly chelate and modified as digging spades. Fine fringes of setae run along their margins, to intercept particles from the water current which are then combed off on the broad maxillipeds. The two posterior pairs of legs form slender struts, supporting the shrimp against the wall of the burrow. Five pairs of flat pleopods create a current through the burrow. The uropods and telson form together a wide tail-fan.

Upogebia major has a similar design and habit to *Callianassa,* but can be distinguished by the two equal sub-chelate claws of the first thoracic limbs. More robustly built than *Callianassa, Upogebia* tends to favour muddy sand, often making horizontal galleries under boulders. A water current is kept up by the pleopods; particles are apparently strained off by a frame of setae on the second thoracic legs, and removed by mobile third maxillipeds.

None of these local species has yet been studied in detail. With a little trouble they can be kept alive in glass tubes, and observation of their feeding and other limb functions could give rewarding results.

Several crabs are very typical, permanently or seasonally, of the Tai Tam lower shore. The portunid *Macropipus corrugatus,* pinkish-brown and finely rugose across the carapace, submerges just below the surface of coarse or silty sand as does *Ovalipes* (p. 158) on cleaner sand. The far more aggressive portunid *Thalamita prymna* (p. 92) turns up in similar places, but prefers the shelter of boulders, particularly upon muddy ground. Beyond low water, young specimens of *Portunus sanguinolentus* can sometimes be found, bluish to grey-green and carrying three large blood-spots on the side-spiked carapace. Commonly taken offshore as a food species, its juveniles may enter the inter-tidal zone.

Towards April, when over-mature *Colpomenia sinuosa* breaks loose and collects at low water, the camouflaging crab *Dromia dormia* appears in some numbers. In low tidal silty sand, the small, hairy, xanthid *Pilumnus verspertilio* is common, with short muddy tomentum, slender legs and eye-stalks reduced to vestiges;

Sheltered Flats

it is evidently a true burrower, normally avoiding the light.

Several types of crab, illustrated in Fig. 10.4, can be collected only by dredging the muddy bottoms at three or four fathoms depth. Two species are thin, attenuate and spider-like. One of these, *Naxoides hystrix*, is a true spider-crab belonging to the Maiidae. The second, *Myra fugax*, distinguished by its long, angled chelipeds, is a much modified leucosiid, belonging, with *Calappa* and *Matuta*, to the Oxystomata.

Dorippe granulata is another offshore oxystomatid, of somewhat novel appearance. The carapace is small and flat, with the first and second walking legs long and strongly clawed. The third and fourth appear lacking at first glance, but will be found short and hooked, securely gripping the margin of the flat grey anemone (*Carcinactis ichikawai*) typically carried for protection on this crab's carapace.

The heavily spinose *Lambrus validus* is a relative of the spider-crab, with the

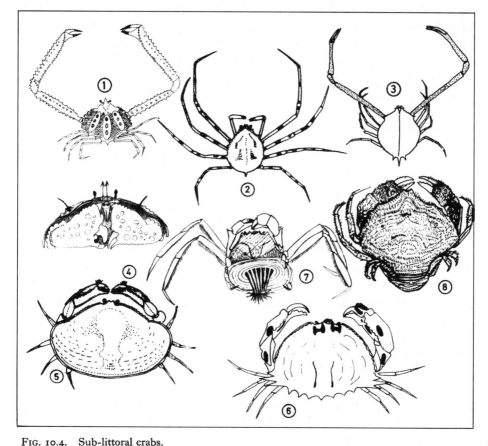

FIG. 10.4. Sub-littoral crabs.
1. *Lambrus validus*; 2. *Naxoides hystrix*; 3. *Myra fugax*; 4. *Calappa hepatica* with front view showing water currents; 5. *Calappa calappa*; 6. *Calappa philargius*; 7. *Dorippe granulata* with anenome *Carcinactis ichikawai*; 8. *Dromia dormia*.

broad triangular carapace typical of the Parthenopidae. It has widely elbowed chelipeds, ornamented with sharp tubercles.

Large hermit crabs, typically *Dardanus* (e.g. *D. haani*) and *Pagurus samuelis*, may also be encountered. At times found attached to the molluscan shell of *Dardanus* is the large commensal anemone *Calliactis japonica*. This collects food left by the crabs messy feeding habits and in turn offers the protection of its stinging tentacles. When changing shells the crab will gently prise the anemone loose and relocate it on its new home.

Echinoderms

The star-fishes, urchins and holothurians all have representatives modified for life in sand. Three species of starfish lie upon the surface or lightly buried in the silty layer. The commonest, *Archaster typicus* is a very regular, five-pointed star, with tapered, rather rigid arms. Spines run along the edges, as in the comb-star *Astropecten,* in many parts of the world the equivalent of *Archaster* on sandy shores. Unlike the carnivorous star-fishes, with more flexible arms, *Archaster* feeds on deposits and minute organisms; the spacious bag of the stomach being partly everted to ingest this food. The upper (aboral) surface is greyish-brown, mottled with black, and carries the anus (lacking in *Astropecten*), the madreporite or sieve plate, and five genital pores. At the breeding season in April and May, *Archaster* forms sex-pairs, a male lying on top, with its arms over the inter-arm angles of the female. The eggs are thus fertilized on exit from the female openings.

The second star-fish of sheltered flats, *Luidia maculosa*, is, like *Astropecten,* a primitive star, of ancient lineage, having a simple gut with no separate anus. The seven arms are much longer than in *Archaster* and slightly curved. Their marginal plates are spineless, but the convex upper surface of the arms is beset with 'paxillae', small mushroom-shaped spines whose flat tops adjoin to form a false integument. A second species of *Luidia* is the smaller, dull brown *L. hardwicki,* which occurs in vast numbers in the sub-littoral deposits of Tolo Harbour and is reported to be a predator of bivalves which it engulfs whole. Closely similar to *L. hardwicki* is *L. longispina*, with longer lateral spines to the arms. It too is carnivorous.

A third and smaller star-fish *Aphelasterias japonica* is occasionally found on the sand surface. With five short, plump arms, covered with tubercles and mottled in purple and brown, it is one of the bivalve-predating stars of the family Asteriidae.

The sea urchins (Echinoidea) have sand species belonging to two groups. First are the 'regular' echinoids, with round, elevated tests. *Temnopleurus reevesi* grows to 3–4 cm across, with the longer spines dark brown, and the shorter spines and test cream-brown. A second species is *T. toreumaticus*. *Salmacis sphaeroides* is broader based and more depressed, with short golden spines darkly cross-barred.

The tube-feet of the upper surface take hold of debris, including algal fragments, small stones and shells. Generally held to be a camouflage, this may also be a method of collecting algal fodder, to be carried by the tube-feet to the mouth.

The heart urchin *Lovenia elongata,* common at Tai Tam, is the best-known local representative of the Spatangoidea. Intimately adapted for life in sand, the test is convex above with numerous long curved spines, and flatter below. At its wide end is a shallow sinus, beneath which lies the mouth, and this, being the advancing end, may be referred to as the 'front'. Ambulacral areas carrying the tube-feet form wide 'petals' on the upper surface. The front petal has long, extensible tube-feet, which pass upwards through a chimney, guarded by a palisade of spines, to the surface of the sand; their viscous terminal papillae collect and haul down fine particles, which are passed by brush-like tube-feet, at the oral ends of the ambulacra, to the mouth. The heart urchin moves forward within a horizontal shaft by the 'rowing' action of long, flattened spines, on the under surface. Food and sand are freely swallowed without trituration and the Aristotle's lantern, the mill of jaw plates found in regular echinoids, is absent. Faeces are discharged into a horizontal shaft behind the urchin. From the muds of Tolo Harbour has been recovered the dull green-brown heart urchin *Schizaster lacunosus* with short, dense-set spines.

Of the Echinoidea the clypeasteroid 'sand-dollars' probably achieve the greatest specialization, some, such as the green *Clypeaster virescens* from Hoi Sing Wan, being remarkably flat (Fig. 10.5). They burrow just below the surface of the sand and collect detritus from the sand-water interface. The larger, green-tinged sand dollar *Peronella lesueuri* has been collected from cleaner sands off-shore.

The worm-like holothurian, *Protankyra bidentata,* also feeds on poorly sorted deposits, and the thin-walled, often transparent, body is usually crammed with silty sand. When empty it appears pinkish-grey, with five longitudinal ambulacral lines, which at once identify it as an echinoderm, even though the locomotor tube-feet are lost, and locomotion is by peristalsis of the body wall. The animal lies mouth downward, pushing material into the mouth by its ten pinately branched oral tentacles, as it burrows, or rather eats, its way through a sandy substrate. The anus keeps surface contact, taking in and discharging respiratory water, and from time to time voiding heavy sand-castings.

CERIANTHUS-ATRINA FLATS

As shelter from the land increases, protected sand silt flats can develop local areas that are virtually mud-flats. These are generally low tidal stretches, lying within a hook of the land, towards one end of the beach. The surface is soft and sticky and may carry drifting or attached vegetation. Large masses of green *Monostroma, Enteromorpha* and *Ulva,* and of pale brown *Ectocarpus* can break away from surrounding rocks in late winter; and at this season migrant herbivores such as sea-hares come in to graze in large numbers. The small monocotyledon *Halophila ovata,* a relative of the sea-grass, can form extensive patches, shallow rooting, and scrambling by runners. Hong Kong sea grasses are described in more detail in Chapter 11.

A good example of a protected shore with the anemone-like *Cerianthus,* and the fan-shell *Atrina* (Pinnidae)[2] is at the sheltered end of the small bay of Hoi Sing Wan (Starfish Bay), near Three Fathoms Cove on the southern shore of the Tolo Harbour. The whole neighbourhood is rich in shore habitats and could be a study centre for schools in the New Territories, complementing Tai Tam, on Hong Kong Island. Beyond the badly polluted Shatin and Tolo Harbour, Tolo Channel is comparatively unspoiled. A YMCA youth village has been established at Wu Kwai Sha, with living and sleeping facilities and a permanent laboratory could easily be provided. In front of the youth village there is a small beach of angular boulders. In Three Fathoms Cove and Long Harbour there are mangroves. Crossing the harbour by boat, the sedimentary rocks already described at Ma Tse Chau (p. 115) can be reached. In the middle of the Channel is Centre Island, widely girded by a small-boulder shore and used in studies by the Chinese University of Hong Kong. Clean, moderately exposed rocky shores are to be found on the points fringing the bays. Corals begin to develop sub-tidally off the rocky shores, coming to their richest at Chek Chau (Port Island) (p. 258) in the channel mouth.

The protected flats at Hoi Sing Wan are depicted in Fig. 10.5. The main beach, crossed by a stream, is divisible into upper, middle and lower shore, with *Anomalocardia, Lingula, Scopimera, Mictyris* and *Thalassema* abundant and typical. At the sheltered west end, the low tidal flats are interspersed with rocky stacks, from which algae break free, to cover the beach with a rich grazing, especially for the sea urchins *Salmacis sphaeroides* and *Temnopleurus reevesi.* Both species lie upon the surface, being 'regular' echinoids only slightly modified for life on sand (p. 188).

At the end of March, the sea-hare *Bursatella leachii* arrives on shore in great numbers, to breed and to browse the loose algae. At times their soft and flaccid bodies are scattered six or more to every square metre, immersing themselves in silt or lying free at the surface. *Bursatella* has a general likeness to the sea-hares *(Aplysia)* described for rocky shores (p. 50); the body is long and flask-shaped, narrower at the head, greyish-green in colour, with dark spots and light green flecks. Small dendritic processes cover the whole upper surface. Respiratory water is taken in through a narrow slit that leads into a wide space beneath the parapodia. These are fused along most of the mid-dorsal line unlike those of

---

[2] For a guide to the Pinnidae of China the reader should consult: Wang Zhen-rui. 1964. Preliminary studies on Chinese Pinnidae. *Studia Marina Sinica* 5: 30–42.

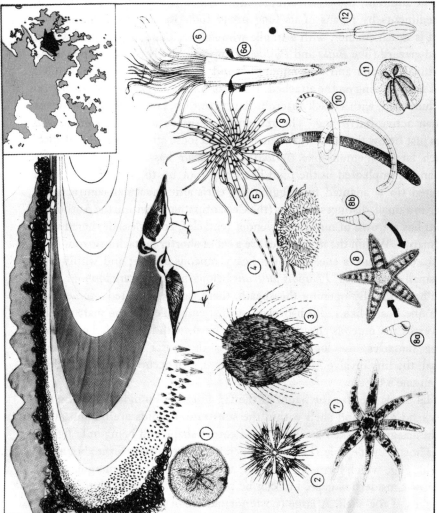

FIG. 10.5. The protected flats at Hoi Sing Wan (Starfish Bay), Tolo Harbour. The three zones of the upper, middle and lower shores are separately shaded. In the foreground are two alexandrine plovers, *Charadrius alexandrinus*.
1. *Clypeaster virescens*; 2. *Temnopleurus toreumaticus*; 3. *Lovenia elongata*; 4. *Salmacis sphaeroides*; 5. *Boloceroides macmurrichi*; 6. *Cerianthus filiformis*, with 6a. the commensal phoronid worm *Phoronis australis*; 7. *Luidia maculosa*; 8. *Archaster typicus*, with 8a. and 8b. two parasitic gastropods *Mucronalia fulvescens* and *Balcis shaplandi*; 9. *Sipunculus nudus*; 10. *Siphonosoma cumanense*; 11. *Schizaster lacunosa*; 12. *Ochetostoma erythrogrammon*.

Aplysia. Like other aplysioids, *Bursatella* forms long copulatory chains, and lays the spawn in coils of long jelly-like yarn. See the paper on Chinese Aplysiidae by Tchang Si and Lin Guang-yu[3] for a guide to this family.

Two bullomorph opisthobranchs found on muddy shores are related distantly to sea-hares. *Philine orientalis,* specialized for permanent burrowing, is a carnivore (p. 162); in addition, the small bubble shell, *Haloa japonica* ploughs through surface sediments by means of its long broad foot. Related to *Bulla ampulla* (p. 162) it is much smaller, with a fragile white shell. Though equipped with a hard-lined gizzard (like *Bulla* and *Philine*) it appears chiefly to graze upon algae and nutritive deposits and is commonly found at the inter-face of embedded rocks and sand-feeding on the attached strands of *Enteromorpha* and *Monostroma*.

Semi-fluid mud, with a black anaerobic layer beneath, favours firm attachment rather than active burrowing. The red and dark-green anemone *Haliplanella luciae* lies just below the surface, fastened to small loose stones by its basal disc. The much larger anemone-like *Cerianthus filiformis* has a tapered muscular column, and is implanted in the mud like a tap-root up to 20 to 22 cm long. The pointed tip is adapted for digging, by peristaltic muscular contractions. Belonging to a small, distinct sub-class (the Ceriantipatharia) within the Anthozoa, *Cerianthus* has a crown of numerous slender tentacles, colourless or tinged with green or mauve. Within the main series are a set of shorter tentacles surrounding the mouth. *Cerianthus* is enclosed by a heavy mucous sheath; and within this lodges a small commensal, *Phoronis australis* belonging to the rare lophophorate phylum Phoronida. By opening the sheath, these can be detected—sometimes several to one host—like reddish-brown worms, embedded in the matrix. The body traverses the mucous sheath, so that the head, with its double horse-shoe of feeding tentacles, the lophophore, reaches the water outside. A second commensal, the tiny bivalve *Montacutona olivacea,* also occurs around the edges of the anemone's tube.

Low tidal muds are also the haunt of the fan-mussels, Pinnidae, fixed by the byssus to sand grains and shell within the substrate. Pinnids are like no other bivalves in design; the shell has a flattened cornet shape and being only lightly calcified is horny and flexible. The valves open not at the lower margin, but along the posterior edge which forms a gaping slit at ground level. The margins of the mantle lie just inside this gape, but can be pulled back for protection into the lower reaches of the shell. A large (posterior) adductor muscle is placed at the lower centre of the shell, and a small anterior adductor survives at the pointed extreme. The foot lies also at this end, secreting the long radiating threads of the byssus.

Two genera of fan-mussel are found in Hong Kong. The genus *Pinna*

[3] Tchang Si and Lin Guang-yu. 1964. A study on Aplyside from China. *Studia Marina Sinica* 5: 25–29.

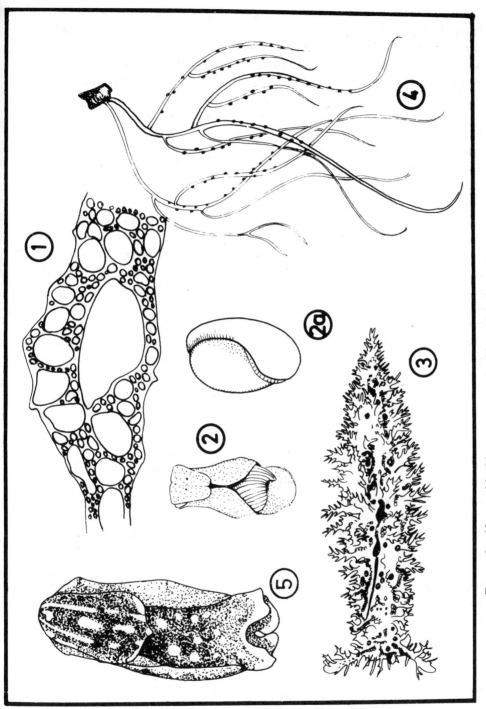

FIG. 10.6. Algae and herbivorous gastropods of low tidal silty sand. 1. *Ulva reticulata*; 2. *Haloa japonica* expanded animal and 2a. shell; 3. *Bursatella leachi leachi*; 4. *Gracilaria chorda*; 5. *Doridium cyaneum*.

comprises long, thin and equally straight-sided mussels. Three species occur here, the common slender, ribbed, light transparent brown *P. muricata;* the deeper water, very slim *P. attenuata* and the most commonly encountered, smooth, slender, pale or transparent *P. atropurpurea*. The genus *Atrina* comprise widely angled species, typically with only one side straight the opposite curved. *A. penna* is very delicate, transparent light brown and delicately ribbed; *A. pectinata* is dark-brown, somewhat inflated and weakly ribbed; *A. inflata* is more inflated, again finely-ribbed with projections while *A. vexillum,* the commonest species, is quite distinctively smooth, dark-brown and smoothy rounded (Fig. 10.7). Attached to the shell of *Atrina* has been found the filter-feeding hipponicid gastropod *Amathina tricarinata*. The pinnid shell is very easily damaged (and regrows very quickly) so that the shells are often badly deformed and thus difficult to differentiate. See the paper by Wang Zhen-Rui[4] on Chinese Pinnidie for a guide to this family.

The chief sedentary polychaetes of these muddy flats are tube-dwelling terebellids and sabellids and deposit-feeding Cirratulidae. The Terebellidae live in thin parchment tubes, strengthened with sand or shell. Already encountered on stable boulder shores (p. 118) they can be recognized by their long extensible tentacles, radiating over the surrounding substrate. Food particles are brought to the mouth by ciliated grooves along the tentacles or wiped off as a retracted tentacle is drawn between the glandular lips. The finest and commonest local burrowing terebellid is *Loimia medusa,* with long, cross-striped tentacles and numerous pairs of branched blood gills. A second terebellid common to these shores is *Lanice conchilega* with a similar but more roughly constructed tube, thickly fringed at the sand surface.

The family Sabellidae is related to the Serpulidae, but is set apart from it by the tough, leathery tubes. Generally projecting up to 2.5 cm from the surface, these close flat when the animal has retracted. *Sabellastarte japonica* carries, like a serpulid, a crown of dark brown or black filtering pinnae located on an orange and yellow body. Common in sandy or muddy shores, the tube is more than 22.5 cm long; the worm is shorter than its tube and can shut down the tentacles and retreat into it on the passing of a vibration or shadow.

The cirratulid *Cirriformia tentaculata* has a great mass of retractile and sticky tentacles around the head which are extended over the mud surface in search of food. When withdrawn they carry detrital particles to the mouth. The body bears numerous, presumably respiratory, filaments.

HEMICHORDATES AND CEPHALOCHORDATES

The yellow acorn-worm *Balanoglossus misakiensis* is typically found in muddy places where the shore has some admixture of sand. Like the king-crab and

[4] Wang, Z. R. 1964. Preliminary studies on Chinese Pinnidae. *Studia Marina Sinica.* 5: 30–42.

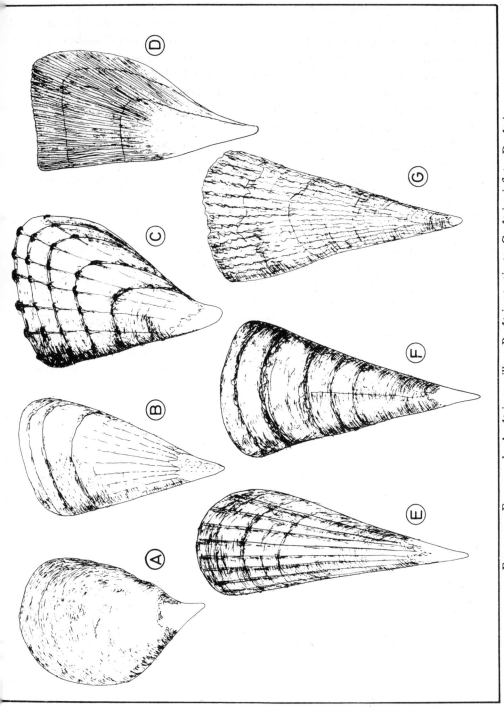

Fig. 10.7. Fan mussels: A. *Atrina vexillum*; B. *Atrina pectinata*; C. *Atrina inflata*; D. *Atrina penna*; E. *Pinna attenuata*; F. *Pinna atropurpurea*; G. *Pinna muricata*.

Lingula, it is to be regarded as a zoologist's prize. Highly primitive, as well as aberrant, the acorn-worms are classed at the base of the phylum Chordata, though very remote even from primitive ascidians and the lancelet *Branchiostoma*. The body is divided into three very unequal segments. The acorn-shaped or conical proboscis, when turgid with water, is used for burrowing, being raised clear of the ground to expose the mouth, which lies within the second segment, a glandular 'collar'. All the rest of the body is morphologically the third segment. This is sub-divided into a branchiogenital region, with two lines of gill pores and bright yellow genital wings, recurving to form a tube, then a region with two rows of small saccules, the 'hepatic diverticula', and a long, featureless 'hind-body' or tail. *Balanoglossus* feeds by ingestion of fine surface deposits, but has also a semi-selective filtering mechanism. From the exit of the prolonged, sometimes spiral burrow, the anus discharges heavy coiled faeces similar in appearance to those of the temperate polychaete *Arenicola*.

The last and most mobile of the early chordates are the lancelets, familiarly called by their old name 'Amphioxus'. In Hong Kong, *Branchiostoma belcheri* (Fig. 10.10), regularly recorded from Hoi Sing Wan, can swim freely through the water but typically burrows in sand. The body is elongate (5 cm long), laterally flattened and possesses no pigment; the muscles can thus be seen as a series of myotome blocks. The body is pointed at both ends, there is no recognizable head, though the head end is marked by a series of buccal cirri which form a sieve around the opening of the oral hood. There is a posterior anus and behind it a short tail. The body is supported by a notochord. In every sense Amphioxus clearly points the way to the higher chordates.

Fishes and Cephalopods

These are the most mobile of all the benthic predators of soft shores. They range widely over beaches and flats when the tide is in, but at low water are seldom seen, retreating beyond the tides, or into the sand.

The fishes have various degrees of structural adaptation for bottom life. Some are intimately camouflaged against the substrate. The sandy grey or cream dragonet, *Callionymus richardsoni,* of low contour and flat beneath, skims about rapidly, half submerged in silty sand. Specialized also for bottom life are the flat-fishes, *Cynoglossus robustus, Paraplagusia blocki,* and the small *Sillago sihama,* slenderly tapered and sand-coloured. The puffer fish *Amblyrhynchotes hypselogeneion* is characteristically shaped, inflating its body to a sphere as it buries within clean sand, and resuming its slender shape when mobile. In the dorso-ventrally flattened *Chelidonichthys kama,* the pectoral fins form wide, laterally located fans, some of the lower pectoral rays being detached from the remainder and modified to form stout finger-like appendages. These are sensory and are used both to search for food and as limbs. In the beautiful sea-robin *Daicocus*

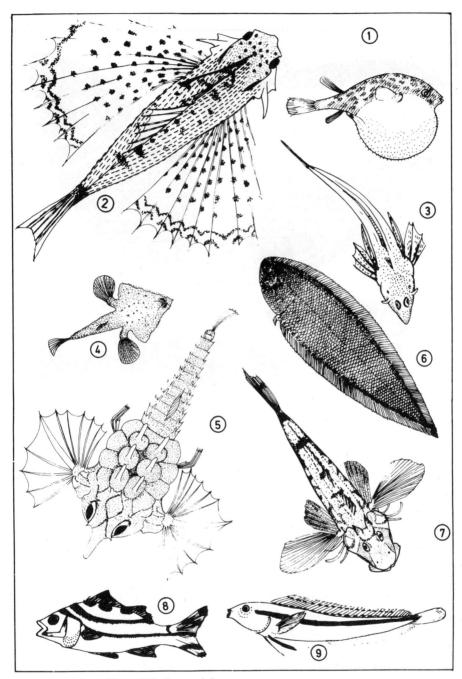

FIG. 10.8. Fishes of low-tidal silty sand-flats.
1. *Amblyrhynchotes hypselongeneion*; 2. *Daicocus peterseni*; 3. *Callionymus richardsoni*; 4. *Malthopsis annulifera*; 5. *Pegasus volitans*; 6. *Cynoglossus lineolatus*; 7. *Chelidonichthys kama*; 8. *Therapon jarbua*; 9. *Dasson variabilis*.

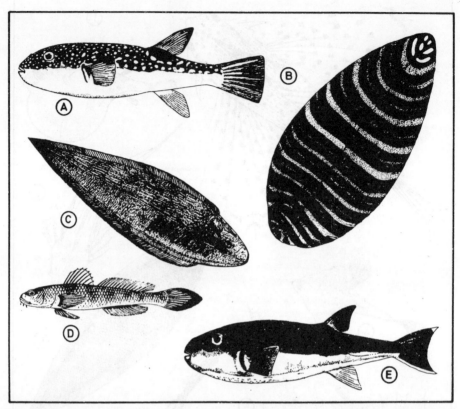

FIG. 10.9. Fishes of sand flats and the shallow subtidal: A. *Fugu vermicularis*; B. *Zebrias zebra*; C. *Cynoglossus robustus*; D. *Chaeturichthys hexanema*; E. *Tetrodon spadiceus*.

peterseni, the pectoral fins form enormous brightly coloured fans used in courtship by the males. Enlarged pectoral fins are also seen in *Pegasus volitans* and *Malthopsis annulifera*.

Working over the surface in small shoals and sifting deposits with the mouth for their nutritive contents is the small grey sand mullet *Mugil engeli*. The obliquely dark-banded *Therapon jarbua* is a voracious predator. All of these species, and many more besides, may be taken in a fine beach seine net dragged over the bottom. Numerous small sand gobies (numerous species of *Gobius* and *Tridentiger*) hide in the sand while the common blenny *Dasson variabilis* will seek refuge in small holes, under stones or shells or even within discarded, beverage cans.

A specialized group of inter-tidal fishes are adapted to a burrowing mode of life. The most important of these is the sand-eel, *Ammodytes*, which lies buried obliquely in the sand. Most strongly adapted for burrowing is the blind, purple-red, goby *Trypauchen vagina*, inhabiting a permanent burrow.

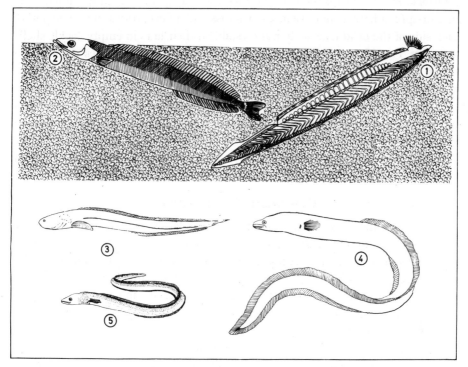

FIG. 10.10. Burrowing fishes and the lancet of sand-flats.
1. *Branchiostoma belcheri*; 2. *Ammodytes personatus*; 3. *Trypauchen vagina*; 4. *Anguilla japonica*; 5. *Ophichthys apicalis*.

Two sorts of eels may be encountered. *Ophichthys apicalis* builds temporary burrows in sand-flats. On more muddy, estuarine shores may be found the holes of the large, principally fresh-water eel, *Anguilla japonica*. This eel, an important Asian food resource, is dark grey dorsally silvery beneath and may reach a length of over 1 metre.

Of the Hong Kong cephalopods, species of three important genera are characteristic of sand beaches and sand-silt flats. The largest are the cuttlefish, *Sepia phaoronis* and *S. lycidas*. Shield-shaped, with an undulating marginal fin, and a buoyant internal shell, cuttlefish hover and manoeuvre close to the ground. *Sepia* has a fine command of camouflaging colour change and flushes shrimps and other benthic crustacea from the sand, by water spurts from its funnel.

The small rounded cuttlefish *Euprymna berryi* is found shallowly submerged in sand on clean flats as at Shiu Hau. The body is stout and thimble-shaped, a little more than 2.5 cm long, and with a broad semi-circular fin at either side. With these flaps, it buries itself lightly in the sand, or emerges to swim about and hunt small crustaceans (cumaceans and amphipods) with its funnel jet .

Octopus membranaceus is one of the smallest of its genus. Small specimens with an arm-spread hardly more than 5 cm across and a false, iridescent 'eye-spot' on each side of the head take shelter and establish their lairs in empty cockle shells, as at Tai Tam. Larger, deeper water individuals cover themselves with piles of shells, each attached to the upturned suckers of the arms. It feeds by pulling apart small bivalves, though also on crabs and shrimps. Egg masses are attached and guarded within the lair. A second species, *Octopus dollfusi* with a coarse brown lace-like patterning, is of similar size and also occupies empty bivalve shells, notably species of *Anadara*. It seems almost exclusively to feed on bivalves. Voss and Williamson[5] have produced a guide to the cephalopods of Hong Kong in which these species are illustrated.

COMMENSALS AND PARASITES

Different animal species can maintain inter-relationships of many kinds and degrees. The commonest are those of a predator with its prey, or the contest between two species in competition for the same resource, whether food or living space. These are both antagonistic relationships: in the first, the predator alone benefits, in competition neither. The relation of parasite and host is like that of predator-prey, with benefit to one, though in parasitism the host resource is conserved, not at once eliminated.

Commensalism is a relationship of permanent association. Usually only one partner benefits, and the other is left unharmed. The partnership is usually obligatory for the commensal species, while the host is not adversely affected, either in its growth or survival. The inter-tidal coast-line of the tropics and sub-tropics is extraordinarily rich in examples of commensalism, and many of these are found on soft shores. Some of the chief examples observed in Hong Kong are given in Table 10.2.

The chief advantage gained by commensals of soft shores seems to be shelter in a place where the host creates a water current, either in the burrow or, for a bivalve, within the mantle cavity. Some of the food introduced by this current is regularly taken, but not in such amounts as to deprive the host.

The thin-shelled and colourless bivalves *Fronsella* and *Nipponomysella* (p. 204) are common commensals in the burrows of sipunculid worms. Species of *Montacutona* live with coelenterates, while species of *Galeomma* probably share a loose relationship with alpheid shrimps, and *Curvemysella paula* lives inside empty shells with *Diogenes edwardsii*. *Pseudopythina subsinuata* attaches to the abdomen of the mantis shrimp *Oratosquilla oratoria*. Other bivalves, e.g. *Crenatula* and *Vulsella*, embed in sponges.

Some of the commonest crustacean commensals belong to the crab family

[5] Voss, G. L. and Williamson, G. 1971. *Cephalopods of Hong Kong*. Hong Kong, Government Printer.

TABLE 10.2 SOME COMMENSAL AND PARASITIC RELATIONSHIPS OF HONG KONG SHORES

Host	Commensal/Parasite	Notes
PORIFERA		
Green, branching sponge (*Sigmadocia symbiotica*)	*Crenatula modiolaris* *Acasta sulcata*	Embedded bivalve. Embedded barnacle (COMMENSAL).
Black, rounded sponge (*Suberites*)	*Vulsella vulsella*	Embedded bivalve (COMMENSAL).
Sponge (*Spirastrella vagabunda*)	*Pyrogopsella stellula*	Embedded barnacle (COMMENSAL).
CRUSTACEA		
Oratosquilla oratoria	*Pseudopythina subsinuata*	Leptonid bivalve, with dwarf male, attached to shrimp (COMMENSAL).
Charybdis bimaculata *Dorippe granulata* *Portunus pelagicus*	*Octolasmis warwicki*	Barnacle attached to carapace and legs (COMMENSAL).
Macrophthalmus convexus *Scylla serrata* *Charybdis truncata*	*Octolasmis tridens*	Sessile barnacle attaching to the mouth-parts (COMMENSAL).
Pollicipes mitella	*Nemertopsis gracile*	Worm securing food in barnacle's mantle cavity (COMMENSAL).
Diogenes edwardsii	*Curvemysella paula*	Leptonid bivalve (COMMENSAL).
Clibanarius infraspinatus	*Syphopatella walshi*	Slipper limpet attaching inside hermit crabs shell (COMMENSAL).
Clibanarius striolatus	*Pseudostegias setoensis*	Bopyrid isopod attached to abdomen (PARASITE).
Panulirus stimpsoni	*Temnaspis amygdalum* *Octolasmis neptuni* *Octolasmis warwicki*	Barnacles attached to mouth-parts and gills (COMMENSAL).
ARACHNIDA		
Tachypleus tridentatus	*Octalasmis warwicki*	Barnacle settled round the mouth (COMMENSAL).

Host	Commensal/Parasite	Notes
BIVALVIA		
Pinna atropurpurea	*Anchistus custos*	A commensal shrimp in a sex pair (Fig. 11.19).
Soletellina diphos	*Mytilicola intestinalis*	Modified copepod of the rectum (PARASITE).
Crassostrea gigas *Pinna atropurpurea* *Geloina erosa*	*Pinnotheres sinensis*	Pea-crabs within the mantle cavity (COMMENSAL).
Laternula truncata *Marcia japonica*	*Pinnotheres affinis*	Pea-crabs within the mantle cavity (COMMENSAL).
Barbatia virescens *Geloina erosa*	*Pinnotheres cyclinus*	Pea-crabs within the mantle cavity (COMMENSAL).
Gafrarium tumidum	*Pinnotheres dilatatus*	Pea-crabs within the mantle cavity (COMMENSAL).
Crassostrea gigas	*Protoeces ostreae*	Flat-worm in digestive system (PARASITE).
Crassostrea gigas	*Echinocephalus sinensis*	3rd stage larval nematode found around gut. Final host is the ray *Aetabates flagellum* (PARASITE).
AMPHINEURA		
Liolophura japonica	*Stylochoplana pusilla*	Turbellarian flat-worm living in the mantle cavity (COMMENSAL).
GASTROPODA		
Nassarius pullus	*Paranthus sociatus*	Anemone attached to shell (COMMENSAL).
Batillaria zonalis *Batillaria multiformis*	*Patelloida lampanicola*	Limpet attached to shell (COMMENSAL).

Host	Commensal/Parasite	Notes
ECHINODERMS		
Anthocidaris crassispina	Athanas dorsalis	Crustacean lying beneath the urchin's base; often in a sex pair (COMMENSAL).
Diadema setosum	Chryseulima philippinarum	Eulimid snail (PARASITE).
	Echineulima mitterei	Eulimid snail, occurring as a sex pair (PARASITE).
Archaster typicus	Podarke angustifrons	A hesionid polychaete in ambulacral grooves (COMMENSAL).
	Balcis shaplandi Mucronalia fulvescens	Two eulimid gastropods ectoparasitic on upper and lower surface (PARASITE).
Macrophiothrix longipeda/ Alpheus hippothoe· Stenopus cf. hispidus	Ephippodonta oedipus	Leptonid bivalve living with hosts in cavities under coral heads (COMMENSAL).
Holothuria leucospilota	Balcis kuronamako	Eulimid snail (PARASITE).
COELENTERATES		
Cerianthus filiformis	Phoronis australis	Embedded in sheath (Fig. 00.0) (COMMENSAL).
	Montacutona olivacea	Leptonid bivalve on rim of tube (COMMENSAL).
Cyanea nozakii	Alepas pacifica	Stalked barnacle (COMMENSAL).
	Caranx malabaricus C. kalla	Fishes travelling with the jellyfish as young (COMMENSAL).
Stoichactis kenti	Amphiprion percula	Clown fish (COMMENSAL).
Parasicyonis actinostoloides	Amphiprion bicinctus	Clown fish (COMMENSAL).
A wide range of scleractinian corals	Armatobalanus allium Cantellius pallidus Creusia indica Nobia conjugatom Savignium crenatum Pyrgoma cancellatum	Barnacles attached to living corals (COMMENSAL).
Pteroeides sparmanni	Porcellana picta	Porcelain crab (COMMENSAL).

Host	Commensal/Parasite	Notes
WORMS		
Sipunculus nudus	*Fronsella oshimai*	Leptonid bivalve (COMMENSAL).
	Asthenognathus inaequipes	Pinnotherid crab (COMMENSAL).
Siphonosoma cumanense	*Nipponomysella subtruncata*	Leptonid bivalve (COMMENSAL).
	Pinnixa rathbuni	Pinnotherid crab; (has entoproct *Loxosomella* sp. attached) (COMMENSAL).
Mesochaetopterus japonicus	*Tritodynamia rathbuni*	Pinnotherid crab within tube (Fig. 9.13) (COMMENSAL).
Loimia medusa	*Tritodynamia horvathi*	Pinnotherid crab (Fig. 9.13) (COMMENSAL).
Ochetostoma erythrogrammon	*Mortensella forceps*	Crab (COMMENSAL).
	Achasmea sp.	Bivalve (COMMENSAL).
	Sigaretotornus plana	Gastropod (COMMENSAL).
	Red flat-worm	
	Scale-worm (Fig. 9.14)	
HEMICHORDATE		
Balanoglossus misakiensis	*Pinnixa balanoglossana*	Pinnotherid crab (Fig. 9.13) (COMMENSAL).

Pinnotheridae. These include pea-crabs (*Pinnotheres* sp.) living in the mantle cavity of various bivalves. Round-bodied and thin-shelled, especially the permanently immured females, they intercept mucous strings of food from the host's gill, by the fine setal fringes of their legs. Other members of this family, such as *Tritodynamia* and *Pinnixa* live in the burrows of polychaete worms or of *Balanoglossus*. Their carapace is sideways elongated, with an enlarged and flattened second pair of legs, enabling them to move nimbly sideways through the burrow. The shrimp *Anchistus custos* lives in a sex pair inside the mantle cavity of the fan-shell *Pinna*. They feed on the mucous food strings of the ctenidia.

A third important group of commensals is provided by the scale-worms Polynoidae, with numerous slender species associated with other burrowing polychaetes. The unusual and rare worm *Phoronis australis* embeds in the leathery tube of the burrowing anemone *Cerianthus* thereby gaining protection.

Tiny, sharp-spired gastropods are frequently found on the integument of worms, echinoderms and molluscs. Not commensals but ectoparasites, they live by sucking blood or body fluids from the host. They may belong to one of two families, the Eulimidae, without hard mouth-parts but entering the host by the action of salivary enzymes, or the Pyramidellidae, piercing the host with minute stylets derived from the jaws. The sand star-fish *Archaster typicus* has two eulimids, *Balcis shaplandi* on its upper surface, and *Mucronalia fulvescens* underneath. The white *Balcis kuronamoka* feeds on the common black sea-cucumber *Holothuria leucospilota* while other eulimid snails parasitize the long-spined sea-urchin, *Diadema*. The barnacles are important commensals and parasites. A variety of genera embed in sponges *(Acasta; Pyrogopsella)* or attach to crabs *(Octolasmis; Temnaspis)* and corals *(Armatobalanus; Cantellius; Creusia; Nobia; Savignium; Pyrgoma)*. Others, e.g. *Sacculina,* are highly modified parasites of crabs.

There are many further examples on Hong Kong shores of a wide range of highly personal relationships between members of a community. Many of these are not yet described and await research.

CHAPTER 11

Enclosed Flats

Broadly defined, enclosed shores lie at the sheltered heads of harbours or in bays land-locked by the surrounding coasts. Rocky surfaces are scarce. The greater part of the enclosed inter-tidal could be designated 'mud-flats', with a virtual lack of sand-sized particles, or such a predominance of silt and clay that they are uncomfortable and sometimes hazardous to walk over.

Wave action is almost lacking; but the stirring up of fine sediment by currents gives the water a permanent turbidity. Almost to the surface, the ground becomes anaerobic and black, with the accumulation of ferrous sulphide.

Sediments brought by streams are deposited with little subsequent shifting; but, as a stream or tidal channel crosses the flats, a mosaic of bedding differences can sometimes be seen, a response to periodic changes of the course of flow.

Those mud-flats so heavily polluted that the normal fauna and flora are unbalanced or impoverished, even extinguished, will be ignored. Bad examples are to be found in the Sha Tin arm of Tolo Harbour, and in some long, narrow inlets such as Junk Bay and, recently, even part of Tai Tam Bay.

Pollution most often comes from the run-off of organic wastes from the land: human sewage or more importantly, in the New Territories, discharge from piggeries and poultry farms. Sewage out-flow results in multiple damage to the environment. Suspended particulate matter leads to accumulation of silt, clogging of respiratory and feeding organs, and reduction of photosynthesis with reduction of light. Organic nutrients—nitrates and phosphates—even in effluents given primary and secondary treatment in purification plants before discharge, can produce over-productivity or eutrophication, with heavy algal blooms, the most noticeable of which are red tides caused by the dinoflagellate *Noctiluca scintillans*, followed by swift regression and decay.

The presence of putrefrying organic matter greatly raises the B.O.D. (biochemical oxygen demand) of water, leading to oxygen lack and so death of many normal species. In such situations, the diversity of the natural population is greatly lowered. A few resistant species survive and prosper, with others eliminated. Ultimately, only anaerobic bacteria may survive; and with low diversity comes also low stability, leading to vast fluctuations in biotic composition.

All these evils are compounded where tidal circulation is sluggish and pollutants slow to clear. At Tai Po Channel, fresh water and sewage were estimated to remain for 23 days before flushing out to sea; at Tide Cove the period was 14 days and in Tolo Channel 10 days. Population increases are expected to raise the suspended pollution load $3\frac{1}{2}$ times at Tai Po, and 11 times at Sha Tin[1]. Anaerobic muds lie over the most of the bottom of Victoria Harbour where removal mechanisms are slow.

Discharge of toxic industrial wastes and of garbage often run close to sewage in their destructive effects. The Advisory Committee on Environmental Pollution on Land and Water, reporting in 1972 noted that 'the sea and especially the harbour, is the city's dumping ground, receiving the bulk of the city's sewage and large quantities of refuse'.

A final accompaniment of pollution, with perilous hazards for mankind, is the rising pathogen count of local waters, generally indicated by soaring levels of faecal coliform bacteria, with the sub-microscopic content of *Salmonella* and viruses of infective hepatitis and other sewage-borne diseases. The contamination of the oyster beds at Lau Fau Shan has become an acknowledged source of concern[2] and have more recently been found to possess high levels of the toxic heavy metal Cadmium.

Large areas of the enclosed shores are still however in a healthy state and could be salvable far into the future; and these maintain today a full and characteristic range of life. Their flora and fauna are in many respects different from those already found in protected silty sand-flats. First, they have a considerable vegetation of rooted angiosperm plants: often the sea-grass *Halophila ovata* at low water, and—far more constant and considerable—mangrove swamps and salt-meadow on the upper shore.

Of bivalve molluscs neither of the 'cockles' *Anomalocardia* or *Fulvia* is found. Absent too are the sand-loving tellinid species, though *Macoma* and the small pink *Moerella juvenilis* are common. Thin-shelled bivalves such as Pinnidae, razor-shells *(Ensis* and *Solen)*, window-shells *(Placuna)*, and the fragile *Laternula* are all typical. So too are the heavier and less mobile 'bloody cockles' *Scapharca* and *Anadara* (Arcidae) and the venerid *Gafrarium*.

Burrowing and surface echinoderms, as well as *Lingula* and the echiuroids are lacking on muddy shores. Of the crabs, the sand-processing *Scopimera* and *Mictyris* will generally be looked for in vain; but there is a long list of mud-flat crabs with *Uca, Macrophthalmus* and *Sesarma* species the most numerous.

---

[1] Watson, J. D. and Watson, D. M. 1971. *Marine Investigation into Sewage Discharges. Brief Report.* Hong Kong, Government Printer.

[2] Morton, B. S. 1975. Pollution of Hong Kong commercial oyster beds. *Marine Pollution Bulletin* 6: 117–22.

The presence of mangroves does not always denote muddy grounds. Several species can grow at the heads of sandy shores, as at Shiu Hau, wherever there is the protection of a berm, or an encroaching bay-head bar as at Pui O. Such upper levels, where silt and clay can accumulate, contrast with the predominantly sandy middle and lower shore. They support not only salt-marsh with mangroves, but also salt-meadow with grasses and succulents, properly flooded only during the highest of spring tides.

SALT-MEADOWS WITH *ZOYSIA* GRASS

Salt-meadows lie to the landward of mangroves, or sometimes in the midst of them, where the ground is appreciably raised above the surrounding areas. Mingling with salt-meadow vegetation are frequently mangrove-associate shrubs, of somewhat higher levels, especially the milky mangrove *Excoecaria*. Though rarely inundated by tides other than springs, salt-meadows come under the constant action of salt spray, and the ground is generally so salty as to lead to physiological drought, and call forth succulence and related adaptations in the plants.

The substratum of salt-meadow is generally firm, sandy or pebbly, rather than the sticky soils where mangroves flourish to seaward. Always compact enough to walk on clean-footed, salt-meadow soils are sufficiently coherent to allow sesarmids and other high-level crabs to establish permanent burrows.

In Hong Kong, a *Zoysia* community always defines the lower limits of more terrestrial vegetation on most shores. By reference to Figure 11.1 it can be seen that on sheltered sand and cobble beaches as are typical of many of the shores of Tolo Harbour, the *Zoysia* community progresses, landwards, into plants of terrestrial affiliation and occurring here in response to factors other than those defined by the sea. With increasing freshwater wetness, as in front of paddifields, this community is replaced by marsh sedges that still, nevertheless, merge seawards into *Zoysia*. A marsh community will be described later. With a greater marine influence the sedges are replaced by mangroves though again a *Zoysia* community often separates the pioneer plants *(Kandelia* and *Aegiceras)* from the higher zoned mangroves *(Lumnitzera, Excoecaria, Acanthus)*.

The *Zoysia* community in Hong Kong is very distinctive, the plants stragglingly sparse on rocky beaches but growing dense and luxuriantly on sand and gravel in comparison to the rather flattened truly terrestrial grasses behind. We take our example of this community from the beach in front of Wu Kwai Sha village on the southern coast of Tolo Harbour (Fig. 11.2). The broad, wide blades of *Axonopus compressus* grow flat on the ground and are sharply replaced by the alternately branching coastal grasses *Cynodon dactylon* and *Eleusine indica,* the latter with rather flat blades arising like a rosette from a common base. The extent of the *Zoysia* here is wide and further seawards, and mixed with it, are

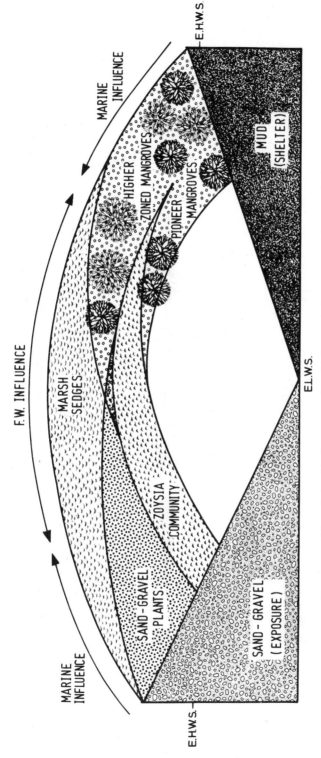

Fig. 11.1. The relationship between various high-zoned plant communities on soft shores in Hong Kong.

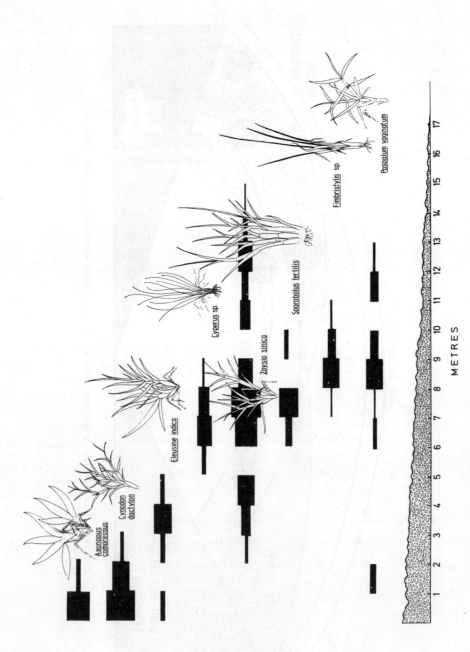

FIG. 11.2. The zonation of grasses and sedges into a *Zoysia* community on a sand-gravel beach at Wu Kwai Sha in Tolo Harbour.

PLATE 17. *(Above)* A view from the hills overlooking the extensive mud flats of Deep Bay towards the Mai Po marshes. In the foreground is the pier at Tsim Bei Tsui with a small natural stand of mangroves. *(Below)* The bunds of the Kei Wai's in the Mai Po marshes looking towards the land.

PLATE 18. *(Above)* The rock pools and terraces of Ping Chau in Mirs Bay. Note the wave-cut notch on the cliff-face behind. *(Below)* Wang Chau or Hole Island seen from the south. Both photographs illustrate different effects, with different rock types, of high exposure to wave action upon the structure of a rocky coast.

Enclosed Flats

clumps of the delicate leaved *Cyperus* sp. and *Fimbristylis* (typical of the Cyperaceae). Also scattered here and there are the sedges *Sporobolus fertilis* and *Paspalum vaginatum*. On this beach the lower limit of the *Zoysia* is sharply defined, elsewhere, on more clean sands, the *Zoysia* rhizomes wage a continual battle to stabilize their habitat and spread widely and sparsely.

Fig. 11.3 shows a typical area of salt-meadow with grass, succulents and clumps of intruding mangroves at Cheung She, in Tolo Harbour, near the marine biological station of the Chinese University of Hong Kong. By comparison with the similar formations of many temperate shores, the list of plant species is rather short. By far the dominant and most ubiquitous species is *Zoysia sinica*, giving a grey-green or brown tinge to wide expanses. Narrow leaves, stiff and pointed, diverge from slender uprights coming off from horizontal runners. *Zoysia* grass keeps the mud-surface moist beneath it, so that a plexus of filamentous green algae can generally establish. *Rhizoclonium* and other algae may form a coherent mat that can be stripped away to reveal small gastropods and crustaceans beneath.

Next in abundance, though much scarcer than *Zoysia*, are separate clumps of *Limonium sinense*. A semi-succulent member of the family Plumbaginaceae, this plant stands out by its rich green, radical leaves and its upright flower-stalks. The inflorescence is a one-sided spike with a row of small crinkled, yellow or white flowers.

Scaevola hainanensis is a creeping succulent with long runners and leaves not unlike *Limonium*. Though wholly unlike its beach shrub relative, *Scaevola frutescens*, both are recognizable as members of the Goodeniaceae by the small white flowers, like asymmetric stars, with their pointed petals all directed to one side.

A fourth succulent species, common on many salt-meadows, is the sea-blight *Suaeda australis* (Chenopodiaceae), with short branches loaded with small succulent leaves, and inconspicuous flowers.

Of the mobile animals of salt-marsh, the crabs are the first to become conspicuous. The fiddler-crabs *(Uca)* of the family Ocypodidae have already been encountered (p. 184) on the protected shores at Tai Tam. Two species are very common at high spring tide, making a warren of close-set burrows in stiff mud. *Uca lactea* has a brown flecked carapace, and the male's enlarged claw is white, flushed with yellow. In *Uca chloropthalmus* the large claw is cherry red and white; the rest is jet black, except for the carapace which can vary from cross-streaked black and greenish-blue to bright scarlet or dull brick-red.

It is on salt-meadow, particularly where stones are strewn, and in mangrove swamps, that the grapsoid crabs of the important sub-family Sesarminae come into their own. Fast-running and with square or rectangular carapace, they have short eye-stalks characteristically widely set, at the corners of the wide 'front'.

FIG. 11.3. Salt-meadow with grass and succulents at Cheung She, Tolo Harbour. The main extent is of *Zoysia* grass, with mangroves *(Avicennia)* and *Limonium* in front. The fringe is of cobbles on mud. In the background is a village with a 'Fung Shui' wood.

SPECIES:
Bird inset is a Common sandpiper *Tringa hypoleucos*.
1. *Assiminea violacea*; 2. *Littorina melanostoma*; 3. *Cassidula plecotrematoides*; 4. *Ellobium chinense*; 5. *Auriculastra duplicata*; 6. *Melampus triticens*; 7. *Cassidula schmackeriana*; 8. *Onchidium* sp.; 9. *Limonium sinense*; 10. *Zoysia sinica*; 11. *Nerita yoldii*; 12. *Scaevola hainanensis*; 13. *Terebralia sulcata*; 14. *Dostia violacea*; 15. *Suaeda australis*.

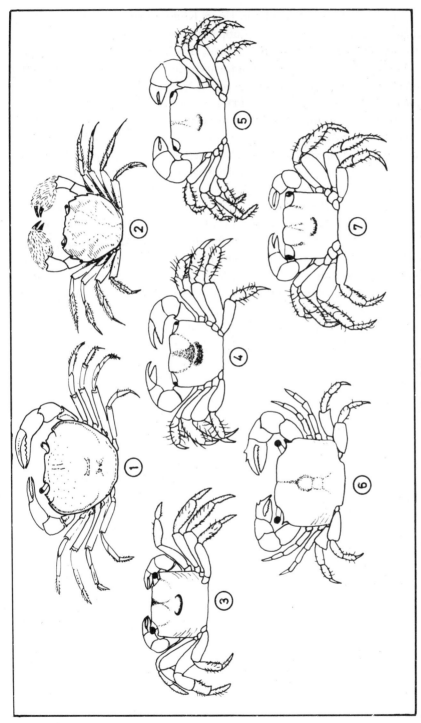

FIG. 11.4. Crabs of Hong Kong's paddy field, mangroves and salt-meadow. 1. *Potamon hongkongensis*; 2. *Eriocheir sinensis*; 3. *Holometopus serenei*; 4. *Holometopus dehaani*; 5. *Chiromanthes fasciatum*; 6. *Neosarmatium punctatum*; 7. *Pseudosesarma patshuni*.

A recent study by C. L. Soh[3] of the Hong Kong sesarmines has revealed no fewer than 15 species on or near the upper shore.

Highest occurring and fresh-water adapted, *Holometopus serenei* lives in streams and paddifields. Two other crabs accompany it in streams, the Chinese mitten crab, *Eriocheir sinense*, a grapsid easily recognized by the hairy tufts on its chelae; and *Potamon hongkongensis* (Potamonidae).

Where streams drain into paddifields, especially where the ground dries out, and in the clay bunds, three more sesarmids appear: *Neosesarma punctatum*, *Pseudosesarma patshuni* and *Chiromanthes fasciatum*. With progression to salt-meadow, under stones and among succulent vegetation, the sesarmid series greatly enriches, as may be seen from the summary diagram (Fig. 11.5).

The gastropods of salt-meadow belong to a few families rather constant in this habitat; they reach their greatest size and variety in Southeast Asia and Malaysia. The species are well distributed among the different micro-habitats available. First come snails not decisively adapted for salt-meadow but found wherever small boulders cover a sheltered shore at high level. The ubiquitous *Monodonta australis* occurs from wave-exposed boulders to deep shelter. Much more confined to shelter is *Nerita yoldii*, with pale yellow exterior, variably black-banded and often eroded and dull yellow mouth.

The periwinkles, Littorinidae, have two members well adapted to saltings and grass-flats. *Littorina melanostoma* is the commoner, generally found climbing above the ground on *Zoysia* grass. The elongate shell is thin and smooth, coloured fawn to grey, and with a conspicuous brown patch on the inner lip. The other is *Littorina scabra*, sometimes present on rocks in salt meadow, but most common on the foliage of mangroves (p. 222).

Two small prosobranch gastropods occur sparingly at this level, under succulent plants or stones, being common only at certain localities. They are the tiny, sharp-spired *Truncatella* sp. (found commonly at Deep Bay) and the conical *Assiminea violacea*.

The Cerithiidae have several species common on mangrove swamps (p. 222) and on mid-tidal mud-flats (p. 245). The large *Terebralia sulcata* is the most important in salt-meadow. Though still equipped with a gill, it uses the mantle cavity as an air-breathing lung; to facilitate this habit the two sides of the lip fuse to leave a narrow canal through which air can pass even with the operculum closed to prevent water loss. Cerithiids, like all these gastropods, feed on detritus, living on rich organic deposits, or scouring the surface of isolated stones. They lay thin gelatinous spawn strings, attached under debris, and like the littorines have a free-swimming larval stage.

The remaining gastropods are pulmonates (Fig. 11.6), not like cerithiids adapted

[3] Soh, C. L. 1978. On a collection of sesarmine crabs (Decapoda, Brachyura, Grapsidae) from Hong Kong. *Memoirs of the Hong Kong Natural History Society* 13: 9–22.

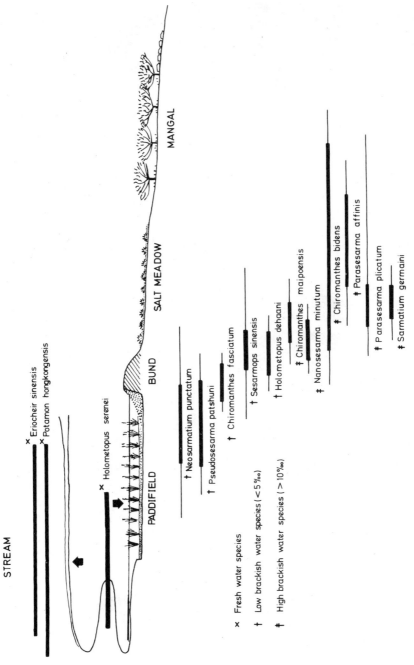

FIG. 11.5. The distribution of fresh-water and salt-meadow/mangrove brackish water crabs.

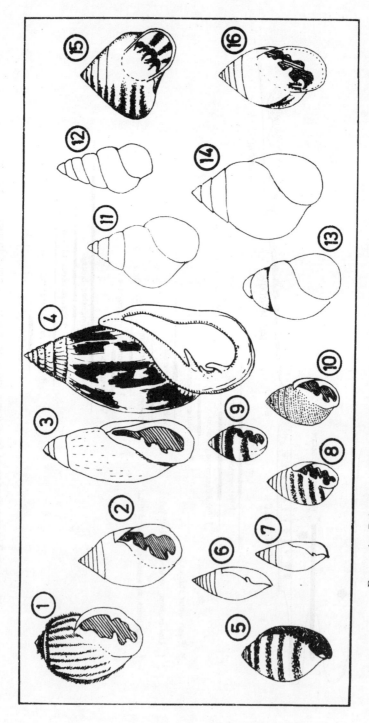

FIG. 11.6. Pulmonate and prosobranch snails of salt-meadows and mangroves.
1. *Cassidula schmackeriana*; 2. *Cassidula plectorematoides*; 3. *Ellobium chinense*; 4. *Ellobium politra*; 5. *Melampus triticens*; 6. *Auriculastra duplicata*; 7. *Auriculastra subula*; 8. *Laemodonta punctatostriata*; 9. *Laemodonta punctigera*; 10. *Laemodonta exaratum*; 11. *Assiminea violacea*; 12. *Assiminea subeffusa*; 13. *Assiminea lutea*; 14. *Assiminea brevicula*; 15. *Pythia fimbriosa*; 16. *Pythia cecellei*.

for trailing over soft ground, but plump and ovoid, concealed under stones, wood and other debris, often in large numbers[4]. The Ellobiidae are the most widespread of all maritime pulmonates, with short, dull-coloured shells, having a toothed aperture. Living in damp, sheltered places, seldom inundated, they avoid the light and breathe atmospheric air. The largest Hong Kong species, *Ellobium polita*, up to 7.5 cm high, is generally found not in saltings but around mangrove trunks in the Deep Bay marshes. *Ellobium chinense*, with long oblique mouth, drawn out behind into a short canal, is more widely distributed. Most abundant of all the ellobiids is the small *Melampus triticens*, the shape and size of a wheat-grain when adult, and darkly cross-banded.

The *Auriculastra* species (*A. duplicata* and *A. subula*) are smooth ellobiids, thin-shelled and sharp-pointed at apex and base. The small *Laemodonta* species are by contrast short and barrel-shaped, with both sides of the aperture strongly toothed. The shells have finely punctate sculpture, being plain in *L. exaratum* and dark-banded in *L. punctatostriata* and *L. punctigera*. The two *Cassidula* species are of larger size and sturdy build, with the aperture surrounded by a strongly rimmed lip. *C. plecotrematoides* is smooth-shelled and *C. schmackeriana* is distinguished by its coat of coarse bristles arranged vertically along the lines of growth.

The genus *Pythia* contains moderately large ellobiids, living more terrestrially, associated with trees or grass clumps. In *Pythia fimbriosa* the ovoid shell is compressed like a plum-stone. *P. cecellei* is more trochoid in shape, and dark-banded. In both the aperture is narrow and heavily toothed.

A second pulmonate family, Onchidiidae, is represented by three or four species of *Onchidium*. *O. verraculatum*, a naked flattened slug, about 2.5 cm long occurs on estuarine rocks. Other, as yet unidentified species characterize salt-meadows and mangroves. The body wall is rubbery, grey and brown mottled and heavily warty on the back. In external symmetry, *Onchidium* is perfectly bilateral, save for the genital aperture at the front right. The anus is in the mid-line behind, and above it the aperture of a small lung rhythmically opens and closes. The shell is entirely lost, the thick skin giving protection from desiccation and possessing numerous secondary pallial eyes. Common on firm surfaces and under stones, *Onchidium* grazes on algae or fine deposits like the ellobiids.

Three bivalve molluscs live on salt-meadows, at a level surprisingly high for ciliary feeders depending on tidal inundation. Small mussels, *Brachidontes atratus*, attach to the bases of *Zoysia* grass, and specimens of *Caecella chinensis* bury themselves in patches of coarse sand (p. 114). The corbiculid *Geloina erosa* may also be found at this level especially around streams, but it is more common in the mangroves.

[4] See the comprehensive treatise by Shen, T. C. 1939. *The Chinese Land and Freshwater Gastropods in the Senckenberg Natural History Museum*. Vittorio Klostermann, Frankfurt-am-Main (in German).

Chan and Trott[5] have described the collembolan insect *Oudemansia esakii* from the upper levels of such beaches. It lives some 36 cm below the sand surface and both adults and juveniles emerge when the tide recedes. The insect completes its life cycle in the sand and is a scavenger feeding mainly on small animal corpses. When in the soil an air-bubble forms around the insect and they can withstand enforced immersion far over 75 hours.

Denser Grass-fields

A much more luxuriant cover of grasses is occasionally encountered close to salt-marsh or mangrove shores, especially in places of continuous fresh-water seepage from the bunds of paddy-fields. The ground is here soft underfoot and a nutrient-rich mud covers the surface. Leaf-fall and organic decay builds up a rich substratum that is rarely covered by the tides, but constantly wetted by fresh seepage water.

In a typical habitat of this kind, close to Sai Kung, on the western shore of Three Fathoms Cove, in Tolo Harbour, the back of the beach is dominated by a sedge and grass community that merges with the paddifields to the rear (Fig. 11.7). Here the grass *Panicum repens*, with *Ischaemum indicum*, defines the landward margin above which even the highest spring tides do not invade. Further seawards the long thin blades of *Schoenus* and *Scirpus* stand erect and up to 1 metre high. The former has a very large rhizome and the leaves are crescentric in section: the leaves of the latter are solidly triangular in section. *Fimbristylis spathacea* and *Cyperus* sp. (both Cyperaceae) with short, narrow, flat blades, rounded at the tips effect the transition to a stubbly *Zoysia* community, earlier defined, but approaching the limits here of its wetness tolerance. At Sai Kung the back of the beach mangroves (*Acanthus, Bruguiera, Excoecaria, Lumnitzera*) are small in stature and widely scattered throughout the sedges and the *Zoysia* grass. Clearly the ground here is too wet, even for them, though further seawards the pioneer mangrove community, largely of *Kandelia*, flourishes as the sea is able to exert its influence. This dense marsh is very rich in amphibious pulmonate snails (Fig. 11.6). The salt-meadow *Ellobium chinense* gives way to *Pythia fimbriosa* and *P. cecellei. Cassidula plecotrematoides*, common on salt-meadows, is here replaced by *C. schmackeriana*. Smaller *Laemodonta* and a range of *Assiminea* species are common beneath the ground-cover or under litter.

Amongst the grasses are scattered shrubs, such as specimens of *Bruguiera conjugata* and *Excoecaria agallocha*. On their stems and leaves are littorines and *Cerithidea ornata. Dostia violacea* and *Nerita lineata* are nerites common at root level.

[5] Chan T. D. and Trott, L. B. 1972. The collembolan, *Oudemansia esakii*, an intertidal marine insect in Hong Kong with emphasis on the floatation method of collection. *Hydrobiologia.* 40: 335–43.

Enclosed Flats

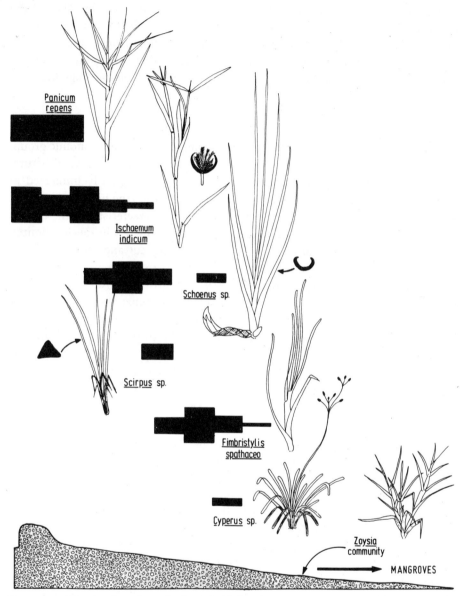

FIG. 11.7. The zonation of marsh plants, behind a pioneer mangrove community at Sai Kung, in Three Fathoms Cove, Tolo Harbour.

A profusion of ants, beetles and gammarids with other Amphipoda, inhabit the litter, along with some important sesarmid crabs. The fresh-water species, *Holometopus serenei*, commonest in paddy fields, is found here, and even the fresh-water crab *Potamon hongkongensis*. More noticeable are the faster running sesarmids *Chiromanthes fasciatum*, and the widespread *C. bidens*, together with *Sesarmops sinensis* (see Fig. 11.4).

Mangrove Swamps

The name 'mangrove' does not denote the plants of any single taxonomic group. It is a description of a habit and a habitat. Scattered among several natural families, mangroves are amphibious shrubs and trees, regularly immersed at trunk level, and even up to the foliage, by high tides. Like the hermatypic corals of clean oceanic waters (Chapter 12) mangroves are characteristic of the tropics and are here in Hong Kong held in delicate balance being few in species, limited in extent and dwarf in stature. They are also being cut and reclaimed at an alarming rate.

Rooted in a water-logged anaerobic soil, mangroves show a complex of structural and physiological adaptations. They may stabilize themselves in mud by strong, radial root systems, as in *Avicennia* (Avicenniaceae), or by prop-roots as in *Kandelia* and *Bruguiera*, both belonging to the tropical family Rhizophoraceae. The horizontal main-roots of *Avicennia* put up breathing roots in the form of erect, twig-like pneumatophores, with a corky waterproof bark, lenticels and a tissue of spongy parenchyma. Buttressed roots occur in the rare 'looking glass' mangrove *Heritiera littoralis* (Sterculiaceae) that elsewhere in S.E. Asia is a dominant mangrove species.

Other anatomical peculiarities among various mangroves are sclerified leaves, often with a polished, waxy cuticle above and a white tomentum beneath, serving both to reflect bright light, especially off the water surface, and to repel water. The water milieu is paradoxically one of physiological drought, with a salty external medium tending to dehydrate the tissues. The correspondingly high osmotic pressure of the cell sap safeguards the plant's ability to take up and retain water. Their dense network of veins and several layers of water-storing epidermis are leaf modifications generally found not in aquatic plants but in xerophytes. Oxygen conduction and storage, with water-logged and anaerobic soil, is provided by underground aeration systems, as well seen in *Avicennia*.

In the more seaward-growing mangroves, reproduction takes place by the germination of viviparous embryos: clusters of torpedo-like 'droppers' in Rhizophoraceae, smaller droppers in *Aegiceras* (Myrsinaceae), and a rounded embryo with folded, fleshy cotyledons in *Avicennia*.

Not all the species in a mangrove swamp or 'mangal' have gone as far as the seaward members in acquiring these special habits. There is a group of regularly occurring shrubs and trees to the landward, never immersed by normal tides, that can be called 'mangrove associates'. Furthest to the landward will generally be the screw-pine *Pandanus tectorius,* with its long and heavy serrate leaves, and the trunk forked at the bottom into strut-roots. Along with *Pandanus* grows the pale yellow or red-flowered *Hibiscus tiliaceus,* a familiar sea-coast tree. The ubiquitous tropical shrub *Lantana camara,* with rough aromatic leaves and red and yellow flower-heads, is also certain to be found. Common also may be the mangrove fern *(Acrostichium aureum),* the scrambling *Clerodendrum inerme* and the St. John's lily *Crinum asiaticum.* We may first approach the study of a 'mangal' by visiting the small, very accessible habitat at Ting Kok, in Tolo Harbour, before going on to the more extensive and spectacular marshes of Deep Bay. Fig. 11.8 shows a mangrove sequence from one of the small but fairly complete stands left at Tolo, in simplified form, proceeding out from land to seaward.

TABLE 11.1 KEY TO THE MANGROVES OF HONG KONG

1. With seeds germinating on the parent plant to produce embryos viviparously 3–6
2. With viviparous embryos lacking . . . . 9 or 10
3. With large elongate torpedo-shaped embryos ('droppers') RHIZOPHORACEAE 7 or 8
4. With clusters of smaller banana-shaped embryos, following small white 5-petaloid flowers: stems red . . . . . . *Aegiceras corniculatus* (MYRSINACEAE)
5. With rounded embryos having convoluted green cotyledons; leaves opposite and bronze green, pale beneath; flowers small, with 4 inconspicuous petals
Avicennia marina (AVICENNIACEAE)
6. Fruit a large, keeled, brown nut. . . *Heritiera littoralis* (STERCULIACEAE)
7. Leaves about 10 cm long, with pointed tips; stems reddish; flowers 11–14 petaloid. . . . . *Bruguiera conjugata* or:
8. Leaves about 5–6 cm long, with rounded tips, stems green; flowers 4–6 petaloid. . . . . *Kandelia candel*
9. Leaves flat and pale green, with rounded tips; small white 5 petaloid flowers.
Lumnitzera racemosa (COMBRETACEAE) or:
10. Leaves angled in section, and stems when broken exuding a poisonous milky latex; reduced flowers in small catkins, unisexual with male and female on different trees.
Excoecaria agallocha (EUPHORBIACEAE)

To the seaward of *Hibiscus,* the mangrove associate *Excoecaria agallocha,* may often grow right out on to the upper beach when true mangroves are lacking. It is recognized by its greenish catkins, carrying reduced flowers of separate sexes, and its copious white latex from broken leaves. Overmature leaves turn dull red. The holly-like shrub, *Acanthus ilicifolius,* with its stiff long-spined leaves, grows commonly at the inner edge of the mangrove swamp. The pale blue flowers are in terminal spikes, supported by leafy bracts. In the denser mangal, as at Deep Bay, the climbing vine *Derris trifoliata* (Papilionaceae) is found, with pinnate

Fig. 11.8. Mangrove flat at Ting Kok, Tolo Harbour.

ZONES:

A. Sandy inter-tidal shore with soldier-crabs and higher up a few succulents; B. Pioneer zone of mangroves with *Kandelia, Aegiceras* and *Avicennia*; C. Back-zone of mangroves with *Lumnitzera, Excoecaria* and *Acanthus*; D. Paddy fields behind with fringe of *Pandanus, Derris* and *Hibiscus*. A generalized cross-section of the mangal *(middle)* (for symbols see Fig. 11.9).

SPECIES:

1. Black-capped kingfisher *Halcyon pileata*; 2. White-eye *Zosterops japonica*; 3. *Nerita yoldii* with rock oyster attached; 4. *Clithon oualaniensis*; 5. *Nerita undata*; 6. *Nerita lineata*; 7. *Dostia violacea*; 8. *Cerithideopsilla djadjariensis* with inset of lip; 9. *Cerithidea ornata* with inset of lip; 10. *Terebralia sulcata* with inset showing respiratory aperture and animal expanded; 11. *Cerithidea rhizophorarum* inset of lip and mode of attachment during hibernation; 12. *Truncatella* sp.; 13. *Assiminea violacea*; 14. *Littorina scabra* also showing animal expanded; 15. *Littorina melanostoma*; 16. *Clypeomorus moniliferum*.

oval leaves formed into 5 leaflets; it produces viviparous fruits, with large, rounded, green cotyledons. The creeper *Canavalia maritima* has pinnate leaves formed into three leaflets and produces long, banana-shaped seed cases. Both of these features easily separate this creeper from *Derris*. In this region of the mangal the grasses *Zoysia sinica, Sporobolus virginicus* or *Fimbristylis dichotoma* form a ground cover.

Of the species properly forming the mangal, five can be recognized belonging to three families. Table 11.1 in addition includes the euphorbiacean *Excoecaria* and the sterculiacean *Heritiera*.

The seaward-lying mangroves can, with a little practice, be recognized at a distance by their colour and shape. Hong Kong is too far north for the mangroves of the genus *Rhizophora* so characteristic of tropical estuaries, with arched strut-roots forming an impenetrable thicket down to the water's edge. On the shores, the outlying mangroves are small trees, generally *Avicennia, Aegiceras* or *Kandelia*. *Avicennia marina* is a low shrub with a flat, spreading canopy, dull green or slightly bronze-tinted. The scatter of upright pneumatophores, as wide as the extent of foliage, identifies an *Avicennia* beyond mistake.

Kandelia candel is more compact in shape, with its foliage brighter yellowish-green; even from a distance, its clusters of terminating droppers stand out clearly. The white flower most generally has four petals. The nearest relative of *Kandelia* is *Bruguiera conjugata,* the many-petalled mangrove, distinguished as a taller tree (up to 2 metres), with the leaf petioles and receptacles of the droppers reddish (green in *Kandelia*). *Lumnitzera racemosa* occurs higher on the shore than the foregoing species; its light green leaves are round-tipped and leathery. *Aegiceras corniculatum* has an inflorescence of small white flowers producing, as fruits, clusters of droppers-like tiny-green bananas. To the seaward of the mangal, the sea-grass *Halophila beccarii,* (a S.E. Asian mangrove associate with more slender leaves than *H. ovata* (p. 249)) is common; and on more consolidated mud grows the succulent *Suaeda australis*.

The leaves of *Kandelia* are colonized by a scale insect *(Ceroplastes rubens,* Coccidae) which sucks sugary sap. Since the supply is in excess of their needs some sugar is excreted as 'honey dew'. This is in turn utilized by a fungus of the family Meliolaceae that forms a sooty powder on the leaves giving a dirty appearance. Mangrove leaves are bored by insect leaf-miners of the family Agromyzidae.

Mollusca

The leading gastropod families of mangrove swamps are the Potamididae and Neritidae. The first are typical of the mud-surface, the second prefer isolated patches of hard substrate. The potamidids are trailing long-spired snails, feeding on surface detritus, like *Batillaria,* sometimes found in more sandy patches, and

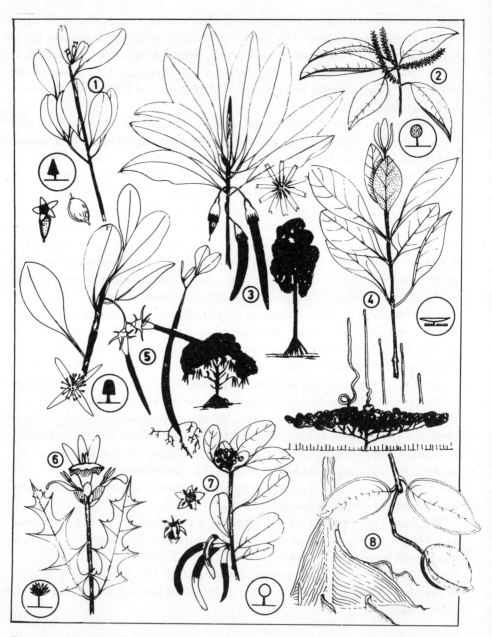

FIG. 11.9. Mangrove plants.
1. *Lumnitzera racemosa* with inset flower and fruit; 2. *Excoecaria agallocha*; 3. *Bruguiera conjugata* with attached droppers and inset flower and tree shape; 4. *Avicennia marina* with inset below pneumatophores and tree shape; 5. *Kandelia candel* with attached droppers and inset flower, germinating dropper and tree shape; 6. *Acanthus ilicifolius*; 7. *Aegiceras corniculatum* with inset flowers; 8. *Heritiera littoralis*. The small circles show the symbols used in the profile in Fig. 11.8.

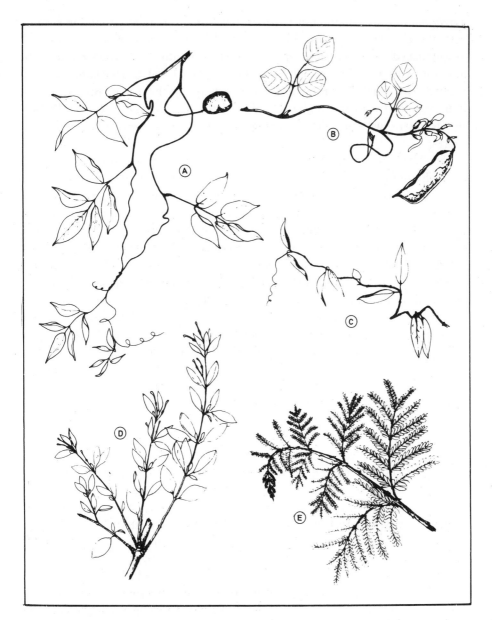

FIG. 11.10. Creepers from the back of the shore mangrove plants.
A. *Derris trifoliata*; B. *Canavalia maritima*; C. *Ecdysanthera micrantha*; D. *Clerodendrum inerme*; E. *Caesalpinea crista*.

Terebralia sulcata, common in salt-meadow, but also regularly found in swamps with the *Cerithidea* species. Three *Cerithidea* are illustrated here (Fig. 11.8), distinguished by details of sculpture and shell mouth. *C. rhizophorarum* is the heaviest, with a strong beaded sculpture. *C. ornata* and *C. cingulata* are ornamented with axial sculpture. Both species may attach to mangrove trunks or branches, held with mucus by the edge of the lip. *Clypeomorus moniliferum* is the darker-brown equivalent in mangroves of *C. humilis* on cobble and sandy shores.

The Neritidae crawl about on the bases of mangrove trunks, or live under small slabs of rock lying in mud. *Dostia violacea* is smooth black on the exposed surface, with the border of the mouth expanded and coloured dull blood-red to mauve. *Nerita lineata* is a larger species, low-pitched and identified from its dark brown, cord-like sculpture. *Nerita yoldii* is generally eroded above, and shows a dull lemon-yellow aperture. The smallest and, where it occurs, most abundant nerite is *Clithon oualaniensis* (Fig. 11.8) found on shell and hard cover where fresh water runs off. *Littorina scabra* and *L. melanostoma* are the familiar periwinkles of mangroves, living chiefly on the upper twigs or foliage. The mucus of the foot is extremely viscid, making it hard to detach, and the shell hangs by dried mucus from the lip when the operculum is closed. *L. melanostoma* either exists in a number of colour and shell form morphs or else comprises a 'species group' in which a number of closely similar 'species' co-exist in the mangal. The elucidation of these 'species' is a matter of high priority.

The long-spired *Truncatella* can also be found in small clusters among the leaves of the mangrove plants.

The common bivalves with mangroves are mostly heavy and only slightly mobile. Some species of the noah's ark-shells (Arcidae) have lost the byssus attachment and lie free in anaerobic muds. By shape and sculpture, the shell is not suited for active movement. Three species are very common in Hong Kong: *Anadara granosa* with heavily rounded ribs, crossed by strong lamellae, and *Scapharca cornea,* dark brown and covered with a hairy stubble. Both lie with the squared-off posterior end close to the surface. Much lower-zoned on muddy shores is the twisted ark-shell *Trisidos semitorta.* This large, highly distinctive species is twisted in the antero-posterior plane and lies embedded at an angle in the mud with the posterior edge of the right value uppermost. The tissues of most arks are dark red and when cut exude blood rich in haemoglobin. This respiratory pigment is rare in molluscs, but serves in arcids as a respiratory store in a habitat where oxygen is physiologically difficult to obtain.

The venerids *Gafrarium tumidum* and *G. pectinatum* are mud-flat bivalves common throughout the tropical West Pacific. Like the arcids, *Gafrarium* has strong sculpture and lies immobile just below the surface, generally in coarse sand at the edge of mangrove flats. *Geloina erosa* is a thick-shelled,

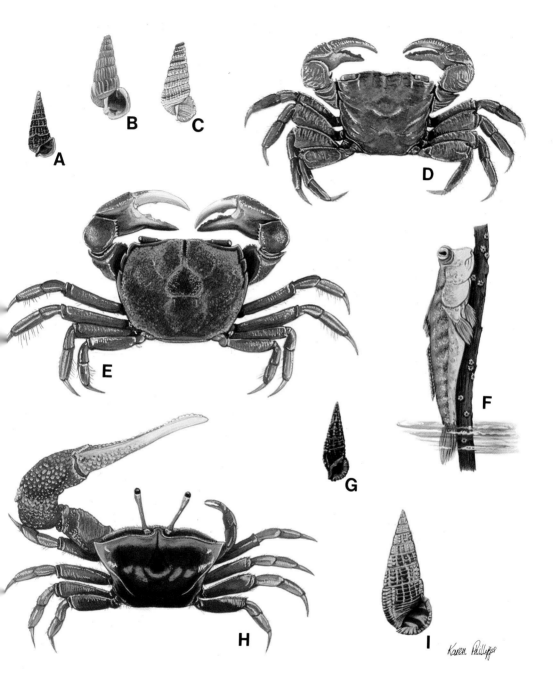

PLATE 19. The fauna of mangroves: A. *Cerithidiopsilla cingulata*; B. *Cerithidea ornata*; C. *Cerithidea rhizophorarum*; D. *Chiromanthes bidens*; E. *Chasmagnathus convexum*; F. The mud-skipper *Periophathalmus cantonensis* (note the barnacle *Euraphia withersi*); G. *Cerithideopsilla djadjariensis*; H. *Uca arcuata*; I. *Terebralia sulcata*.

PLATE 20. Birds of soft shores and marshes: A. Kentish Plover, *Charadrius alexandrinus*; B. Asiatic Golden Plover, *Pluvialis dominicus*; C. Common Sandpiper, *Actitis hypoleucos*; D. Greenshank, *Tringa nebularia*; E. Whimbrel, *Numenius phaeopus*; F. Long-toed Stint, *Calidris subminuta*; G. Bar-tailed Godwit, *Limosa lapponica*.

orbicular bivalve, living in mud or amongst grass and in pools formed between mangrove roots. The heavy periostracum is greenish, often eroded near the umbones by the acid mangal soil. Belonging to the fresh-water family Corbiculidae, *Geloina* can withstand long periods of exposure and filters from the subterranean waters in the burrows of sesarmid crabs. Currents are taken in not only by the siphons, only immersed at spring tides, but by the pedal gape, submerged deeper in the ground. During drought, *Geloina* can respire by the exposed mantle edges. A strangely specialized bivalve, *Enigmonia aenigmatica,* related to the saddle oysters (p. 90), lives on the leaves and trunks of mangroves attached by a stout byssus. Common in Malaysia, it is very occasionally found in Hong Kong. The left (upper) valve is purple, streamlined and limpet-like, with the right valve below functionless.

Crustacea

All the Crustacea of mangroves and mud-flats are burrowers, but it is only on the firmer ground up-shore that permanent home-sites can be established. The crabs are extraordinarily numerous in individuals and species. Like the gastropods they can be broken up into separate habitat groups according to tidal position, and the solidity or fluidity of the ground. It is unlikely that all the Hong Kong species have yet been found, and rewards still await the student willing to do patient collecting, and to make records of zone and habitat data.

The principal mud-flat crabs are shown in Fig. 11.11 with details of the structure of the chela for most of the species. Species of Grapsidae are, as usual, commonest at upper levels. Near high-tide *Helice tridens* may be dug out, with squarish carapace, three-toothed (including orbital spine), and mud-grey to olive-green. The eye-stalks, a little longer than in sesarmids are only of moderate length. In the same patches of stiff clay, some scarcer grapsids may be found, including *Varuna litterata* and *Clistocoeloma merguiense* and *Sarmatium germaini*. The most familiar sesarmid is *Chiromanthes bidens,* leaden grey to dark green, with legs and sides of the carapace marked with thinly traced cross-lines, and the chelae yellow to tan-brown. This is one of the commonest crabs in burrows, especially around the roots of mangroves, and where the substrate is too soft for *Helice* and other grapsids. High-zoned with *C. bidens* is the large *Chasmagnathus convexum*. A heavy green carapace is offset by brilliantly purple, red and blue, approximately equal chelae. Under stones and loose cover, as on mud-flats with oyster beds, *Metopograpsus messor* is particularly common, running about rapidly when disturbed. The carapace is grey-green, finely black-lined and stippled, and the chelae are purple with white tips.

At the lowest shore level, where the mud is stickiest and most fluid, *Macrophthalmus latreille* and *M. definitus* are generally the commonest crabs. The generic distinguishing marks are the very broad carapace (twice as wide as long), and the

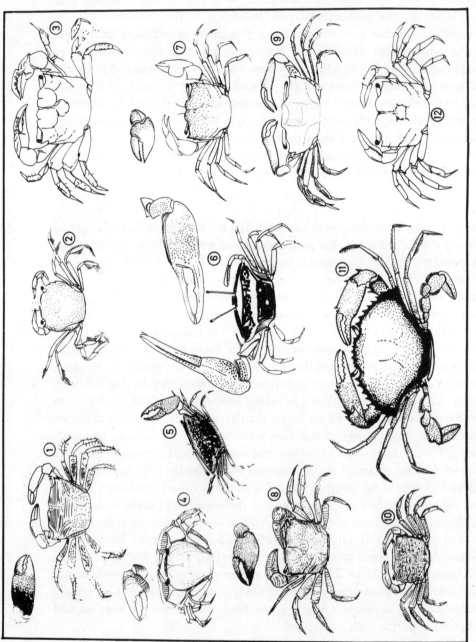

Fig. 11.11. The crabs of salt-meadow, mud-flats and mangrove shores.
1. *Metopograpsus messor*; 2. *Varuna litterata*; 3. *Chiromanthes maipoensis*; 4. *Sarmatium germaini*; 5. *Uca vocans*; 6. *Uca arcuata*; 7. *Helice tridens*; 8. *Chiromanthes bidens*; 9. *Macrophthalmus definitus*; 10. *Clistocoeloma merguiense*; 11. *Scylla serrata*; 12. *Sesarmops sinensis*.

slender chelae with strong, down-curved fingers in the adult male. Slower than sesarmids, these crabs are well-adapted, being drawn out laterally and almost cylindrical in antero-posterior section, to slip sideways into temporary burrows in semi-fluid mud. Two other species, *M. convexus* and *M. boteltobagoe* may be found, the former with bright red-purple chelae, the latter with orange chelae. Both are higher zoned and more common on sandy beaches.

The mud-flat fiddler-crab *Uca arcuata* is the largest and most vivid of the Hong Kong species. An adult male, with claw details, is illustrated in Fig. 11.11. The carapace is up to 2.5 cm wide, and variously black mottled often with cross-bands of turquoise. The large chela has a bulbous base and is brick red on top, spreading into mauve at the sides, with the finger tips white. In Fig. 11.11 we also illustrate the much smaller *U. vocans* which is a low-zoned species commonly occurring on softer muds beyond the mangroves.

The swimming crabs (Portunidae) are represented in mangrove swamps by their largest and heaviest species, *Scylla serrata*, a valuable edible crab. The body is dull greenish-black, with the carapace bow-fronted and strongly toothed. The chelae, as in all portunids, are powerful and sharp, but the walking legs are slender. The paddles on the fourth pair are rather small. *Scylla* makes deep oblique burrows, and is seldom seen at the surface during low tides.

On more solid terrain among mangroves, conical mounds 7–8 cm high, with a hole at the summit, attract attention. From the burrow beneath them can be dug the slow-moving shrimp *Laomedia astacina*, pale pink with blue markings. Belonging to the Anomura, and a close relative of *Callianassa* and *Upogebia*, this shrimp is ill-adapted for movement out of its burrow. The enlarged, sub-chelate claws are the second thoracic limbs, and the frail walking legs (3, 4 and 5) are modified to stabilize the animal against its burrow wall. The mode of feeding is still not understood, and would well repay investigation. The pleopods are vestigial, and evidently do not create a current as in *Callianassa* and *Upogebia*. The third maxillipeds and the first pereiopods have fringes of setae, and these may perhaps be deployed outside the burrow for microphagous feeding. The same method has been suggested for a larger relative, the 'mud-lobster', *Thalassina anomala*, of tropical mud-flats.

Less apparently specialized for burrowing life are the common and widespread shrimps *Crangon affinis*, *Palaemon* (illustrated is *P. serrifer* with a rather straight rostrum upcurving at the tip) and *Palaemonetes*. With slender, sub-chelate claws, depressed body, and pepper-and-salt camouflage colouring, they lie in loose sediments just beneath the surface. These shrimps are harvested by man, and suffer the great toll levied by wading birds between tides.

The mantis shrimps or squillas (Stomatopoda) are burrowing crustaceans of the open surfaces of mud-flat and silty sand. They are seldom found near the root systems of mangroves and prefer semi-fluid low-tidal sediments. Stomatopods are long flat-bodied and loose-jointed Crustacea. Not only do they burrow actively, like *Callianassa*, but they have considerable agility outside, somersaulting, doubling-up, and swimming with large plate-like pleopods.

The Stomatopoda are distinguished from the rest of the higher Crustacea by the very short carapace, leaving four thoracic segments uncovered and the long, abdomen, widening towards the large tail-fan. The most prominent limbs are the second thoracic pair forming two great chelae, with which the shrimp impales active living prey. The last chelar joint has sharp curved teeth, and hinges upon the joint behind it like a mantis' grasping claw. The chelae retract nearly into the smooth contour of the carapace, or can be swiftly extended to grasp prey. They may deliver a painful wound.

The first thoracic limbs are small maxillipeds, and legs 3, 5 and 4 are blunt and heavy-tipped for digging. Behind them are three slender splints, with whip-like exopodites, making in all eight thoracic pairs. The broad pleopods are for swimming, or impelling water through the burrow.

Common mud-flat stomatopods are *Oratosquilla oratoria* and *Harpiosquilla japonica*, both cream to greyish-white in colour, and 15 cm or more long. Telson details of each, with the whole body of *Oratosquilla* are shown in Fig. 11.12.

The Deep Bay Marshes

These are the last considerable mangrove swamps left in the Hong Kong territories, and they are among the most northerly in the world. By comparison with mangals of Malaysia, their trees are dwarfed; and they have none of the arched and strut-rooted *Rhizophora* species, so typical of the tropics. As compared with the complete sequences seen from the shore to the outer edge of Taipo marshes, they have been robbed of a number of their landward species by reclamation and fish-pond conversion.

But their several kinds of mangrove cover large areas, and these, along with the fauna of the mud, and the wonderful diversity of birds, make the Deep Bay Marshes without challenge the greatest wild-life resource of Hong Kong.

The Deep Bay marshes, as approached from Mai Po, have a wide-ranging international importance. The Hong Kong–Chinese border cuts through them to the north; and long before man's political jurisdiction, they were a stopping place of many species of migrant wader, such as the bar-tailed Godwit *(Limosa lapponica)* on its journey from New Zealand to its Siberian northern summer breeding grounds or the Curlew sandpiper *(Calidris ferruginea)* similarly travelling regularly between its northern breeding grounds and its winter residence in Australia. The marshes are today the favourite resort for the growing company of Hong Kong ornithologists, and they are regularly the first place an overseas naturalist will be brought to see. With the coasts of China

Enclosed Flats

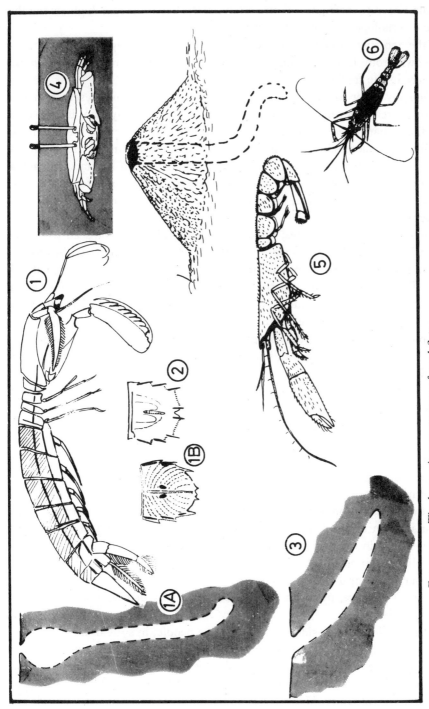

FIG. 11.12. The burrowing crustaceans of mud-flats. 1. *Oratosquilla oratorio* with 1a. burrow and 1b. telson detail; 2. *Harpiosquilla japonica*; 3. Burrow of *Scylla serrata*; 4. Posture of *Macrophthalmus latreille*; beneath the surface; 5. *Laomedia astacina* with mound and shape of body; 6. *Palaemon serrifer*.

virtually closed to international bird study, Deep Bay must rank high among the choice ornithological areas of the world.

Of the 370 species of birds presently recorded from Hong Kong, about 230 are known from Deep Bay and of these over 100 are rarely, if ever, found elsewhere in the territory.

Before any other part of Hong Kong, these marshes call for preservation as a wilderness area. They have already been profoundly altered by man. Until recent times these changes could be said to have enhanced the variety of bird habitats, as well as their accessibility. This stage has been passed: further development could put at risk the survival of the marshes as independent habitats. Already a large acreage of their inner reaches has been cleared of mangroves for fish-ponds, and to the landward of these, green paddy fields have been reclaimed into high productivity.

The three sections, marshes, ponds and paddy fields, can be seen from the map in Fig. 11.13. The truly marine section, ponds and mangroves, lies to the seaward side of the inner bund, a firm mud-bank running parallel to the natural shore. To walk this path, and to look across the ponds to the heavy green mangrove cover, 0.6 km wide, and beyond these to the great mud-flats, is to realize the full extent of the wilderness.

An outer bund separates the fish-ponds, themselves studded with mangroves, from the natural mangal; and this can be approached at intervals by cross-paths from the inner bund.

The pond area, like the paddy fields, is today the habitat of a bare-footed people to whom soft mud and turbid water are elements like the sky and air. The few secure buildings are on the bunds or raised above the fish-ponds on stilts. These waters are used extensively for fattening fish, which are regularly caught as the tide goes out through the sluice gates of the ponds. Within the large ponds, a dredge is regularly at work, shifting and accumulating loads of bottom mud.

Further reclamation is an imminent threat. Part of the marsh territory is in the jurisdiction of the People's Republic of China. Where enhanced primary productivity is in question, modern China is for the foreseeable future likely to be largely indifferent to the claims of wilderness preservation. Such arguments for conservation, or a policy that seemed to set the survival of birds against the people's interest, might in today's China be regarded with somewhat less than enthusiasm.

From the Hong Kong side too, the Chinese people, in spite of an impressive history of appreciation for environment, are hardly able today to give conservation a high place. With present overcrowding, they have not yet learned to think of natural wilderness as something to be safe-guarded for the sake of ecological and psychological sanity. Mangrove swamps and their adjacent flat-lands all too

Enclosed Flats

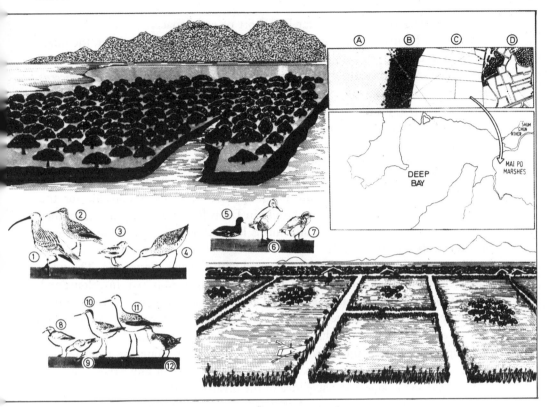

FIG. 11.13. The mangrove shore and fish-poinds near Mai Po, Deep Bay. The sketch map shows the following regions from the seaward in:
A. Soft inter-tidal mud-flat; B. The mangal with the back shore species obliterated by; C. Fish-ponds; D. Paddy fields with some residual upper shore mangroves.
(*Upper left,* B) Sketch of the mangal intersected by mud-banked streams and with bare soft mud beyond. (*Lower right,* C) The fish-ponds with residual mangrove clumps and the inner bund (*foreground*) with cross-paths and outer bund, beyond which lies the mangal (B). The pond borders comprise *Kandelia, Acanthus, Phragmites* and *Ischaemum*. Birds shown inset:
1. Curlew *Numenius arquata*; 2. Whimbrel *Numenius phaeopus*; 3. Fantail snipe *Gallinago gallinago*; 4. Eastern bar-tailed godwit *Limosa lapponica*; 5. Coot *Fulica atra*; 6. Shoveller duck *Anas clypeata*; 7. Teal *Anas crecca*; 8. Alexandrine plover *Charadrius alexandrinus*; 9. Common sandpiper *Actitis hypoleucos*; 10. Redshank *Tringa totanus*; 11. Greenshank *Tringa nebularia*; 12. Chinese water rail *Rallus aquaticus*.

often seem fair places for reclaiming, and what is being done today to the mangroves of Tolo Harbour, Government may soon be pushed to repeat at Mai Po. The extension of Yuen Long satellite town and the construction of a 'garden housing estate' for 30,000 people at Tai Shang Wai could endanger the marshes by pollution from run-off and proximity of urban crowding. Neither Government nor local people properly understand the priority that wilderness preservation should have. Yet for the social well-being of the most important animal in this book, man, shore conservation is even more necessary in Hong Kong than in places far less short of open space.

The Mai Po mangroves can be reached by the cross-paths between the inner and outer bunds. Already in the fish-ponds are clumps of *Kandelia candel,* picked out from a distance by their bunches of droppers. *Avicennia* and *Aegiceras* are also found here; and these continue as the main species, together with a few taller *Bruguiera,* on the mangal covering the soft muds beyond. *Canavalia* and *Derris* cover the back of the shore mangroves, making the beach a virtually inpenetrable tangle. The landward 'mangrove associate' species are scarce at Mai Po as compared with Tolo, having been largely lost to fish-ponds.

The holly-leafed *Acanthus ilicifolius,* is, however, much more in evidence than at Tolo and here constitutes the major 'pioneer' species responsible for the rapid forword growth of the community in Deep Bay in recent years. Together with the tall reed *Phragmites* it also forms long straight fringes along the edges of the ponds and the bunds.

The seaward mud is, in patches, covered with the alga *Bostrychia.* Though a member of the Rhodophyceae, the flat thalli are green to brown in colour, and in addition to coating the mud surface, it also forms a green band on the stems of the more seaward mangrove trees just below a mass of encrusting barnacles, in this case, in the extreme shelter of Deep Bay, *Balanus amphitrite amphitrite.*

Heaps of dredged mud removed from the ponds lie along the main bunds, often studded with white shells. From these, better than any other way, the student can get a comprehensive idea of the bivalves and gastropods of the fluid and inaccessible muds beyond the mangroves. The shell heaps can be sampled to yield good specimens of *Scapharca, Anadara, Heteromacoma, Dosinorbis, Glauconome, Pharella,* and *Placuna,* also cerithiids and *Turritella.*

The muds of the mangal beyond the outer bund are deep and semi-fluid, and are dangerous to walk over unaided. No one should venture off the paths into the mangroves alone. Accompanied by a party, the swamp can best be negotiated attaching small boards to the feet as mud-shoes. On more open stretches a mud-scooter is useful. For emergencies (though with care these ought not to happen) a coil of rope should be carried by the party. Within the mangal itself the trees and their shallow horizontal root systems give a measure of protection from sinking.

But far from encouraging unpractised students to enter the mangroves and their difficult substrate, the best advice to the newcomer is to stay out. Most of the birds and surface species can be seen with field glasses from the edge, or secured with little departure from firm ground.

For example, field glasses will show up the bright-coloured fiddler-crabs (chiefly *Uca arcuata*) as well as sesarmids, *Helice, Chasmagnathus* and *Macrophthalmus.* Glasses will also bring into view the two species of mud-skippers, *Periophthalmus cantonensis* and *Boleophthalmus boddaerti,* and small gobioid fishes prolifically abundant on the mud-surface and mangrove branches. Mud-skippers have numerous modifications for life and progression out of water. By adaptations to the mouth and the provision of a rich blood supply to the walls of the buccal cavity mudskippers are quite amphibious. The eyes are large and spherical, with all-round periscopic vision. The head and chest are raised on strut-like pectoral fins, mobile at the wrist and splayed like hands. These are used for step-wise progression and for climbing mangrove branches, the forward-placed pelvic fins making a cup-like attachment sucker; and with flicks of the tapered muscular tail, the mud-skipper can make parabolic leaps across the mud. In the larger and more lower-zoned of the two species, *Boleophthalmus boddaerti,* the sail-like dorsal fin, spotted with blue, can be raised in prominent display. Males of both species are bigger and, scattered over the mud flats, they protect their territory, fighting both with each other and with crabs. In the breeding season, males attract females by raising their fan-like dorsal fins and raising themselves up on their pectoral fins. Once a female is seduced she accompanies him to the burrow where the eggs are laid. The pair remain together, the female usually occupying the burrow and the male defending it.

Unlike other mangroves in Hong Kong, the muds of Mai Po are so sloshy that even the most hardy cerithiid snails are precluded. Instead, here the mud is speckled by the tiny (2–3 mm), brown, plump rissoacean snail *Assiminea dohrniana*.

The heteropteran 'sea-skater' *Asclepios shiranui coreanus* is an unusually salt-tolerant insect, darting seemingly at random, often in large numbers, on the surface of mangrove waters. The canacid mangrove fly *Trichocanace sinensis* forms unpleasant swarms deep within the mangrove, and seems to feed particularly on *Avicennia*. Biting midges of genus *Culicoides* (Ceratopogonidae) breed in mangrove muds and greatly contribute to the unpleasant side of these coastal areas—the female needing a blood meal before reproduction can proceed; the males, however, feed on nectar or plant sap.

The Fish-ponds or 'Tam-baks'

The *'tam-baks'* or *'kei-wais'* of the Mai Po marshes utilize the ebb and flow of the tide to bring into shallow tidal ponds, especially dug out, prawns, shrimps, fish and crabs. These animals enter the ponds on the flowing tide and during the subsequent ebbing tide the water is let out through netted sluice gates, thereby entrapping the catch. Such ponds are a long-established and characteristic method of fishing throughout Asia and are typically found behind mangroves. The fishermen capitalize upon the natural high productivity of the mangrove environment cropping the species of prawns that breed in the sea, with the post-larvae entering the estuaries to mature. Sometimes, during the breeding season, the prawns are left in the pond (Fig. 11.14) to grow and to mature—a practice

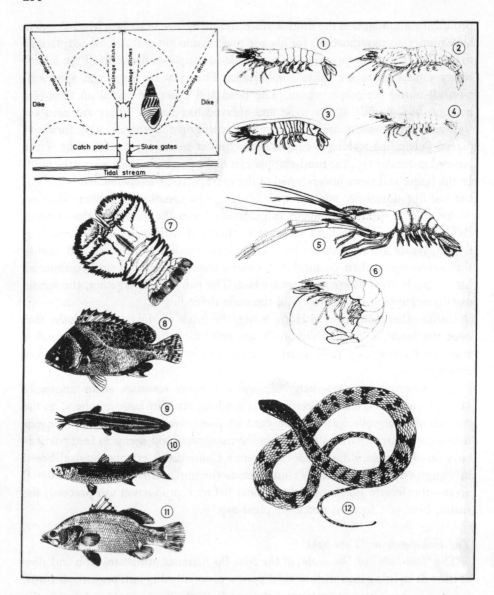

FIG. 11.14. A South-east Asian prawn-pond with the marshland snail *Sermyla tornatella*. Common prawns and shrimps caught in the Kei Wai's of the Mai Po marshes are identified identified below together with the mangrove snake.
1. *Penaeus latisulcatus*; 2. *Metapenaeus intermedius*; 3. *Penaeus japonica*; 4. *Penaeus monodon*; 5. *Macrobrachium nipponense*; 6. *Penaeus semisulcatus*; 7. *Ibacus ciliatus*; 8. *Mylio latius*; 9. *Plotosus anguillaris*; 10. *Mugil cephalus*; 11. *Lates calcarifer*; 12. *Enhydris bennetti*.

that has undoubtedly led to the development, in modern times, of prawn culture in permanent ponds.

Wear and Stirling[6] have produced a useful key to the penaeid crustaceans of Hong Kong. The large prawns *Penaeus monodon* and *P. semisulcatus* are carnivores whilst the smaller species of *Metapenaeus* feed on the algal surface film. During spates of low salinity the fresh-water prawn *Macrobrachium nipponense* (typically found in streams, but breeding in the estuaries) is also captured and can be recognized by the very long chelae held out squarely from the body. Also encountered rarely, are the highly specialized shovel lobsters *(Ibacus ciliatus)* (Scyllaridae) which are highly modified palinurans in which the antennae are laterally flattened.

In recent years there has been a change from this traditional form of shrimp and fish fishing to the culture of fishes and shrimps in the ponds, in which case the fry have to be artificially fed. In addition, oysters are now grown in the ponds, enhancing their productivity. The salinity of the *kei-wai* waters varies considerably (between 2.5–29.5‰) enabling survival and stocking of only a relatively few fish species.

The common fish caught in the tam-baks of Hong Kong are *Tilapia mossambica* (introduced from East Africa), the grey mullet *(Mugil cephalus)*, the marine catfish *(Plotosus anguillaris)* the giant perch *(Lates calcarifer)*, the sea perch *(Lateolabrax japonicus)* and the black and yellow-finned sea breams *(Mylio macrocephalus* and *Mylio latius)*. The mangrove crab *Scylla serrata* can also be caught in the ponds. At Mai Po some of the traditional *tam-baks* have been converted into permanent fish-ponds, with a corresponding destruction of the estuarine animals and plants. In salinities greater than 5°/oo only *Mugil cephalus* can be reared with the young being caught in fine mesh nets in the sea. At lower salinities a range of carp varieties are reared, exploiting a diversity of niches in the pond environment. Nutrients in the form of fertilizers and food additives, augmented by the droppings of ducks and pigeons reared above the ponds, maintain high productivity levels in the ponds and the fish rapidly attain marketable size. Such ponds are now a major source of farming in the Yuen Long plain.

A characteristic snail of the fish-ponds and marshes is *Sermyla tornatella*, elongate and delicately vertically ribbed.

Birds of the Marshes and Flats

The bird-life of the Mai Po marshes is only briefly mentioned here, and for

---

[6] Wear, R. G. and Stirling, H. P. 1974. A guide to the genera and species of prawns of the family Penaeidae (Crustacea: Decapoda) from Hong Kong waters. *Hong Kong Fisheries Bulletin* 4: 97–108.

more detail the reader should consult more authoritative references.[7]

The Hong Kong Bird Watching Society organizes frequent trips, observing common species and keeping watch for new records. This aspect of the natural history of Hong Kong is actively maintained by expert amateur groups.

Of the natural assemblages of birds, the waders must come foremost. The largest species, the curlew *(Numenius arquata)* is relatively frequent, to be seen in flocks at a distance near the water's edge. The related, slightly smaller whimbrel *(Numenius phaeopus)* is one of the commonest waders of the marshes and flats, being a passage migrant in August–October and late April to early May. It prefers to feed in the mud around the mangrove roots. The eastern bar-tailed Godwit *(Limosa lapponica)* is another common bird in passage, in spring and autumn. The common sandpiper *(Tringa hypoleucos)* is seen almost everywhere, both on sandy shores, and on mud-flats, paddy fields, streams and fish-ponds. At Deep Bay it frequents particularly the bunds. The Kentish or Alexandrine plover *(Charadrius alexandrinus)* already seen in pairs or small groups on sand-flats *(see* Fig. 10.5), is also a bird of the paddy fields and mud-flats. The longer-legged redshahk *(Tringa totanus)* and Dunlin *(Calidris alpina)* belong in or about the deeper marshes. The Greenshank *(Tringa nebularia)* tends more frequently to fish the banks of streams where they intersect the mangrove thickets. Of the three species of snipe, the fantail *(Gallinago gallinago)* and pintail *(Gallinago stenura)* are both found in paddy-fields, while Swinhoe's *(Gallinago megala)* prefers marshy ground and the borders of streams.

The heron and egret tribe have their typical habitat on the flats beyond the mangrove; they also commonly perch upon the mangroves and breed in the marshes. Present at most times of the year are the grey heron, *(Ardea cinerea)* the purple heron, *(Ardea purpurea)* and the little green heron *(Butorides striatus)*. The Chinese pond heron *(Ardeola bacchus)* is dusky and difficult to see until it rises and spreads its white wings. The little egret *(Egretta garzetta)*, which is entirely white, is a swamp bird whereas the reef egret *(Egretta sacra)*, slate-grey and yellow-legged, is confined to rocky shores. Two other commonly encountered marsh egrets are the great egret *(Egretta alba)*, and cattle egret *(Bubulcus ibis)*.

The bittern *(Botaurus stellaris)*, found along the edges of the ponds, freezes to look like a stump or reek clump; it relies on posture, and its speckled and streaked coloration for camouflage. The small bitterns *(Ixobrychus)* nest off the ground, perching and shyly concealed in mangroves. They include von Schrenck's little bittern *(I. eurhythmus)*, the yellow bittern *(I. sinensis)* and the chestnut bittern *(I. cinnamomeus)* present from April to October.

Much larger than the herons or bitterns is the black stork *(Ciconia nigra)*

---

[7] Herklots, G. A. 1967. *Birds of Hong Kong*. Hong Kong, South China Morning Post; and Webster, M. and Phillipps, K. 1976. *A New Guide to the Birds of Hong Kong*. Hong Kong, Sino-American Publishing Co.

Enclosed Flats

FIG. 11.15. Some birds of the mangrove shore.
1. Grey heron *Ardea cinerea*; 2. Purple heron *Ardea purpurea*; 3. Little egret *Egretta garzetta*; 4. Yellow bittern *Ixobrychus sinensis*; 5. Bittern *Botaurus stellaris*; 6. Spotted-billed pelican *Pelecanus philippensis*.

which is a winter visitor. So too are the white ibis *(Threskiornis melanocephalus)* and the two species of spoonbill *(Platalea leucorodia* and *P. minor)*. These feed by trapping surface-dwelling organisms in the 'spoon' of the bill as they sweep it from side to side in loose mud.

Of the rail tribe, the secretive banded rail *(Rallus striatus)* and the Chinese water rail *(Rallus aquaticus)* can generally be glimpsed darting beneath the mangroves. The coot *(Fulica atra)* and the little grebe *(Podiceps ruficollis)* swim and dive in the ponds. The latter species and the pheasant-tailed jacana *(Hydrophasianus chirurgus)* took advantage in 1968 of large areas of emergent vegetation to breed in Deep Bay; numbers have since declined owing to conversion of these ponds.

Of the five kingfishers found in Hong Kong, at least four can be regularly seen in the Deep Bay marshes. The small vivid blue, red-breasted *Alcedo atthis*, familiar in Britain, is the common kingfisher throughout Eurasia. The pied kingfisher *(Ceryle rudis)* is another fishing species flying along the water-courses and dropping like a stone. Two crab-eating species are the white-breasted kingfisher *(Halcyon smyrnensis)* commoner in winter near the sea and the black-capped kingfisher *(Halcyon pileata)*.

Marsh-harriers *(Circus aeruginosus)* can commonly be seen flying low in search of their prey of birds and rats. Ospreys *(Pandion haliaetus)* plunge

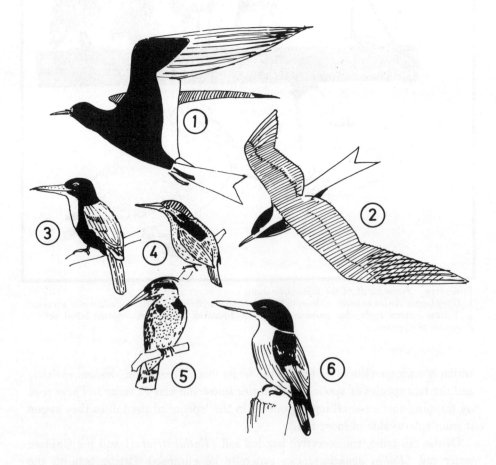

FIG. 11.16. Terns and kingfishers of the marshes.
1. White-winged black tern *Childonias leucoptera*, summer plumage; 2. Whiskered tern *Childonias hybrida*, summer plumage; 3. White-breasted kingfisher *Halycon smyrnensis*; 4. Common kingfisher *Alcedo atthis*; 5. Pied kingfisher *Ceryle rudis*; 6. Black-capped kingfisher *Halcyon pileata*.

dramatically, feet first, into fish-ponds to catch fish of considerable size. Carrion is the main food of the collared crow *(Corvus torquatus)*.

The duck tribe are represented at Deep Bay by the teal *(Anas crecca)* and garganey *(A. querquedula)* in small streams, along with the widgeon *(A. penelope)*, yellownib duck *(A. poecilorhyncha)*, pintail *(A. acuta)*, and the shoveller *(A. clypeata)*. The last is a broad-billed filtering species, commonest upon the ponds. On the mud-flats, flocks of conspicuous white shelduck *(Tadorna tadorna)* can sometimes be seen.

Gulls and terns (Laridae) occur in Hong Kong in winter only (p. 54) but can then be common around the mangroves. The black-headed gull *(Larus ridibundus)*, which loses its dark crown in winter, and the smaller Saunders' gull *(Larus saundersi)* can often be seen 'puddling' with their feet, to stir up small crustaceans from the mud beyond the mangal. The whiskered tern *(Chlidonias hybrida)*, the little tern *(Sterna albifrons)* and the gull-billed tern *(Gelochelidon nilotica)* are all graceful fliers, plunging for food just below the surface. They are birds of passage in the marsh, in spring and autumn.

Numerous passerines or song-birds play a part in the mangal community. Nesting among the mangroves will be found the white-eye *(Zosterops japonica)*, the chestnut munia *(Lonchura malacca)*, Chinese bulbuls *(Pycnonotus sinensis)* and coucal *(Centropus sinensis)*.

The yellow-bellied and brown wren-warblers *(Prinia flaviventris and P. subflava)* are commonly found among the *Phragmites;* they are joined in spring and autumn by von Schrenck's reed-warbler *(Acrocephalus bistrigiceps)* and the great reed-warbler *(A. arundinaceus)*. In winter the bluethroat *(Erithacus svecicas)* can be seen on the bunds whilst yellow wagtails *(Motacilla flava)* dance over floating vegetation or drained fish-ponds in search of insects. Swallows *(Hirundo rustica)* are frequently observed manoeuvering for flies close to the water.

The largest of all marsh birds is the Dalmatian pelican *(Pelecanus crispus)* with its smaller relative the spotted-billed pelican *(P. philippensis)*. These graceful birds are often seen in their V-shaped flying formation, or fishing at the edge of the mangroves. When fishing a group of ten or so birds will form a wide circle on the water and gradually close on a shoal of small fish. By raising the wings to create a shadow the alarmed fish are concentrated into a cluster and scooped up by the bag-like lower beak of the pelicans.

The Mangrove Snake

A regular member of the Deep Bay mangrove community is Bennett's water snake *(Enhydris bennetti)*, about 60 cm long and grey with black bands. Not wholly aquatic, as is the colourful and venomous *Hydrophis cyanocinctus* found in Hong Kong amongst coral, *Enhydris* can be found warming itself either on the fish-pond bunds but more commonly in the mangal itself, either in the trees of slithering over the mud. Belonging to the viper family (Colubridae) *Enhydris* is characterized by nostrils located on the top of the snout, enabling it to breathe whilst almost totally immersed. It probably feeds on mud-skippers and on the fledglings of mangrove-nesting birds.

MUD-FLATS WITH OYSTER BEDS

Oysters have been cultivated in Hong Kong for perhaps 700 years. The industry is today located around Deep Bay, at the north-west of the New Territories, on part of the border between Hong Kong and China. Oyster beds lie on both sides of the boundary. The cultivated species is the Pacific cup-oyster, *Crassostrea gigas*, though wild populations of the more rounded *C. rivularis* also occur here and may co-occur on the oyster tiles. As a cultivated oyster *C. gigas* has rapidly displaced the smaller rock oyster *Saccostrea cucullata* that still, however, settles and grows along with it.

Deep Bay is an area of broad inter tidal mud-flats of about 112 km^2 fringed to the landward by a mangrove swamp, the largest and best developed in Hong Kong. Into the Bay drain numerous streams and small water-courses, notable among which are the Shum Chun River and Yuen Long Creek; Deep Bay itself opens into the estuary, about 35 km wide, of the Pearl River, the major system draining the hinterland of southern China.

The Bay has an estuarine hydrology affected by the interaction of the north-easterly monsoon in winter and the south-easterly monsoon in summer. The cool dry north-easterly monsoon lowers the water temperature to 10–15°C from November to February, when the salinity is maintained at the comparatively high level of 26–32°/oo. In summer, the warm, rain-bearing south-easterly monsoon raises the water temperature to 28–32°C, typically from June to August. The seasonal cooling and warming are hastened and enhanced by the shallowness of the water. The large rainfall associated with south-easterly winds causes flooding of the rivers and streams leading into Deep Bay, reducing the salinity to as low as 5–15°/oo. At the height of the July monsoons, the salinity may fall even further, almost to fresh-water conditions.

The foreshore of Deep Bay is divided into about six family or T'ong plots which are farmed by the owner or are often rented out. The industry is thus made up of a number of family businesses.

A natural distribution centre has grown up at the village of Lau Fau Shan, where—still with relatively primitive methods and dubious hygiene—oyster culture has transformed the whole aspect of the inter-tidal flats. Workers from the village are constantly out on the flats, laying out, tending or harvesting. The journey across the mud, though seldom dangerous as in mangrove swamps, is irksome and unpleasant. The oyster people make light of it by the use of the mud-scooter, upon which a man can glide over the semi-liquid surface, with the ease and speed of a cyclist.

Spat is collected further up the flats, near the upper shore. The spatting season is from May to August; and as the traditional flat stones became gradually used up, or removed from the foreshore, large, six-year-old oyster shells came to be used as collectors. When oyster shells began to be pre-empted for lime-making they were in turn replaced by rectangular tiles or concrete posts. After settlement, the spat collectors are taken off-shore and placed in neat rows within the borders of a farmer's plot. Because of high sedimentation they have periodically to be lifted out of the mud and relocated to prevent the spat from suffocating. The oysters are tended in this way until three or four years of age and are then cropped. Occasionally they are left longer, six-year-old oysters growing to almost 30 cm long. The age of cropping varies with demand, and at times even younger individuals can be marketed, and most spat-collectors are then put down. With reduced demand, only older individuals are marketed.

Except at low spring tides, tiles or 'cultch' are raised from a boat by a pair of tongs, about 4 m long, used to feel for the tiles, grasp and recover them. The same tongs are used for periodic relocation of silt-covered cultch.

The oysters are opened or 'shucked' with an oyster hammer (Fig. 11.17); its short strong point punches a hole in the upper valve, into which the curved, sharp point is inserted to cut the adductor muscle. The soft parts can then be easily removed. Oysters are marketed fresh in the village of Lau Fau Shan, and are also in large numbers dried or made into oyster sauce. For a more detailed account of the oyster industry, the reader is referred to a paper by B. S. Morton and P. S. Wong[8].

The Fauna of Oyster Beds

The texture of the mud-flat is much softer than even the most protected of silty sand-flats, as at Tai Tam Harbour. A sticky sediment of silt and clay lies from the surface to 15 cm or more deep. The ground beneath is usually firmer, with a higher content of sand or stiffening clay, though anaerobic and grey or black in appearance.

The flats laid bare at low tide can differ in sediments and in fauna from place to place. Essentially unsorted and generally without significant effect from wave-action or along-shore currents, the sediments are chiefly modified by variations in land run-off. Frequently the course of a stream across the beach will remove part of the finest grades, and leave more walkable stretches, either with sand or shell fragments.

On walking out across the flats at Lau Fau Shan, the first habitat to be

[8] Morton, B. S. and Wong, P. S. 1975. The Pacific oyster industry in Hong Kong. *Journal of the Hong Kong Branch of the Royal Asiatic Society* 15: 139–49.

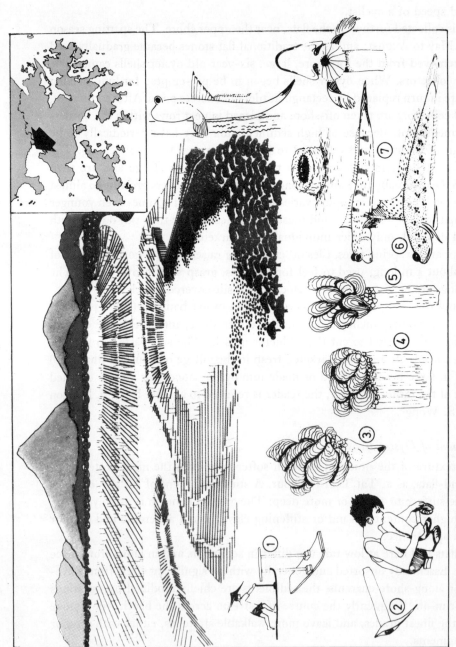

FIG. 11.17. Mud-flats with mangroves and oyster cultivation beyond at Deep Bay. Stakes and batons are shown in their assemblies on the inter-tidal flats and in the foreground the mangal below the Tsim Bei Tsui Police Post.
1. Mud sledge; 2. Oyster hammer and its use; 3. Shell cultch; 4. Tile cultch; 5. Pillar cultch; 6. *Boleophthalmus boddaerti*; 7. *Periophthalmus cantonensis* with detail of the burrow, the pelvic sucker and the animal twig-climbing.

Enclosed Flats

encountered on the level shore lies in 5–6 cm of standing water. *Enteromorpha* species accumulate thickly (especially where there is fresh-water seepage) and form a meshwork with the small oval-leaved angiosperm *Halophila beccarii*; another species *(H. ovata)* has already been met with at low tide on *Cerianthus* flats (p. 190). Where sheets of fresh or brackish water spread over the flats from stream outfalls, the small greenish nerite *Clithon oualaniensis* is incredibly abundant. A second, more ovoid brackish-water species *Clithon retropictus*, is often found together with the red or mauve-mouthed nerite *Dostia violacea*, common also among mangroves.

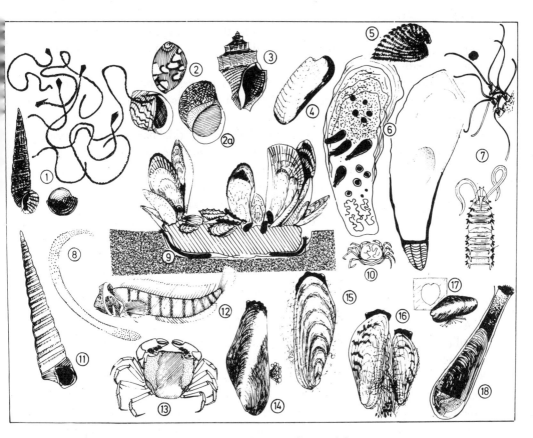

FIG. 11.18. The fauna of oyster clumps and surrounding mud-flats.
1. *Cerithideopsilla djadjariensis* with meandering trails and end view; 2. *Clithon oualaniensis*, 2a. *Clithon faba*; 3. *Thais carinifera*; 4. *Trapezium liratum*; 5. *Capulus yokoyami*; 6. Shell of *Crassostrea (left)* eroded exterior with burrows of *Polydora*, *Clione* and *Aspidopholas* and *(right)* interior; 7. *Polydora* sp. head detail with *(above)* palps emergent from tubes; 8. *Phascolosoma scolops*; 9. Settling plate with *Crassostrea gigas*, *S. cucullata*, *Trapezium liratum* and *Brachidontes* and *Polycheira* beneath; 10. The commensal pea-crab *Pinnotheres sinense*; 11. *Turritella terebra cerea*; 12. The blenny *Omobranchus uekii*; 13. *Hemigrapsus penicillatus*; 14. *Modiolus metcalfei*; 15. *Arcuatula elegans*; 16. *Musculista senhausia*; 17. *Botula silicula*; 18. *Aspidopholas obtecta*.

A flat stone with oysters attached has been illustrated in Fig. 11.18, introducing an island of hard benthos in an expanse of mud. Both oysters, *Crassostrea gigas* and *Saccostrea cucullata*, with a third bivalve, *Trapezium liratum* are attached to the top. Gastropods are much fewer than on clean rocky shores, but two species are universally common on Hong Kong oyster flats: the strongly ribbed *Nerita undata* and the oyster predator *Thais carinifera*, distinguished from the other thaids by its lightly built brown shell ornamented with spinose tubercles around the shoulders of the whorls.

Beneath the stone, the fauna is generally sparse, especially where the underlying sediments are black and anaerobic. But sipunculid worms *Phascolosoma scolops* are often abundant, with bodies reaching out radially, to extend the introvert, with its tentacle fringe, just beyond the covering rock. Alpheid shrimps burrow in the mud; and *Harmothoe* and flat-worms are also found in suitable places beneath the boulders.

The large and handsome blennioid *Omobranchus uekii* is often found amongst oyster cultch; and should be carefully handled, for its sharp teeth can give a painful bite.

Heaps of tiles and stone slabs, put out higher on the tidal flat as spat collectors, provide further hard islands. The two oysters, *Saccostrea cucullata* and spat-sized *Crassostrea gigas*, cover the tops. Underneath, the bivalves are byssus-fixed: the small finely ribbed mussel *Brachidontes atratus*, *Trapezium liratum* and the saddle oyster *Anomia ephippium*, distinguished by its wrinkled upper valve. Two sorts of crab are abundant and active. The faster is *Metapograpsus messor*, attractively coloured when washed clean of silt, with the carapace eau-de-nil green, with fine transverse lines and stippling, and the chelae purple with white tips. The second crab is the dull-coloured *Hemigrapsus penicillatus*, with the carapace typically brown and white blotched, as in the related *Gaetice depressa*, from which this crab differs in having two post orbital spines, and dense setal tufts on the chelae.

At the same level of the flats lie occasional sheets of *Musculista senhausia*, a mytilid characteristic of muddy ground. These mussels sit vertically upright, compacted side to side in great numbers with pointed anterior end down. By attaching byssus threads to each other and to sand grains they make a coherent mat, with much sediment lodged between them. Two other mytilids are common infaunal residents of the mud-flats: the black triangular *Modiolus metcalfei*, strongly attached by its byssus to broken branches and stones, and the greatly elongate and shiny green *Arcuatula elegans*, similarly embedded in the mud and cocooned in a jelly-like blob of byssal threads and mucilage. Seawards before the oyster holdings are reached is a broad mid-tidal expanse of semi-fluid mud. Several sorts of gastropods are very typical of these sediments. The long and trailing cerithiids, *Cerithideopsilla djadjariensis* and *C. cingulata*, form meandering tracks, everywhere intersecting and crossing. A feeder on fine surface muds,

Enclosed Flats

these snails replace the two *Batillaria* species on the softest ground. *Nassarius* is common and very mobile, as already seen on protected flats. The speckled *Natica tigrina*, is often found in localized sandy patches and can be detected by its spawn coils.

Turritella terebra cerea is a trailing gastropod with a long tapered spire, its shells being common at low water, though not always occupied. Unlike the superficially similar cerithioids, *Turritella* is a highly specialized ciliary feeder. The mantle cavity is protected at its entrance against coarse sediment by a screen of pinnate tentacles. The finer material let through is strained from the pallial cavity water current by the elongated gill filaments, by a mechanism similar to that of the bivalves, but structurally simpler. The particles collected are carried forward to the head in a mucous cord, seized and passed inward by the radula. The *Umbonium* species are other burrowing gastropods showing this habit.

The mature shells of *Crassostrea gigas* form a hard substrate, with their own sessile and penetrating fauna. A small pholad bivalve *Aspidopholas obtecta*, found only in this habitat, can be detected by its short, bottle-shaped burrows. Rounded siphon holes are visible at the surface and the pholad sometimes penetrates right through the shell. *A. obtecta* burrows using the abrasive anterior end of the shell; much rarer, the unrelated oyster borer *Botula silicula* is an acid borer and is allied to the date-mussels *(Lithophaga)* encountered in greater diversity as coral borers (p. 269).

The outer layer of oyster shells is eroded by the tiny spionid polychaete worm *Polydora*. The tube is bent double like a hairpin, running horizontally within the shell, with its flexible openings standing clear. From the head end (Fig. 11.18) two tentacular palps are rhythmically waved, to collect food from the water overhead.

The eroding sponge *Cliona celata* has far more destructive effects than pholad or polychaete. *Cliona* first settles and penetrates the shell as a scatter of small yellow pustules. These join up by excavating a mesh-work of galleries, calcite being dissolved by enzyme-secreting cells of the sponge. The mature sponge finally grows over the flimsy and eroded shell in a soft external crust.

As well as *Thais carinifera,* the oyster whelk (p. 246), *Crassostrea* has a second gastropod *Capulus yokoyami* living at its expense, attaching like a cap-shaped limpet near the outer shell margin. The animal is a sedentary gleaner of food particles travelling across the oyster's mantle edge. With its long and extensible proboscis, *Capulus* foreshadows the suctorial ectoparasitic habit of the minute pyramidellid gastropods that live on body fluids at the mantle edge of the bivalve.

Bivalves of Mud-flats

Bivalve molluscs are as important on muddy shores as in sand-flats. A few species, such as, for example, *Cyclina orientalis,* may be common to both. But as the illustrations (Fig. 11.19) make clear, mud-flat bivalves have a trend and style of their own. First, there are thin and light-shelled species actively burrowing. *Heteromacoma irus* is typical of these, being a long-siphoned deposit-feeder belonging to the Tellinidae. This family is also represented by the small, pink and thin-shelled *Moerella juvenilis*. Of similar habits is *Abra maxima*, belonging to the related family Semelidae. On the other hand, the common venerid, *Dosinorbis japonica* is a suspension-feeder, subcircular and compressed, but with the shell heavier and concentrically sculptured. A second venerid of mud-flats, *Glauconome chinense,* has a uniformly dark-green periostracum, and is adapted for deep-burrowing, with long fused siphons and an active foot.

The two sorts of razor-shells are stream-lined and lightened to a high degree. *Pharella acuminata* is long and parallel-sided, laterally compressed like a narrow

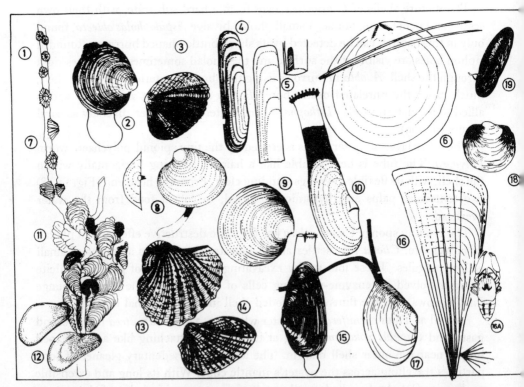

FIG. 11.19. The bivalves and other fauna of mangroves and mud-flats. To the left a group of barnacles and bivalves is shown attached to a young mangrove stem.
1. *Euraphia withersi*; 2. *Geloina erosa*; 3. *Gafrarium tumidum*; 4. *Pharella acuminata*; 5. *Solen corneus*; 6. *Placuna placenta* with profile of the valves; 7. *Balanus albicostatus albicostatus*; 8. *Cycladicama cumingii*; 9. *Dosinorbis japonica*; 10. *Laternula truncata*; 11. *Saccostrea cucullata*; 12. *Brachidontes atratus*; 13. *Anadara granosa*; 14. *Scapharca cornea*; 15. *Glauconome chinense*; 16. *Pinna atropurpurea* with 16a. *Anchistus custos*; 17. *Heteromacoma irus*; 18. *Anodontia stearnsiana*; 19. *Enigmonia aenigmatica*.

Enclosed Flats 249

pod. The plug of the foot emerges from the anterior end and draws the shell obliquely or upright through the soft end. *Solen corneus* has achieved its razor-form by a great extension of the posterior end with the umbones at the extreme anterior, and the shell here truncated and squared off.

Cycladicama cumingii and *Anodontia stearnsiana* belong to the Lucinacea (p. 165 for sand-beach relatives). The shells are thin, smooth and globular and the foot is long and cylindrical, fixing like an anchor and hauling the bivalve down into the mud. The foot also makes separate holes for inhalant and exhalant currents and, as in the lucinid *Codakia,* there are no true siphons. The shell valves can be distinguished not only by their lightness, but by the double tooth ('diplodont') condition. *Cycladicauma* is slightly less circular than *Anodontia*.

Some highly specialized mud-flat bivalves remain to be mentioned. *Laternula truncata* has a thin translucent shell, with a silvery sheen, gaping widely at either end. From the posterior extends a thick tube made up of fused siphons enclosed in a heavy sheath. The foot is very small, the adult *Laternula* lying up to 15 cm deep, and unable to move or to re-embed if disturbed. Among many anatomical peculiarities, *Laternula* has a ring of small but remarkably complex eyes around the siphonal openings, its only site of contact with the light.

Pinna atropurpurea, one of the fan mussels (Pinnidae), is common at low water and sub-tidally. Like *Atrina vexillum* (p. 165), *Pinna* lives immobile, like a flattened cornet open at the top, and fixed by a byssus at the pointed lower end. *Pinna* is distinguished from *Atrina* by its pale colour, more slender form and median ridge along each valve.

The window-shell, *Placuna placenta,* is a much modified relative of the saddle oysters, Anomiidae. The valves are almost perfectly rounded, thin and transparent and the space between them is narrowly compressed. *Placuna* lies on the surface, or obliquely part-buried in mud. The mantle cavity is open to inhalant currents nearly all round the shell, and as in scallops and oysters, the 'quick' adductor muscle can produce fast-clapping movements of the valves to expel inborne sediment but there are no resultant swimming movements.

Sea Grasses

A number of monocotyledonous flowering plants grow under marine estuarine conditions on the lower shore. They are collectively referred to as the 'sea grasses'. The only species common in Hong Kong, with small oval leaves and hardly grass-like, is *Halophila ovata* (of the family Hydrocharitaceae), widely distributed on sheltered flats at Tai Tam, at Three Fathoms Cove and Hoi Sing Wan in Tolo Harbour, and in Deep Bay. It grows as a thin prostrate rhizome, rooted at the nodes and putting out lateral shoots, each carrying a pair of leaves and non-sheathing petioles. The leaf-blade is translucent green, purple tinged and with cross-veins between midrib and marginal vein.

A second species, *H. beccarii,* has recently been reported,[9] forming dense mats amongst pioneer mangroves at Deep Bay. It differs from *H. ovata* in having the leaves in pseudo-whorls of 6 to 10, the lack of cross-veins and the broadly sheathing petioles.

Halophila has solitary flowers of separate sexes borne on a lanceolate spathe. Both sexes have three oblong to elliptical tepals. The male flower has three stamens with linear anthers, releasing chains of pollen that disintegrate in the water, becoming entangled among the two or three filiform styles of the female flower. The fertile seeds sink when liberated.

In 1977 the first discovery in Hong Kong of eel grass, *Zostera,* a cosmopolitan, chiefly temperate plant, was made at Lai Chi Wo in Crooked Harbour, N.W. New Territories.[10] Covering the middle shore in a brilliant green mat, it puts up from its rhizomes groups of three narrow, linear leaves. Much of the reproduction of *Zostera* (in some parts of the world exclusively) takes place by vegetative advance of the rhizomes. The inflorescence—when it occurs—is a flat spadix (1.5 cm long) at the base of a leaf. Reduced male and female flowers alternate in two rows along one side. Pollination takes place under water from thread-like floating pollen, caught by chance on the stigmas of other flowers.

The bed of *Zostera nana* at Lai Chi Wo has been awarded protection by the Agriculture and Fisheries Department of Government, thereby recognizing its importance in the overall picture of Hong Kong's shore ecology. Exclusively from this sea grass bed has been recorded the small shrimp *Periclimenes demani*.

A fourth sea grass, *Ruppia maritima,* once recorded from Shek-O Lagoon,[11] has never subsequently been found.

---

[9] Hodgkiss, I. J. and Morton, B. S. 1978. *Halophila beccarii* Ascherson (Hydrocharitaceae)—a new record for Hong Kong together with notes on other *Halophila* species. *Memoirs of the Hong Kong Natural History Society* 13: 28-32.

[10] Hodgkiss, I. J. and Morton, B. S. 1978. *Zostera nana* Roth (Potamogetonaceae)—a new record for Hong Kong. *Memoirs of the Hong Kong Natural History Society* 13: 23-27.

[11] Dunn, S. T. and Tutcher, W. J. 1912. Flora of Kwantung and Hong Kong (China). *Royal Botanic Gardens, Kew Bulletin of Miscellaneous Information, Additional Series X.* pp. 370.

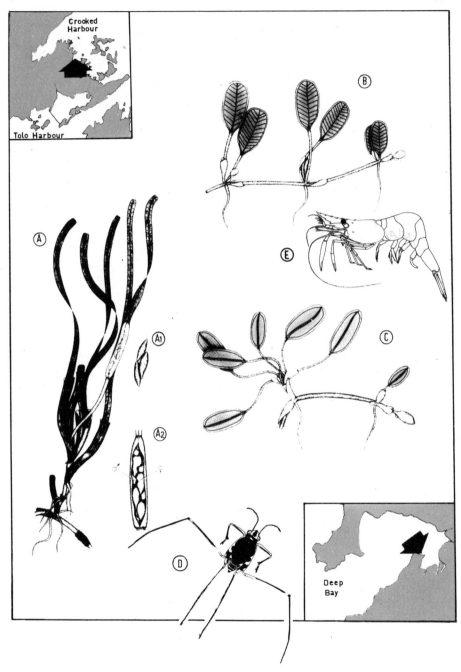

FIG. 11.20. Hong Kong sea grasses.
A. *Zostera nana* with A₁ details of the fruit and A₂ the inflorescence; B. *Halophila ovata*; C. *Halophila beccarii*; D. *Asclepios shiranui coreanus*; E. *Periclimenes demani*.

PART FOUR
Eastern Coral Shores

CHAPTER 12

The Coral Sub-littoral

INTRODUCTION

The sub-littoral fringe and the sub-tidal reach their zenith of diversity and complexity with the coral communities of the Indo-Pacific and Caribbean seas. To the east, in clear oceanic waters of high salinity, though almost always sub-tidally, the reef corals touch Hong Kong.

Two great groups of the coelenterate class Anthozoa give the distinctive character to coral reefs. Within the Subclass Alcyonaria, the order Alcyonacea contains the soft corals, with a fleshy gelatinous coenenchyme, from which simple polyps protrude, recognizable by their eight pinnate tentacles. Within the Subclass Zoantharia belongs the much larger order Scleractinia, containing the true or stony corals. These, though related to the large naked and solitary anemones, have for the most part small and colonial polyps, responsible for the building up of coral reefs. As their name suggests, they are characterized by their ability to produce an exoskeleton of calcium carbonate, the corallum. This, except in a few solitary forms, is a colony of individual corallites. In the open end of each corallite, forming a concave cup or 'calice', sits the living polyp.

Scleractinian corals can be divided into two major ecological groups. The first are the 'hermatypic corals' or the reef-builders, which are noted for the presence of vast numbers of symbiotic unicellular algae (zooxanthellae) within their endodermal lining. Now known to be the resting phase of a planktonic dinoflagellate *Gymnodinium microadriaticum*[1], these are responsible for the metabolic well-being and vigorous growth of the corals, aiding both calcium binding and the removal of wastes. For their nutrition however all corals are basically micro-carnivores, feeding on microscopic zooplankton.

---

[1] The only other dinoflagellate of importance in the sea shore ecology of Hong Kong is the large (1 mm diameter) *Noctiluca scintillans* that is responsible for the luminescence in the coastal waters especially in Spring and Autumn, but when occurring in supra-abundance can form a red tide. Morton, B. S. and Twentyman, P. R. 1971. The occurrence and toxicity of a red tide caused by *Noctiluca scintillans* (Macartney) Ehrenb., in the coastal waters of Hong Kong. *Environmental Research* 4: 544–57.

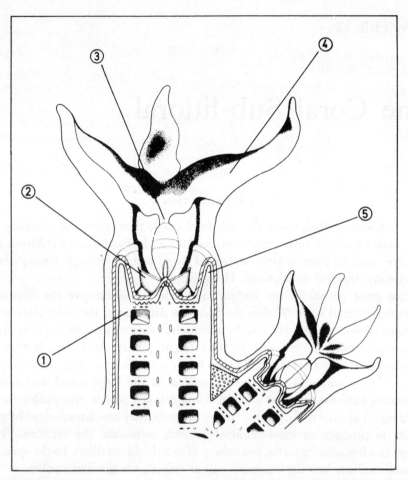

FIG. 12.1. Structure of a coral polyp, with the soft parts cut away to show the calcareous basal plate and radial septa of the calice.
1. Basal plate; 2. Septa; 3. Stomodaeum; 4. Tentacle; 5. Calice.

'Ahermatypic', or non-reef-building corals are often solitary and lack zooxanthellae. They have a wider, almost global, distribution and occur at much greater depths. It should be mentioned that these two divisions are not taxonomic, but functional; within the single family Dendrophylliidae, indeed, are genera of both types.

The ecology of the hermatypic corals is largely controlled by the requirements of their photosynthesizing zooxanthellae. They thus require strong light, and occur in the tropics, rarely below a depth of 50 m, with their ultimate limit at about 90 m. Ahermatypic corals on the other hand require no light and can exist at depths of up to 6000 m. Most hermatypic corals prefer temperatures of above 18°C, though they can tolerate somewhat lower ones. This accounts for their marginal presence in Hong Kong where summer temperatures are adequate, but winter temperatures almost limiting. The slightly warmer waters of the Kuroshio, which exert an influence on Hong Kong's cold water masses in winter, probably help maintain the local coral community in delicate balance. But hermatypic corals flourish well only in sea temperatures between 25°C and 29°C. Hong Kong is clearly at their natural edge of distribution. Ahermatypic corals grow maximally at water temperatures of between 10° and 25°C, but can survive very low temperatures (-1.0°). All Scleractinia are marine, and are adversely affected by reduced salinity, with which increased turbidity is often associated.

In general terms then hermatypic corals are found in shallow oceanic waters, from about 35°N Lat. to 32°S Lat. They tend to be lacking from the eastern sides of the great oceans, along the western coasts of continents with their cold currents. In East Africa, north-west, north and north-east Australia, as in east Malaysia, they occur away from deltaic coasts, with access of fresh water, and favour areas where oceanic waters impinge coastally. Most notably they form atolls upon submerged oceanic peaks: our nearest atoll is Pratas Reef, about 220 km south of Hong Kong.

The richest of all coral areas is the Indo-West-Pacific, with more than 500 species of stony corals representing about 80 genera. In the heart of the tropics there may be over 200 species on a single reef. In Hong Kong, approaching the northern limit for hermatypic corals, there are probably a few more than 50 species, from about 25 genera. Of these the Faviidae are well-represented, accounting for almost half of the total number of corals. This well-adapted group of corals is able to tolerate a wide range of hydrographic conditions, including the somewhat turbid waters of the islands to the south-west of Hong Kong where they scatter the subtidal boulders.

The coral communities of these shores have not yet been fully explored, and much ecological work remains for organized divers to undertake. In Hong Kong the corals are assembled into small sub-tidal reefs surrounding offshore islands and rarely extending to a depth of more than -10 m. C.D. On the eastern mainland, their distribution is more patchy and they are generally restricted to headlands away from the mouths of rivers and streams. In winter the upper coral levels are hidden by the broad swathes of *Sargassum* that characterize such shores (see Part 2). In winter also very cold surface temperatures, coinciding with low tides, can kill off the least resistant corals forcing them to release their zooxanthellae and to whiten.

Mirs Bay and the Eastern Islands

The small islands outside the Bay, and to the east of the New Territories are bathed by waters of full oceanic salinity, free from the periodic dilution and turbidity brought to the western reaches by the Pearl River; and protected from the extreme wave exposure of the coasts to the south-east, as at Cape D'Aguilar and Waglan Island. Their waters are thus not only saline and clear, but warm and sheltered: all these conditions promote the growth of hermatypic corals.

These islands are the richest and most nearly tropical shores of Hong Kong. Along with their corals they have a host of gastropod, bivalve, crustacean and echinoderm species not found elsewhere in Hong Kong.

The corals of Hong Kong are mostly sub-tidal. Only in one locality, the far eastward island of Ping Chau, are corals found high up in permanent pools on the inter-tidal benches formed by the sedimentary strata. Ping Chau is the most remote of all the Hong Kong Territories, 11 km from the entrance to Tolo Harbour, and only 3 km from mainland China to the east.

The Coral Sub-tidal

Though mostly hermatypic, the Hong Kong corals do not build consolidated reefs. But to a growing generation of snorkel and SCUBA divers, their patches and fringes are now familiar collecting grounds. The patterns of coral communities can readily be seen even without diving. A glass-bottomed viewing cylinder is almost indispensable in making a trip round the eastern islands.

The patterns and combinations of species vary widely. The particular association shown (Fig. 12.3) is primarily held together by strong upright flanges of *Pavona decussata*. Radially arranged, they can reach a metre high, and are interconnected by smaller bifacial fronds.

Around and between the *Pavona,* are the heads of other species. First, up to half a metre across, will be the solid rounded heads of the Faviidae, often with large, green-centred polyps. There are also smaller clustered cauliflower-shaped heads of *Stylophora* and the rounded tips of vertical clubs and branched columns of *Goniopora* and *Alveopora*. In thick hummocky crusts, or convex heads, is one of the commonest of the stony corals, *Porites,* with small and rather featureless calices requiring a lens for clear observation.

Usually conspicuous are some of the branched corals, chiefly *Acropora,* both in short-fingered clumps, or in wide-spreading horizontal brackets. Smaller brackets are formed by dull brown *Montipora,* which may also encrust the bottom and put up foliose branches. Other thin and foliose forms include the fragile *Echinophyllia* and *Lithophyllon*. Much thicker and more massive are the flanges of the hermatypic *Turbinaria peltata,* purple-brown in colour and with wide-spaced calices.

From contemplating such an assemblage from 2–3 m below low water, at Port Island (Chek Chau) closer observation may be made by diving and handling the separate entities. Carefully selected pieces can be brought up to the boat deck or the shore. The Hong Kong coral beds will not stand indiscriminate collecting and despoiling, and as a dried object or ornament coral soon loses its attraction.

PLATE 21. Some Hong Kong corals: A. *Psammocora superficialis* (Thamnasteriidae); B. *Stylocoeniella guentheri* (Astrocoeniidae); C. *Acropora pruinosa* (Acroporidae); D. *Montipora venosa* (Acroporidae); E. *Pavona decussata* (Agariciidae); F. *Leptoseris mycetoseroides* (Agariciidae).

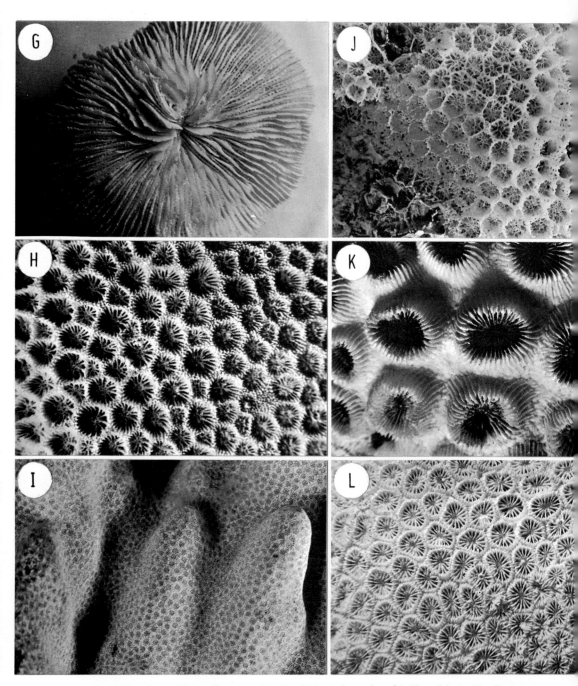

PLATE 22. Some Hong Kong corals: G. *Lithophyllon edwardsi* (Fungiidae); H. *Goniopora columna* (Poritiidae); I. *Porites lobata* (Poritiidae); J. *Alveopora irregularis* (Poritiidae); K. *Barabattoia mirabilis* (Faviidae); L. *Plesiastrea versipora* (Faviidae).

It is usually enough for a morning's work to bring up one or two largely dead pieces, and to study the associated interstitial animals by breaking them up with a light hammer. Students and divers should avoid damage to living corals.

THE FAMILIES OF STONY CORALS

The principal families of Scleractinia found in Hong Kong are described here. For reliable identification, carefully cleaned and bleached SMALL specimens are needed, and suitable fragments can be brought back for this purpose. After sun-drying these can be steeped for a short time in a weak solution of a hypochlorite. All traces of this chemical should be finally washed away from the bleached specimen. For most purposes, however, the collection of dead coral heads or washed up skeletons is preferable to the indiscriminate gathering of living colonies.

The scleractinian families are summarized in the following list:

Order SCLERACTINIA

Sub-order ASTROCOENIINA
Family ACROPORIDAE

Less important in Hong Kong than the full tropics, *Acropora* has various branching growth forms: stagshorns, compact thickets, bottle-brushes, or wide brackets and plates, with shortened, closely massed branches. The calices are long or short tubes, and the terminal one of each branchlet is always enlarged, and often pink, mauve, blue or yellow, according to species.

Hong Kong species: *Acropora pruinosa, A. tumida; A. candelabrum* is distinguished from the other two species by its more plate-like form with short tapering branches.

In the same family, *Montipora* has variable growth, from brackets and spreading crusts to short upright branches or small knobs. The corallum is dull brown or grey, with identification features somewhat difficult to find. The calices are small and simple, and the surface hispid, or roughened with small tubercles or granules.

Hong Kong species: *Montipora informis, M. striata, M. venosa*

Family ASTROCOENIIDAE

Represented by only one species in Hong Kong, *Stylocoeniella guentheri* which is common in many coral areas. Usually massive colonies with short club-shaped knobs and light brown corallites surrounded by darker coenosarc. It is distinguished from *Porites* by the stylliform columella and stellular septal arrangement.

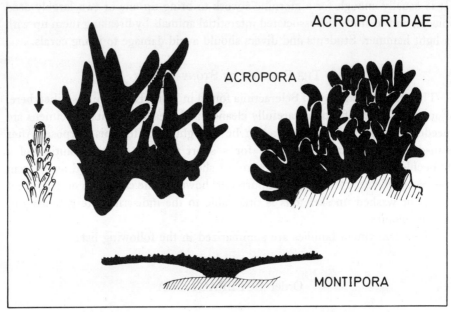
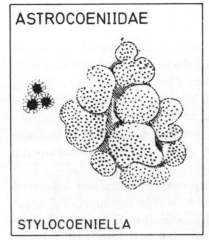
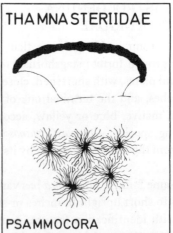

FIG. 12.2. Growth forms of the families of the Astrocoeniina.

Family THAMNASTERIIDAE

Represented in Hong Kong by the encrusting form *Psammocora*. The corallite walls are absent, and individual groupings may include several centres (polyp mouths), united by radiating septa. Hong Kong species: *Psammocora superficialis*, *P. contigua*, *P. haimeana*.

Sub-order FUNGIINA
Family FUNGIIDAE

Very large corallites, typically solitary and unattached. *Fungia* itself (not found in Hong Kong) breaks free from its juvenile stalk to lie on the surface, like an inverted mushroom, convex and with strong radial septa. A délicate colonial fungiid common in Hong Kong is *Lithophyllon edwardsi*. Expanded into a fragile plate it has several mouths or 'centres', with the upper surface radially traversed by long, delicate septa. The margin buds off irregular small discs, and the centre of the colony base is fast-attached, but the growth form later becomes encrusting and finally foliate with large rounded lobes projecting upwards.

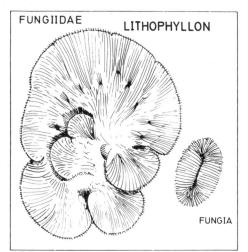

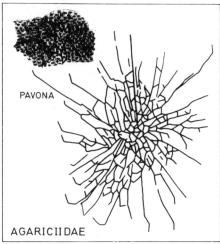

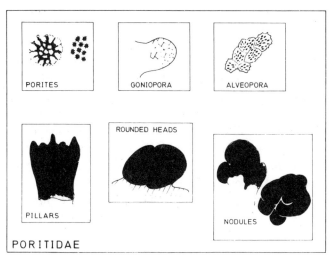

FIG. 12.3. Growth forms of the families of the Fungiina.

Family AGARICIIDAE

With highly modified calyx structure, the cup centres being opened out flat, and the surface covered with radial or parallel colline systems. May be elaborately foliose with high vertical plates or almost flat with plates reduced to mere ridges in more turbid waters. The family is represented in Hong Kong by two genera and species. *Pavona decussata,* which commonly grows with all the abundance and luxuriance of more tropical waters, is instantly recognizable by its high, vertical plates. *Leptoseris mycetoseroides* has an encrusting growth form.

Family SIDERASTREIDAE

The family is represented in Hong Kong by a single genus and species, *Coscinaraea columna,* with a submassive to encrusting growth form and very similar in appearance to *Psammocora.*

Family PORITIDAE

Familiar fast-growing corals of light, porous build, forming pillars, small nodules, rounded heads or crusts. A mature colony of the small-polyped *Porites* can form a flat-topped 'micro-atoll', with living growth, as in island atolls, confined to the outside. Other members of the family have larger calices, with open fragile lace work in *Alveopora*. Polyps of *Goniopora* and *Alveopora* are long and villiform, expanded up to 10 cm in daytime. Common Hong Kong species: *Porites lobata, Goniopora columna, G. stutchburyi, Alveopora irregularis.*

Sub-order FAVIINA
Family FAVIIDAE

The most diverse and important family of Hong Kong Scleractinia. Convex heads with the largest calices (up to 2 cm across) in *Favia* (rounded 'plocoid') and in *Favites* (hexagonal 'cerioid'). In *Goniastrea,* hexagonal calices (less than 1 cm across) form a honeycomb. *Cyphastrea* has small, circular calices, with intervening spaces. In *Platygyra,* sometimes called 'brain corals', the calices are narrowly stretched out and linear, with several centres, often meandering and branching. *Hydnophora* is highly distinctive, with calices everted to form small upstanding cones (monticules) ridged with septa. *Oulastrea* forms very small colonies, the skeleton being very dense, metallic-sounding when tapped, and retaining a grey-black colour when bleached.

Common Hong Kong species: *Favia speciosa, F. pallida, F. favus, F. lizardensis, Favites pentagona, F. abdita, F. flexuosa, Goniastrea aspera, Cyphastrea serailia, C. microphthalma, Platygyra sinensis, P. pini, P. daedalea, Hydnophora exesa, Montastrea curta, Plesiastrea versipora, Leptastrea purpurea, L. pruinosa, Oulastrea crispata.*

The Coral Sub-littoral

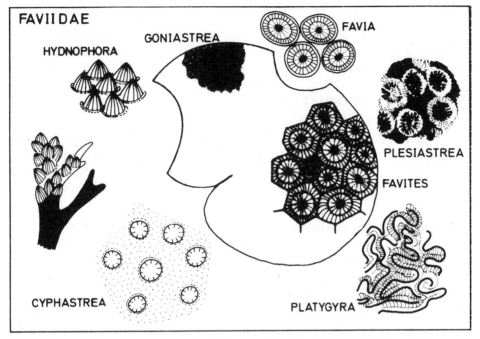

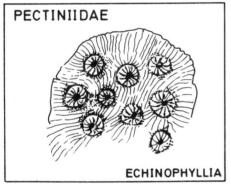

FIG. 12.4. Growth forms of the families of the Faviina.

Family OCULINIDAE

This family is exceedingly rare in Hong Kong being respresented by a single genus and species, *Galaxea astreata* with small, stunted colonies.

Family PECTINIIDAE

Corallum broad and fragile, spread out in thin flanges. Represented in Hong Kong by *Echinophyllia aspera,* with corallite walls reduced, and septal systems running from the calices right over the face of the lamella.

Family MUSSIDAE

Large corallites, simple or more often colonial, with sharp, dentate septa. Commonly represented in Hong Kong by *Acanthastrea bowerbanki*, with the corallites monocentric.

Sub-order DENDROPHYLLIIDA
Family DENDROPHYLLIIDAE

Large corallites, solitary or colonial, usually orange, yellow or brown. *Dendrophyllia* is ahermatypic, with large cups at the angles of long branches. *Tubastrea* is also ahermatypic, forming small clumps with the corallites separate but united by feeble costae; found in deeper water, or under shaded overhangs. *Turbinaria* is hermatypic, with a heavy growth form of solid, contorted or foliaceous branches. *Turbinaria peltata* has much larger polyps than *T. mesenteria*. Common Hong Kong species: *Dendrophyllia gracilis, Turbinaria peltata, T. mesenteria, Tubastrea aurea, Tubastrea diaphana*.

Zou Ren-lin and his co-workers[2] have produced a comprehensive guide to the scleractinian corals of Hainan Island and although some of the specific names used here differ from those ascribed by Zou and his colleagues, the guide is nevertheless of particular value, particularly for the identification of the coral genera.

CORAL RELATIVES

A further group of colonial Anthozoans is placed beside the hard corals (Scleractinia) in the order Zoanthidea of the Zoantharia. They are in fact small branching or colonial anemones, loosely packed together in clusters. The tentacles are short, and in two circlets and the mouth disc often brightly coloured. The zoanthid *Palythoa* (Fig. 12.6) is pale brown, compacted into stiff pads or spreading crusts.

As well as the Zoantharia, the class Anthozoa possesses two other sub-classes—the Ceriantipatharia and the Alcyonaria (or Octocorallia).

The first of these, the Ceriantipatharia are primitive anthozoans possessing a simple arrangement of septa and tentacles. The sub-class comprises two orders. The first, the Ceriantharia, was encountered on soft shores and is represented in Hong Kong by the burrowing anemone *Cerianthus*, without a pedal disc and encased within a mucilaginous sheath. The second order is the Antipatharia, the 'black corals'. These, again are a colonial form with a thick black or brown axial skeleton bearing polyps arranged laterally. In Tolo Harbour *Antipathes* cf. *densa* forms a tree-like colony up to 1 m in height. The branches are ensheathed with the fleshy yellow polyps. *Cirrhipathes anguina* is a long whip-like colony

---

[2] Zou, R. L., Sung, S. M., and Ma, K. F. 1975. *The Shallow Water Scleractinian Corals of Hainan*. Peking, Scientific Publisher, pp. 66, pl. I–XV.

The Coral Sub-littoral 265

perhaps up to 1 m in length, while *C. spiralis* has a spirally coiled axial skeleton. All are attached to a hard substrate by a thick base.

The third class of the Anthozoa, the Alcyonaria, comprises colonial anthozoans with eight pinnate tentacles. There are three major orders, the Alcyonacea (soft corals), Gorgonacea (horny corals) and the Pennatulacea (sea pens).

The surface of the representatives of the Alcyonacea is yielding and rubbery, and often grey or buff. They are of simple construction, cushions or lobed branches, sometimes profusely sub-divided. The fleshy colony is strengthened throughout with scattered limy spicules. The surface is beset with small star-shaped polyps, that open out to reveal eight pinnate tentacles typical of all Alcyonaria.

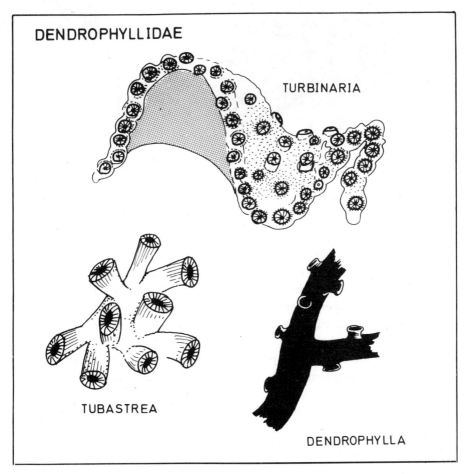

FIG. 12.5. Growth forms of Hong Kong representatives of the Dendrophyllidae.

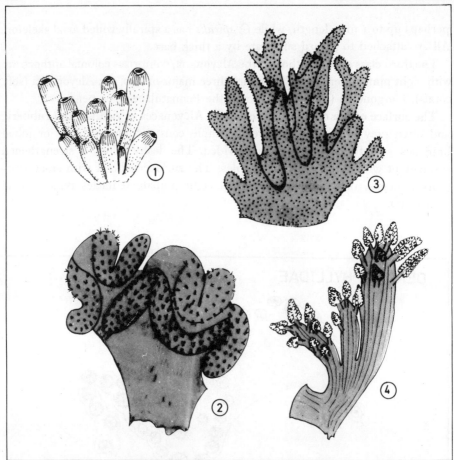

FIG. 12.6. Characteristic growth forms of the soft corals of the Zoathidea and Octocorallia. 1. *Palythoa* polyps; 2. *Sarcophyton elegans*; 3. *Sinularia polydactyla*; 4. *Dendronephthya gigantea*.

Lobophytum is a soft coral found in rather heavy surge, forming rubbery brown sheets with strong corrugations. The *Sarcophyton* species are more elevated, forming cushion shapes, generally with thick, undulating edges. *Cladiella digitulata* is a soft grey with black flecks to the stubby 'fingers' of the soft-coral colony. In *Sinularia* the colony is drawn out into bunches of long or short whip-like branches. *Nephthya* species have a softer texture of more watery consistency and a different build, with the fleshy stems and terminal polyps arranged in upright bunches like cauliflowers.

The commonest sub-tidal species is the highly distinctive, large, brilliant red-orange *Dendronephthya gigantea* forming a fleshy colony up to 60 cm in height. This alcyoniid is fed upon by *Diminovula punctata* a brilliant orange-red ovulid gastropod with a reflected mantle beautifully patterned to match the prey host.

In deeper waters, the ahermatypic scleractinian corals members of the

The Coral Sub-littoral

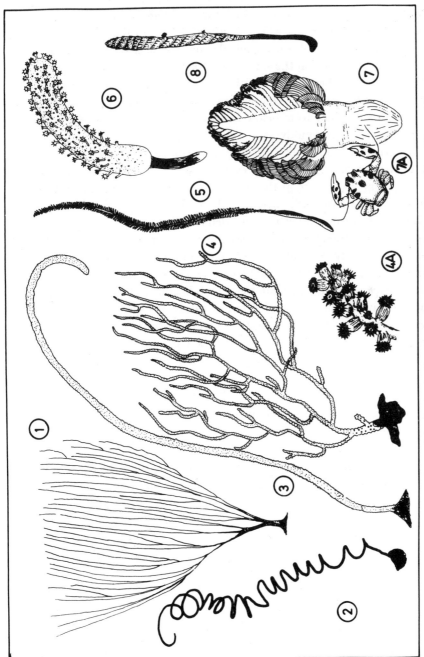

FIG. 12.7. Members of the Antipatharia, Gorgonacea and Pennatulacea from Hong Kong. All the species are sub-littoral.
1. *Antipathes densa*; 2. *Cirrhipathes spiralis*; 3. *Junceella juncea*; 4. *Echinogorgia*, with 4a. detail of polyp; 5. *Virgularia*; 6. *Cavernularia obesa*; 7. *Pteroeides sparmanni* with 7a. the commensal crab *Porcellana picta*; 8. *Stachyptilum dofleini*.

Gorgonacea or 'horny corals' are found. The best known gorgonid is the precious red coral belonging to the genus *Corallium*. This is not found in Hong Kong but representatives of the flat sea-fans *Villogorgia, Echinogorgia,* and the short *Euplexaura* or long slender sea-whips *(Junceella)* are. In the Gorgonacea the skeleton comprises either calcareous spicules or it is of a horny material. The polyps are small and borne on the sides of the skeletal axis.

On the soft sea bed, beyond the coral reef fringe, and in general outside the scope of this book, the Anthozoa complete their story. In Hong Kong hermatypic coral reaches to a depth of 10 m below low water. Thereafter, the sea bed is of soft deposits of sand or muddy sand. Burrowing bivalves, such as hammer-oysters *(Malleus)* and fan shells *(Atrina)* are accompanied by large burrowing anemones *(Actinodendron)* and many others beside.

Projecting from the surface of the sand are sea-pens, of the anthozoan order Pennatulacea. These form fleshy colonies where the primary polyp is elongated, with its lower part forming a fleshy embedded stalk. From its upper part project two rows of lateral branches, bearing secondary polyps along their axes. The resultant structure may resemble a short, thick quill pen *(Pteroeides),* or a long thin quill *(Virgularia)* or in *Cavernularia* it may be stout and clavate, with secondary polyps arranged without trace of bilaterality. In all sea pens, the fleshy polyps have an embedded skeleton of calcareous spicules.

Pteroeides sparmanni, the commonest sea pen and occasionally to be collected by snorkelling over sand flats, often possesses a pair of the porcellain crabs, *Porcellana picta,* white with large black spots, commensally nestling among the lateral branches.

Interstitial Communities of Coral

Coral limestone is markedly different from most hard substrates because it can be tunnelled and excavated. It offers thus a new habitat for numerous molluscs, worms, crustacea and echinoderms. Some species make borings directly into the soft calcium carbonate and others, more numerous, nestle in the galleries and cavities that pioneer species have vacated. No other shore micro-habitat is so diversely rich; and the 'infauna' of Hong Kong corals is just as extensive and complete as it is in the tropics, despite the relative paucity of the corals themselves.

Borers

The most numerous agents in the 'bioerosion' of the massive coral heads are bivalve molluscs. The pholad borer *Barnea* was found in soft sedimentary rocks in Tolo Harbour, excavating its circular shafts mechanically by the abrasive action of the shell sculpture. Exclusively in dead coral rock, similar mechanical boring is performed by *Gastrochaena cuneiformis,* a bivalve resembling a pholad in its wide anterior shell-gape. The sculpture is not, however, file-like and each valve is prolonged in front into a sharp 'rostrum' or beak. The borings can be recognized by the narrow exit shaft like a bottle neck and a flask-shaped cavity, circular in section. Two other common species of *Gastrochaena* are *G. interrupta* and *G. laevigata*.

The mussels of the family Mytilidae are the most common of all coral borers, often excavating deeply and in large numbers. Two genera are recognized; the first of these, *Botula,* we have seen boring into the oysters of Deep Bay *(B. silicula)*. Widely distributed in the Pacific but not yet recorded from the coral of Hong Kong is the related *B. cinnamomea*. P. J. B. Scott[3] has made a detailed study of the mussel borers of the genus *Lithophaga* and demonstrated an unusual

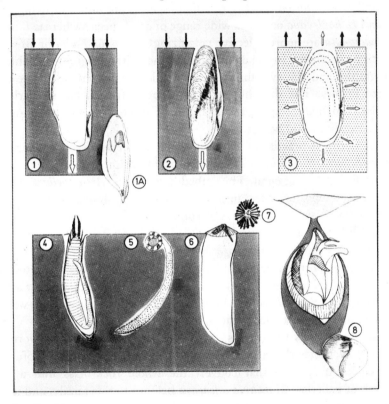

FIG. 12.8. Coral borers.
1. *Gastrochaena laevigata,* with 1a. *G. cuneiformis* internal view of the left shell valve; 2. *Lithophaga teres,* the dead coral borer; 3. *Lithophaga simplex,* the live coral borer. In each of the above cases the direction of boring is indicated by open arrows with posterior areas of niche erosion or deposition represented by closed arrows. 4. *Stylariodes parmata;* 5. *Phascolosoma scolops;* 6. *Pyrgoma* sp.; 7. *Creusia spinulosa;* 8. *Magilus striatus* shown in a flask shaped burrow within a faviid coral head. The shell is also shown inset.

[3] Scott, P. J. B. 1980. Associations between scleractinian and coral-boring molluscs in Hong Kong. In *The Malacofauna of Hong Kong and Southern China,* ed. B. S. Morton, pp. 121–38. Hong Kong, Hong Kong University Press.

diversity of species relative to the proportionately poorer diversity of local corals. The shell of *Lithophaga* is not rotated but is attached by widely radiating byssal threads to the floor of the boring. Boring takes place by the secretion of mucus carrying acid phosphatase from the thickened edges of the mantle extended both ventrally and dorsally beyond the anterior end of the shell.

The *Lithophaga* community in Hong Kong can be divided into two major groups; those boring into dead coral and those boring into live. In the latter category, the siphons also possess glands that produce a secretion to prevent the coral from growing over the aperture to the burrow.

In the former category, *Lithophaga malaccana* is most common, with the black *L. teres* and *L. antillarum* much rarer. Four species are found in living corals; *L. lima* and *L. hanleyana* ocur in a wide range of coral species whereas *L. simplex* and *L. mucronata* are more selective colonizing species of *Favia* or *Montipora* and *Porites* respectively. In the distribution of these lithophagines a combination of hydrography and the availability of a suitable site for settlement are clearly interlinked factors. The distribution of shipworms (p. 84) has similarly been found related to hydrography, the availability of wood and competition from the crustacean limnoriid borers.

Abrasive boring is practised somewhat by ark-shells that freely colonize coral. The unspecialized *Barbatia helblingi* is to be found in large clusters, byssus-attached to the surface of coral. The embedding species, *Arca avellana*, lies with its flatly truncated posterior end flush with the surface, showing the two spirally coiled umbones. A yet third species of coral-associated ark is *B. obliquata*.

The mytilid mussel *Gregariella coralliophaga* is probably also a borer but possibly enlarges empty *Lithophaga* burrows in which it settles. It can be recognized by the posterior fringe of long periostrical fibres that separate it from *Lithophaga*. A final nesting bivalve, though often regarded as a coral borer, is *Coralliophaga coralliophaga*, a close ally of *Trapezium* earlier met with as a nestler of boulder shores and crevices.

Increasingly towards the tropics, coral rock becomes eroded by worms. Of the boring polychaetes the largest family is the Euniciidae, with a profusion of brightly coloured species, that convert the whole rock-mass into a fragile honeycomb of galleries. *Marphysa adenensis* belongs to this family, and its close relative *M. sanguinea* has already been described from sedimentary rock crevices in Hong Kong. The most common coral euniciid is *Palola siciliensis*.

Sipunculids are the commonest worms in tropical coral reefs. They seem less abundant in Hong Kong, but there are several species to be found, notably of *Phascolosoma* (e.g. *P. scolops*), with a cylindrical, muscular and unsegmented body. The proboscis can be extruded to collect surface detritus by its crown of branched tentacles. The mechanism of boring is hardly yet understood; it has generally been attributed to the hard denticles scattered over the integument.

A common boring polychaete certain to be found in dead coral is *Stylarioides parmata*. From the surface, it is to be detected by the pointed wisp of head bristles, emerging from its bored gallery, in which the worm lies bent double, with the head and front half larger. The head bristles can be opened out into a broad fan, intercepting larger particles. Smaller particles are collected by the cilia of a clump of tentacular gills; there are as well two large prostomial palps.

Two barnacles are worthy of mention. The first of these is *Creusia spinulosa* which attaches to the surface of coral, especially *Favites* and *Hydnophora* (Plate 25). Its strongly ribbed shell progressively becomes overlain with coral growth leaving only the aperture free. Less obvious is the small aperture of the deep burrowing barnacle *Pyrgoma* sp. in which the shell plates are deeply embedded within and are fused to the surrounding coral skeleton. Breaking up *Alveopora*, *Hydnophora* or *Platygyra* heads will often reveal a number of these barnacles. Regrettably little is known of these borers in Hong Kong and there are certainly other species, possibly in more intimate relationship with the coral than is at present understood.

Crustacea

The most mobile and apparent animals, on turning over or breaking up a corallum, are generally the small Crustacea: crabs (both brachyuran and anomuran) and shrimps, especially snapping alpheids. Most of them are intimately adapted to the substrate and some are highly camouflaged. Many of them are modified for slow-moving, secretive life in crevices and many are commensal.

In Fig. 12.9 most of the characteristic crustacean life-forms are illustrated, but the number of species could be multiplied. Alpheid shrimps are generally the most numerous, both in species and individuals. In this family (p. 276) one of the chelae is greatly enlarged, with a peg-like projection on the movable finger, that fits into a socket on the fixed finger. The sudden closing of this claw, under strong muscular power, produces a snapping sound as the peg goes into the socket and the heavily calcified finger-tips meet. The sharp sound is probably defensive, as could also be the jet of water usually displaced by the peg.

Three *Alpheus* species are commonly collected with Hong Kong coral. The largest is *A. hippothoe*, greenish and purple-brown banded across the abdomen; a second *(A. malleodigitus)*, somewhat smaller, is a uniform transparent orange-pink; and a third *(A. parvirostris)*, still smaller, is transparent orange with transverse dark bands. More retiring than the alpheids are the *Synalpheus* shrimps, of which one species. *S. coutierei* is illustrated. The carapace and abdomen are broad and depressed, yellowish-orange, with the chelae ovoid and flattened. These synalpheids are found as sex-pairs within hollowed concavities of sponges. The almost white *S. gravieri* lives among soft corals such as *Nephthya*.

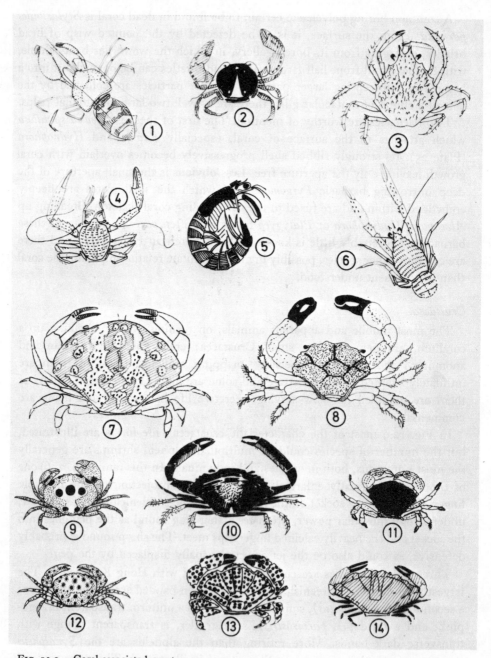

FIG. 12.9. Coral associated crustaceans.
1. *Synalpheus coutierei*; 2. *Pachycheles sculptus*; 3. *Leptomithrax* sp.; 4. *Porcellana ornata*; 5. *Gonodactylus falcatus*; 6. *Galathea elegans*; 7. *Atergatis floridus*; 8. *Chlorodiella nigra*; 9. *Carpilius maculatus*; 10. *Etisus laevimanus*; 11. *Xantho reynaudi*; 12. *Liagore rubromaculata*; 13. *Daira perlata*; 14. *Liomera venosa*.

The banded coral shrimps (Stenopodidae) are also found on Hong Kong corals. *Stenopus hispidus* is slender and transparent, marked with pink highpoints, and known from its habits as the 'cleaner shrimp'. Slender chelae are carried upon the first three pairs of legs, and the prominent, white antennae wave conspicuously; three branches are displayed, (two from the antennule and one from antenna proper on either side). These are used to attract the client fish that come to be cleaned. The toilet operation involves removal of parasites, injured tissue and fungal growth with the sets of small chelae. As will be seen later (Fig. 12.10) a second species of *Stenopus* seems to be more than casually associated with the brittle star *Macrophiothrix*.

The small mantis shrimps (*Gonodactylus* spp.) commonly nestle in coral, being generally shiny black, dark green or brown. These are narrower and more cylindrical than the larger squillids of sand and mud-flats, and they differ interestingly in their chelae and food-catching. Where squillids have serrated claws, used for striking and impaling the prey, *Gonodactylus falcatus* has heavy, smooth-edged claws delivering smashing blows to small crabs, or a stabbing strike to soft-bodied prey (*see* Fig. 12.9). A second species, the bottle-green *G. mutatus*, also occurs in Hong Kong.

The leading brachyuran crabs in coral are the slow-moving Xanthidae and Maiidae. Their commonest forms are *Daira perlata* and several *Actaea* species, the commonest being *A. pura*, compact reddish or brown crabs, covered with ridges or tubercles and with a short hairy tomentum. The smaller *Liomera venosa*, 1.2 cm, is smooth and wine-red, its carapace incised with strong sculpture. Distinctively red-spotted xanthids include the large *Carpilius maculatus*, sometimes found juvenile on corals, and *Liagore rubomaculata*. Large, bulky xanthids nestle in the spaces under the coral heads, or walk about freely. Among these are the smooth, finely patterned shawl-crab *Atergatis floridus*, and the red-eyed crab, *Eriphia sebana*, a genus already noted on exposed shores (p. 43). *Actumnus setifer* is purple red with dark-tipped chelae. *Xantho reynaudi* is laterally expanded with smoothly scalloped margins and with strong tubercules on the chelae.

Some xanthid crabs are so localized in habit they cling throughout life to one sort of coral. The *Trapezia* species, with wide spreading chelae and strong leg grasp, live on *Pocillopora*, and *Tetralia* is confined to certain species of *Acropora*. The commonest of these coral-clinging xanthids in Hong Kong is *Chlorodiella nigra*, with its carapace and long chelae polished black and the legs short and hairy brown. The tips of the chelae are characteristically formed into crushing plates.

The pilumnid crabs are covered in a dense mat of 'hair', local coral gallery species including *Pilumnus longicornis* and *P. minutus*, with, in *Heteropilumnus ciliatus* and *Parapilumnus trispinosus*, a carapace devoid of hair, this being largely confined to the anterior regions of the limbs.

In the decorator-crabs, Maiidae, the carapace is typically pointed in front and the back convex; the legs carry special curled or hooked hairs, by which plant or animal camouflage, such as sponges, ascidians and algae, can be attached. The chelae are long and slender, and the claws give a close attachment to the substrate. Two species, *Micippa philyra* and *Schizophrys aspera* occur in coral galleries.

Two anomuran families, the Porcellanidae or stone-crabs and the Galatheidae or squat-lobsters, are important in the coral community. Porcellanids are numerous both in species and individuals, and usually very prettily marked. The chelae are large, flat and wide-elbowed, the antennae long and whip-like. The third maxillipeds are fringed with straining setae and used for filter-feeding. The anomuran package includes also the long and still mobile abdomen, with tail-fan complete, and the vestigial fourth peraeopods, tucked into the branchial chambers as gill brushes.

The commonest stone-crab is the rose-red *Petrolisthes coccineus*. The smaller *Porcellana ornata* is marked in greenish-grey. Porcellanids also have commensal species (p. 267). *Pachycheles sculptus* lives permanently in concavities within sponges (Fig. 12.9); it is round-bodied, and attractively coloured in coffee brown, with a cream streak down the carapace mid-line. *P. pisum* occurs in coral galleries.

The ancestral derivation of the porcellanids is suggested by their relatives, the Galatheidae. Here the abdomen is longer and fully visible from above, extending out behind the flat carapace. The legs are longer and more slender than in stone-crabs, and the chelae straighter and narrow. The carapace comes in front to a pointed rostrum. Aptly named 'squat lobsters' the galatheids have hardly yet attained the shape of a crab. They feed with setose maxillipeds like porcellanids. *Galathea elegans,* longitudinally brown and white striped, is commensally associated with crinoids, while in coral galleries commonly occurs *G. orientalis*.

Echinoderms

A diversity of echinoderms live in tropical coral formations. Two papers, by Mortensen[4] and Heding[5] deal with some of the local species, but the list is far from complete. Most of the species belong to deeper waters though within the coral galleries and under coral boulders numerous brittle stars or snake stars (Ophiuroidea) can be found. Brittle stars consist of a small central disc, carrying five flexibly articulated arms, by whose sinuous flexion the animal moves. Food is collected by scraping organisms from the surface or by screening particles with mucus spread between the arm spines and carrying them by tube-feet to the mouth. The latter mode of life is practised by the small green

---

[4] Mortensen Th. 1934. Echinoderms of Hong Kong. *The Hong Kong Naturalist Supplement* 3: 1–14.

[5] Heding, S. 1934. On some holothurians from Hong Kong. *The Hong Kong Naturalist Supplement* 3: 15–25.

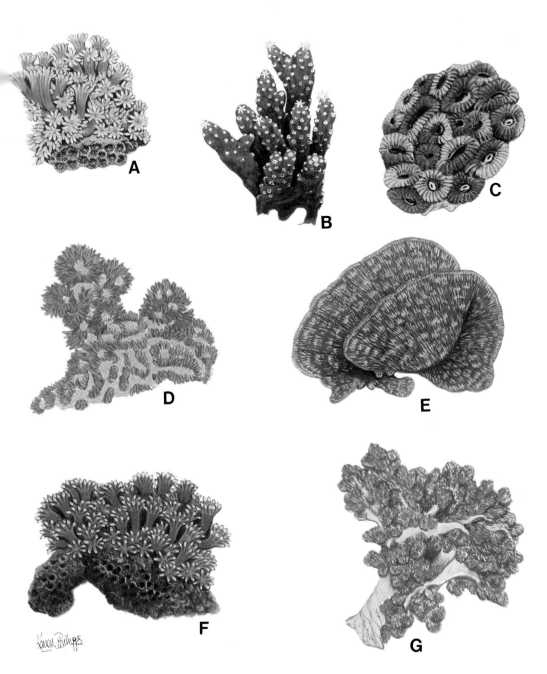

PLATE 23. Some Hong Kong corals: A. *Goniopora columna*; B. *Acropora pruinosa*; C. *Favia speciosa*; D. *Hydnophora exesa*; E. *Pavona decussata*; F. *Alveopora irregularis*; G. *Dendronephthya gigantea*.

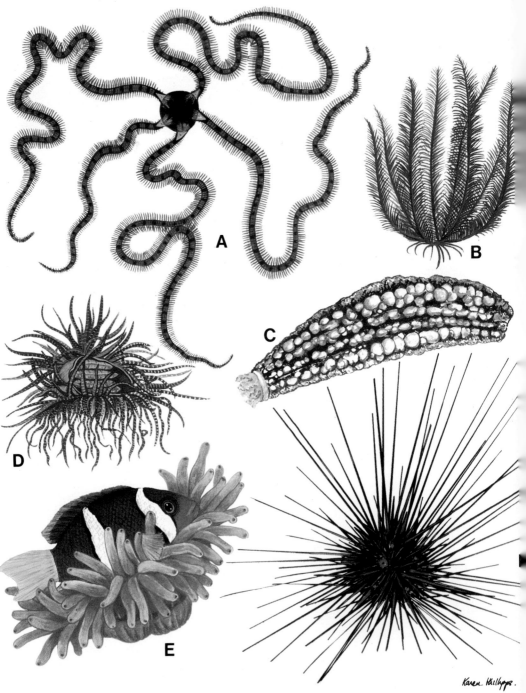

PLATE 24. Coral associated fauna: A. *Macrophiothrix longipeda*; B. *Comanthus parvicirrus*; C. A dendrochirote holothurian (*Thyone* sp.?); D. *Limaria hongkongensis*; E. The clown-fish *Amphiprion bicinctus* with the anemone *Parasicyonis actinostoloides*; F. *Diadema setosum*.

brittle star *Ophiactis savignyi* which typically occupies empty lithophagine holes in dead coral blocks. From the aperture projects the fine tips of the five arms to collect food, their remainder and the disc being located within the hole. *O. modesta* is blue and has the same habit, both being especially common in the coral blocks of Tolo Harbour or sometimes in sponges.

Two other ophiuroids can be found hiding under or within coral blocks. The first of these, *Ophiarachnella gorgonia,* has a perfectly round disc and slender arms, transversely banded with light and dark brown. Superficially very similar is *Ophioplocus japonica,* but in this species, the inter-radii of the disc are lobed. *Ophiotrichoides gratilla* has a brown disc and arms turning to white at the tips. The arms bear numerous long spines.

The highly distinctive, large, brittle-star *Macrophiothrix longipeda* is illustrated in Fig. 12.10. This characteristic species and the very similar *M. variabilis* lives deeply within large coral galleries. It is becoming clear that a number of other coral inhabitants are more than casually associated with it (see also Fig. 12.10). The first of these is the small leptonid bivalve *Ephippodonta oedipus,* with valves widely, almost horizontally, parted and covered by reflected mantle. Large individuals are female, often with dwarf males located within the mantle. A sex pair of *Alpheus hippothoe,* the large pistol shrimp up to 3.5 cm in length with transverse purple-brown bands, one or two *Stenopus* sp. and a scale-worm *Iphione muricata* also seem part of the assemblage. A detailed study of this 'core' grouping could be profitably undertaken.

The sea urchins or echinoids are typified by the long-spined *Diadema setosum,* the common urchin of sheltered rocks and coral patches. The black needle-sharp spines are very mobile. Positively sensitive to shading, the spines, beset with poison glands, can be actively pointed in any direction. *D. setosum* has an orange anal ring which its less common relative *D. savignyi* lacks, the latter also possessing banded spines. On most Indo-Pacific reefs *Diadema* is a secretive, crevice dwelling echinoid but in Hong Kong, possibly because of a lack of mollusc and fish predators, it densely covers the rock surfaces. Related to *Diadema* is *Echinothrix calamaris,* with strong quill-like spines striped in brown and white. Two other coral urchins may be rarely encountered in Hong Kong: the olive green *Parasalenia gratiosa* characterized by a white ring at the base of each short, stout, spine, and *Tripneustes gratilla,* of more standard form with short, fine spines.

Amongst the coral rubble can be found the black *Holothuria leucospilota,* typical of the rocky sub-littoral fringe on many shores in Hong Kong. A second, species, *H. arenicola,* tan with two rows of darker blotches, also co-occurs with *H. leucospilota* on clean coral sands. Amongst coral boulders is a species of *Stichopus,* with strong pointed papillae and up to 30 cm long. Like *Holothuria, Stichopus* is a deposit-feeder, picking up and shovelling large quantities of sediments into the mouth by the adhesive discs of the ten oral podia.

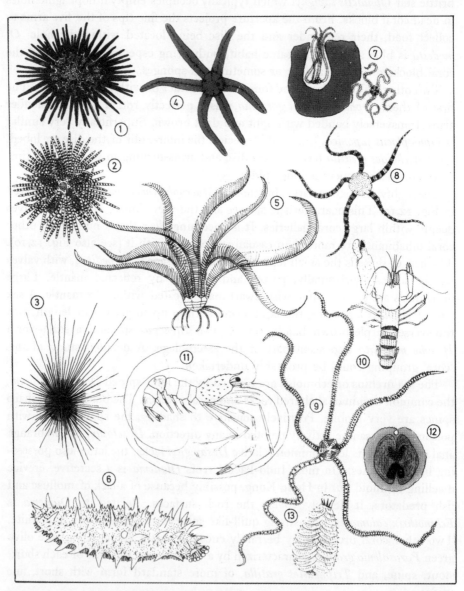

FIG. 12.10. Echinoderms of the coral assemblage.
1. *Parasalenia gratiosa*; 2. *Echinothrix calamaris*; 3. *Diadema setosum*; 4. *Echinaster luzonicus*; 5. *Tropiometra afra macrodiscus*; 6. *Stichopus* sp.; 7. *Ophiactis savignyi* in its cavity in the coral; 8. *Ophiarachnella gorgonia*; 9. *Macrophiothrix longipeda* with its associates; 10. *Alpheus hippothoe*; 11. *Stenopus* sp.; 12. *Ephippodonta oedipus*; 13. *Iphione muricata*.

The Crinoidea (sometimes called sea-lilies) are the last surviving members of the ancient sessile Echinodermata. They are especially numerous in the Indo-Pacific sub-tidal. The feather stars, as they are alternatively called are typified by *Comatula salaris* and the either black or brilliant yellow *Tropiometra afra macrodiscus*. The latter is especially numerous in Hong Kong, occurring from just below the sub-tidal to depths of many metres. The brittle star *Ophiomaza cacaotica* is commensal on *Tropiometra* and *Zygometra comata*, winding its arms amongst the pinnules of the crinoids tentacles. Other, more deep water crinoids include *Oligometra serripinna serripinna* and the reddish brown *Antedon* cf. *bifida* with relatively short arms and stout cirri comprising up to 17 short segments.

Not permanently affixed, these stars can move about or re-attach by the recurved basal cirri. The viscera lie within a small cup, from the rim of which springs a fringe of feather-like tentacles, ten or more in number. These carry delicate side-branches or pinnules, sensitively curling and extending at the lightest stimulus. The pinnules, like the main arms channels, have ciliary tracts carrying fine particles to the mouth. Food is not however secured by filtering, but is caught directly by the waving arms and narcotized by poison glands on the pinnules.

The Asteroidea are extremely poorly represented on corals in Hong Kong. The rust-red, 6-armed, *Echinaster luzonicus,* is sometimes found, but much rarer are the pincushion star *Culcita novaeguinae,* the dark olive *Anthenea aspera* and the large spiny star *Pentaceraster regulus*.

Polychaetes

The galleries of the living and dead coral heads form an ideal habitat for a wide variety of polychaetes. In addition some sedentary polychaetes characteristically associate themselves with coral, their limy tube becoming overgrown by polyps so that they seem to arise from the coral itself. Representatives of many families are found, reflecting the diversity of niches that characterizes the coral community. Many of the polychaete lines of evolution have been mentioned earlier (Chapter 9). Here only a few of these groups that seem to specifically associate themselves with coral are included. Many species also occur fortuitously in coral crevices just as in those crevices of sedimentary rock formations.

Of the errant polychaetes, representatives of the Aphroditidae are common. The polynoid scale-worm *Iphione muricata*, about 2.5 cm long, has already been noted to occur with the brittle-star *Macrophiothrix*. Two other species, *Halosydna brevisetosa* and *Lepidonotus squamatus*, are also very common. The scale-worm's back is covered with two rows of strong, overlapping, greyish-brown elytra, like a flat pine cone. The proboscis is eversible and very muscular and all are carnivores.

With so many tiny creatures inhabiting the coral galleries it is not surprising that the predatory Nereidae are well represented in any collection of coral polychaetes. Most of these occupy semi-permanent burrows in the mud or sand that accumulates in the coral crevices. They roam these galleries striking out, with the proboscis typically equipped with formidable incurving jaws, at soft bodied prey. *Ceratonereis mirabilis* and *Neanthes unifasciata* are illustrated in Fig. 12.11. *Leonnates nipponicus* is also common.

In the Eunicidae the proboscis is characteristically armed. These worms may be algal-eaters, voracious predators or, even microphagous tube-dwellers. *Lysidice ninetta* and *Eunice australis* are illustrated.

The other important polychaetes are tube-dwellers and typically possess a crown of tentacles that collect fine particles held in suspension in the sea above. Of the two familes—the Serpulidae and the Sabellidae—the first is by far the most important in this habitat.

The Serpulidae build calcareous tubes that are typically overgrown by fresh coral growth. The first is the beautiful *Spirobranchus tricornis* in which the tentacular fan, varying in colour from white to yellow, orange or red, is formed into a characteristic spiral. In *Serpula vermicularis* and *S. watsoni,* the fan is a simple crown, often brilliant red. The tube of *S. vermicularis* displays characteristic growth rings. One of the tentacles is modified to form a mushroom-shaped operculum. *Hydroides dirampha,* also associates with coral though this genus is more characteristic of rocky shores (p. 88).

The Sabellidae are occasionally found attached to the under-surface of dead coral heads. They build a flexible, parchment-like tube, often embedded with sand grains. The apical crown of *Myxicola infundibulum* is shown in Fig. 12.11.

The Echiura too are represented in the coral community with *Protobonella* commonly being found and using its proboscis to collect detritus like its more common relatives on soft flats.

Molluscs of the Coral Sub-littoral

Around the islands of Mirs Bay and at the mouth of the Tolo Harbour, free from turbidity or fresh-water dilution, the potentially tropical character of the Hong Kong fauna is most fully realized.

The Mollusca still show only a fraction of the tropical diversity, for example present in the Philippines; but they offer a rich introduction to most of the families of tropical gastropods and bivalves. Only the main groups can be mentioned here, occurring as they do beyond this book's principal territory of the inter-tidal.

Bivalvia

Unlike the burrowing bivalves of soft flats, most of those on coral shores belong to the 'Heteromyaria', a wide assemblage representing a more ancient, though

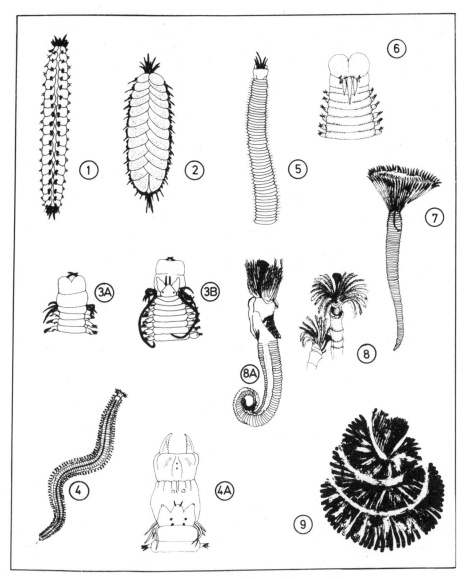

FIG. 12.11. Errant and sedentary polychaetes of the coral community.
1. *Malosydna brevisetosa*; 2. *Lepidonotus squamosa*; 3. *Ceratonereis mirabilis,* with A. head retracted and B. extended; 4. *Neanthes unifasciata,* with 4A. detail of the head of *N. oxypoda*; 5. *Eunice australis*; 6. *Lysidice ninetta*; 7. *Myxicola infundibulum* removed from its tube; 8. *Serpula vermicularis,* with 8A. the animal removed from its tube; 9. The branchial crown of *Spirobranchus tricornis*.

highly specialized, fashion in bivalve evolution. Where most burrowing bivalves retain two adductor mussels and a strong foot, with the shell approximately symmetrical in front and behind, the Heteromyaria have suffered great modification (p. 101). They have retained and greatly developed the byssus (not usually found in burrowing bivalves) and are generally anchored to the substrate by the anterior end. Changes in body form and symmetry follow from this basic fact (*see* Fig. 7.7).

Among the scallops and their allies (Pectinacea), the most primitive are the small, byssus-attached species of the genus *Chlamys*. At the hinge side there are two squared 'lugs', the anterior one larger, where the byssus emerges by a notch.

One other scallop, the flat *Pedum spondyloideum*, retains its byssus through life, attached in deep clefts of *Porites*, kept narrowly open as the coral continues to grow. This interesting wide spread, Indo-Pacific, bivalve does not appear to occur in Hong Kong.

Further evolution follows from a *Chlamys*-like stage. One trend is to lose the byssus and break free. The shell becomes more rounded, as in *Pecten*. With different species, the left valve may be flat and the right cupped, or both shallowly concave. With the strong central adductor muscle, clapping the valves and expelling water from the mantle cavity, *Pecten* is able to swim. According to the control of the mantle edge or velum, water makes exit either at the lugs, when the shell moves with its rounded edge forward, or round the circumference, when the animal proceeds hinge-first.

On soft sublittoral deposits is found *Chlamys pyxidatus* with a flat upper valve, rust red and blotched with large white, black bordered spots and a cup-shaped lower valve. This species swims only weakly.

The greatest freedom is attained by the light, almost perfectly circular scallops, *Amusium japonicum* and the smaller *A. pleuronectes*, frequently gathered offshore on sandy ground. The fragile shell is almost flat, and shining red-brown above and white below. A parallel achievement of light circular shape has been shown in the window-shell *Placuna*, among the byssus-freed anomiids (p. 248). Common features of all scallops are the well-developed velum and the small but elaborate eyes among the pallial fringing tentacles.

An alternative trend among the Pectinacea is towards firm fixation by the right valve, as in the Spondylidae, or 'thorny oysters'. Cemented only by the older part of the valve, *Spondylus barbatus* forms bracket attachments to the edge of coral boulders. The byssus is lacking and the small foot concerned with cleansing the mantle cavity. The shell is heavy and deeply cupped, ornamented with a cover of strong spines.

Some of the Limidae or file shells are thick-shelled with strong, spinose ribs (*Lima lima*) and are permanantly byssally attached. Pallial tentacles, interspersed with eye spots, can be completely withdrawn between the shell valves. As in the Pectinacea however, some limids such as *Limaria fragilis* (found only in coral heads), are free-moving with a thin and translucent shell. The equal, and somewhat elongated valves, rounded at the growing edge, have small lugs beside the

The Coral Sub-littoral

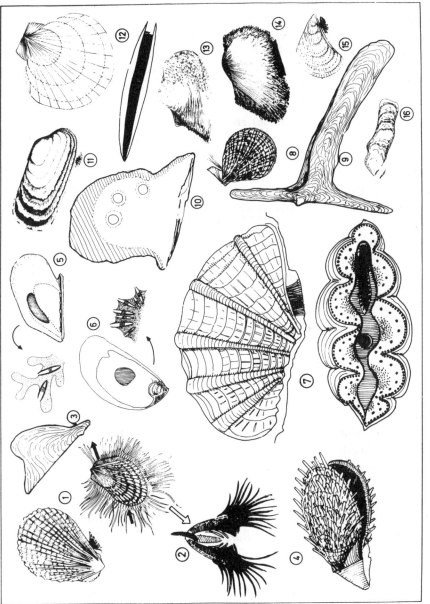

FIG. 12.12. Bivalves of hard sub-tidal shores.
1. *Lima lima*; 2. *Limaria fragilis* showing (*below*) valve gape and velum in end-on view; and (*above*) animal while swimming; 3. *Pteria brevialata*; 4. *Spondylus barbatus*; 5. *Crenatula modiolaris* embedding in the green branching sponge *Sigmadocia symbiotica*; 6. *Vulsella vulsella* embedding in the rounded, black sponge *Suberites*; 7. *Tridacna maxima* with (*below*) a dorsal view of the expanded siphons; 8. *Chlamys irregularis*; 9. *Malleus malleus*; 10. *Pteria penguin* with blister pearls; 11. *Coralliophaga coralliophaga*; 12. *Amusium pleuronectes*; 13. *Arca avellana*; 14. *Barbatia helblingi*; 15. *Septifer virgatus*; 16. *Isognomon legumen*.

hinge. The byssus gland in *L. fragilis* spins a loose nest of threads with which the sedentary animal surrounds itself. But these bivalves can also move freely about, both swimming like a scallop, and also crawling by the tongue-shaped foot in mussel fashion. At rest, the valves can diverge widely to expose the broad velum and, outside it, dense fringes of cream and purple pallial tentacles. As in the Pectinacea, swimming is always with the hinge hindmost, water being expelled across the two lugs. Crawling is performed in upright posture, with the hinge end behind. The purple tentacled *L. fragilis* is easily separable from an otherwise very similar species, *L. hongkongensis*, with red tentacles.

Two distinctive mussels (Mytilidae) are found as coral associates: *Septifer virgatus*, bright blue-green and wedged into the shelter of crevices, and the smooth plump, mussel *Ryenella cuprea*, embedded—like most of its genus—in the massed tests of ascidians.

The wafer-thin Isognomonidae are in some ways the most primitive of heteromyarian bivalves. All byssus-attached, they show various adaptations of form and habit. In *Isognomon legumen*, the shell is long and narrow, to allow for habitation of deep crevices. In *Crenatula modiolaris*, living buried in *Sigmadocia symbiotica*, a green branching sponge, the hinge axis is elongate, straight and winged (the barnacle *Acasta dofleini* is similarly buried in the sponge tissues). The hammer oysters, *Malleus*, have become elongated in the dorso-ventral axis, from hinge to growing margin. The simplest design is that of *Malleus regulus*, compressed like a flat pod and attached at the straight hinge end. In the black hammer oyster, *Malleus malleus*, the hinge line is also long drawn-out, giving the shell its hammer shape. This species lives in clean sand, generally near rock; the hinge line, with byssus attachment is buried, with the long 'handle' emerging at the surface, with the inhalant and exhalant apertures. A third hammer oyster, *Vulsella vulsella*, lies deeply embedded in the massive rounded heads of the black sponge *Suberites*. *Vulsella* has lost the long handles of other malleids and is strongly radially ribbed.

Dead coral especially is often heavily colonised by byssally attached ark-shells. The first of these with its posterior end splayed out to lie flush with the surface is *Arca avellana*, recorded earlier from the lower reaches of boulder shores (p. 99). Other species amongst coral are the large, relatively unmodified *Barbatia helblingi* (p. 270), and the very similar *Barbatia obliquata*.

With the giant clams, *Tridacna*, the bivalves reach their zenith of size and adaptive attainment. Derived from the same stock as cockles (Cardiidae) the Tridacnidae are transformed in structure to shift the umbones ventrally, alongside the byssus and bring the toothed margins dorsally to accommodate the expanded and highly coloured inhalant siphon. These are the site of vast populations of zooxanthellae, or immobile unicellular algae. *Tridacna* shares with the hermatypic corals a close symbiotic dependence on these algae. In the corals the algae assimilate nutrients and dispose of wastes from the host's metabolism. They confer benefit on the coral by facilitating the deposition of calcium in the skeleton.

In the giant clams, the zooxanthellae—exposed to strong illumination in the outspread siphonal margins—carry out photosynthesis, products of which may be available to the host. Algae are also engulfed and intra-cellularly digested by wandering blood cells (amoebocytes) and then transferred either to the kidney or the digestive diverticula for disposal. Though still filtering planktonic food, like other bivalves, by the ctenidium, the tridacnids thus have a supplementary food supply that must have contributed to their high success and achievement of giant size.

Hong Kong comes within the natural range of the tridacnids, but only one species *Tridacna maxima* has been definitely recorded from the coral shores of Mirs Bay. Highly prized by shell-collectors it is almost certainly the first species to have suffered from the enlarged acquisitiveness now possible to SCUBA divers.

Prosobranchia

The Gastropoda of rocky sub-tropical shores have already been introduced, by their inter-tidal members. Their diversity greatly increases beyond low-water mark, especially in sheltered, relatively clean sites.

Common under boulders or coral heads are the Cypraeidae or cowries, some of the most highly prized shells for collectors, renowned for their fine patterns and smooth glazing. The brightly coloured, sometimes papillose folds of the mantle, extended over the shell may however, entirely obscure its character during life. Two species are illustrated in Fig. 12.13, *Cypraea erronea* and the small, dark-striped *Cypraea asellus,* with jet-black integument. The larger *Cypraea arabica,* is perhaps the commonest cowry in harbours and estuaries throughout Hong Kong (p. 99). Other species include *C. onyx* some 3 cm long with a lilac-brown shell, with a blackish margin and often a median darker stripe, *C. gracilis* is only 1–2 cm in length with a blue-green shell often tranversed by one or two dark stripes and with a number of brown spots ventrally.

Cowries graze on a variety of attached animal cover, but probably chiefly on hydroids. Their specialized relatives, in the family Ovulidae, have a narrower diet, particularly of coelenterates. The large white egg cowrie, *Ovula ovum,* lives always on the soft corals, *Sarcophyton*. The small, elongate, spindle-shaped *Volva* species, red, yellowish white or orange, occur plentifully, and feed on gorgonian sea-fans at several fathoms depth.

The conch-shells (family Strombidae) are herbivores, despite the often bizarre appearance of shell and animal. They belong like the cowries to the Mesogastropoda, where their ancestors were probably narrow-spired, somewhat like the

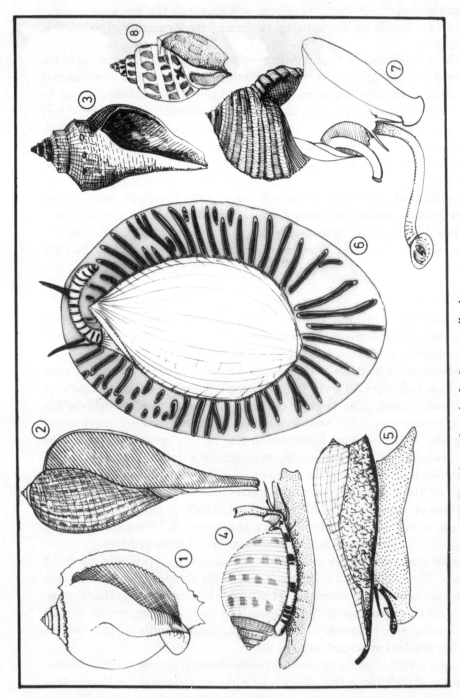

FIG. 12.13. Sub-littoral prosobranch of soft or gravelly shores. 1. *Phalium glaucum*; 2. *Ficus gracilis*; 3. *Hemifusus ternatana*; 4. *Semicassis persimilis*; 5. *Ficus gracilis* extended animal; 6. *Cymbium melo*; 7. *Tonna fasciata* with foot expanded and proboscis extended; 8. *Babylonia areolata*.

present Cerithiidae in form and habit. The lip of the aperture has become wing-shaped or variously expanded, forming a stabilizing device as the animal browses over the surface. The long-spired *Strombus vittatus* shows a simple example of a winged lip. In the spider shell, *Lambis lambis,* the lip is far more complex, with long channelled spines extending from the coral-pink aperture. In *Strombus luhuanus,* the shell has become greatly simplified, resembling a cone shell at first sight.

All the strombids graze on soft deposits or crop fine algae, by the small strong radula, at the tip of the thick proboscis. The locomotion is highly specialized. The sole is reduced, and the foot is compressed and strongly muscular. The mobile anterior edge can feel forward, but the posterior end is armed with a strong operculum, scimitar-shaped and smooth in *Lambis,* serrated in *Strombus luhuanus* and *Strombus vittatus.* The opercular tip is thrust into the ground to give purchase, and the shell is heaved forward about half its length by the strong contraction of the foot. It can also be righted from the overturned position. Unlike other gastropods, strombids have greatly enlarged eyes mounted on stout tentacles; these peep out comically from under the shell lip through special squints, one formed as a posterior notch, the other by the inhalant canal.

Where the offshore bottom is of sand or mud, down to 4 or 5 fathoms depth, other distinctive carnivorous gastropods are found. Of the sand-burrowing Mesogastropoda, the largest are the helmets and bonnets (Cassididae) *(Semicassis persimilis* and *Phalium glaucum)* and the tuns (Doliidae) (typified by *Tonna fasciata).* These prey upon echinoderms or bivalves, either smothering them with the foot, or plunging the proboscis into the soft parts, as tun-shells do with large holothurians. The fig-shells *(Ficus gracilis)* are specialized derivatives of the tun family. The shell is drawn out into a long spout-like anterior canal; the operculum is lost and the foot is too large to withdraw fully in the shell, as in the tuns.

The related trumpets (Cymatiidae) and frog-shells (Bursidae), live on rocky shores, preying chiefly upon urchins or ascidians.

Of the stenoglossan carnivores, a wide assemblage broadly classified with the whelks, are of normal size in relation to the shell, such as *Hemifusus ternatana,* and *Babylonia areolata,* the ovoid, dark-blotched whelk, common in the markets as a food species, and abundant upon offshore bivalve beds.

Among the Volutacea, however, the foot becomes expanded for sand-burrowing and prey-capture, and enlarges beyond the shell, as with the figs and tuns. These snails include the Volutidae, with a wide foot and smoothly elongate shell; the Olividae, highly modified sand-burrowers (p. 161); and the harp-shells (Harpidae). Largest of the Volutacea, and indeed of all the gastropods, are the melon or baler shells (Meloidae). The south-east Asian species, *Melo melo,* is smooth and inflated, up to 30 cm long, orange-brown in colour and wide-mouthed. The foot is extremely broad, and chocolate-brown, with radial yellow stripes.

Though abundant in sand beaches in the full tropics, both the olives (Olividae, p. 161) and the auger-shells (Terebridae) are sparse in Hong Kong. The same is true of the Mitridae, though several species are to be found offshore. Mitres are spindle-shaped and generally smooth, with a rather small foot and a long proboscis able to be inserted into narrow apertures and borings in search of worms and bivalves. The largest mitrids are sand-burrowers, though many smaller species live upon rocky ground. Local species include *Turricula nelliae spurius* and *Lophiotoma leucotropis*.

The family Conidae have a uniquely specialized carnivorous habit. They belong to the group of Toxoglossa, or 'arrow-tongues', having the radular teeth formed into minute harpoons, channelled within a groove conveying a neurotoxin from the salivary glands. Cones stalk and capture many sorts of moving prey by the sudden extrusion of a harpoon.

Each species is food-specific. The narrow-mouthed species, such as *Conus sulcatus* (Fig. 12.14) generally take polychaete worms (nereids, euniciids and terebellids). Others feed on bulloid opisthobranchs and various shelled gastropods, even including other cones. The largest and most toxic cones, with the widest mouths (including *Conus textile,* sometimes lethal even to man) catch live blennies and gobies. Prey is located by testing the incoming water current with the sensitive chemoreceptor, the osphradium. The cone then crawls towards it and covers it with the poised proboscis. During a successful strike, the prey is harpooned by the ejection of a single tooth, and is anaesthetized by injection of saliva. Struggles stilled, it is massively engulfed by the distended proboscis and digested piecemeal. R. Luther and Nee Chung have prepared a small, useful illustrated booklet on the cones and cowries of Hong Kong.[6]

Coelenterate Predators

Those prosobranch snails that prey upon corals, soft corals and other coelenterates can be mentioned separately. The family Magilidae forms a coral-feeding offshoot of the thaids. The shell form becomes rounded and smooth and often thin. The radula is lost, and feeding is suctorial, evidently with the extrusion of salivary digestive enzymes and the use of the muscular pharynx to ingest semifluid food. Two members of the family are illustrated both becoming progressively enclosed by the tissues of the growing host-food. *Rapa rapa* is the larger, with a thin, strongly sculptured shell (Fig. 12.15). It lives with soft corals, maintaining contact with the water outside through a small opening in the host tissues. The smaller species, *Magilus striatus,* is imprisoned in stony corals of the family Faviidae (Fig. 12.8). The shell is white and translucent, about the size of a cherry,

[6] Luther, R. and Nee Chung. 1976. *Cones and Cowries of Hong Kong.* Hong Kong, Sing Cheong Printing Co. Ltd.

The Coral Sub-littoral

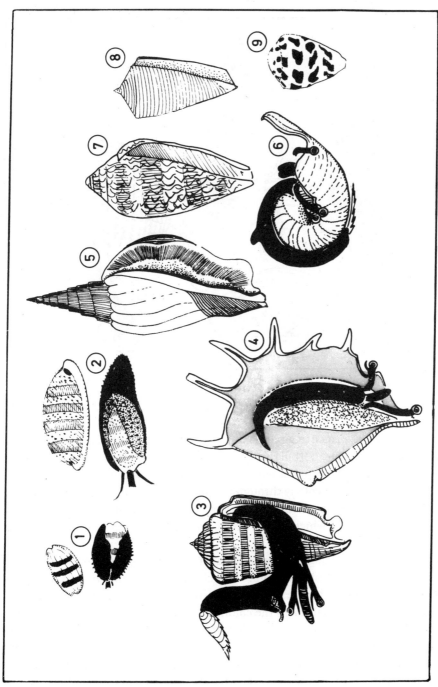

FIG. 12.14. Gastropods of hard sub-tidal shores.
1. *Cyprea asellus*; 2. *Cyprea erronea*; 3. *Strombus luhuanus*; 4. *Lambis lambis*; 5. *Strombus vittatus*; 6. The same species showing the righting action of the foot; 7. *Conus textile*; 8. *Conus sulcatus*; 9. *Conus ebraeus*.

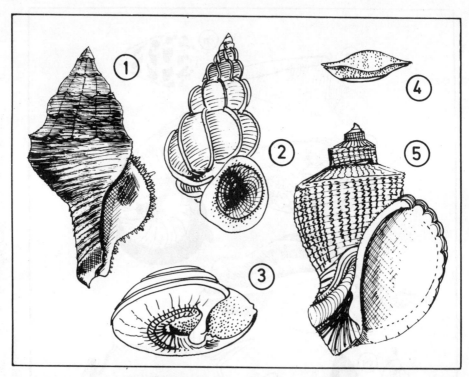

FIG. 12.15. Gastropod predators of coelenterates.
1. *Coralliophila costularis*; 2. *Epitonium scalare*; 3. *Architectonica perspectiva*; 4. *Volva brevirostris rosea*; 5. *Rapa rapa*.

and lies in a flask-shaped cavity opening narrowly to the exterior. A thread-like proboscis can be extruded from this aperture; it may forage from the mucous sheet of collected food on the coral surface, or possibly could enter the communal coelenteric space of the colony through the mouth of one of the adjacent polyps.

A third, much less specialized and more common magilid is the free-living *Coraliophila costularis* which grazes over coral heads leaving in its trail a broad swathe of white coral skeleton.

The Clathridae and the Architectonicidae are chiefly tropical families, each with its special elegance of shell design. The clathrids, or wentletraps *(Epitonium scalare)*, have rounded whorls supported by strong axial ribs; they feed by thrusting the proboscis through the column of an anemone and sucking the body fluids. The circular and flat-based Architectonicidae or sundial shells live by grazing the tissues, or penetrating the coelenteron of corals, soft corals and zoanthids. The common Hong Kong species is *Architectonica perspectiva*.

A few species of muricids (Fig. 7.8) also attack coral including *Cronia margariticola* that more typically feeds on bivalves and *Drupella rugosa* which seems to exclusively feed on corals.

Opisthobranchia

The sea-slugs and their relatives increase markedly in diversity and beauty of colour towards the tropics. One group, the aplysioids or sea-hares, already observed on rocky shores and on soft flats (p. 50 and p. 193) is untypical in its herbivorous browsing habit and cryptic camouflage. The majority of opisthobranchs, including the shell-less true 'nudibranchs' are vividly coloured, warning predators of their distasteful properties; in diet they may best be called 'grazing carnivores', often with a specific reliance on a particular sessile animal: sponge, polyzoan or ascidian.

John Orr[7] has produced a checklist of Hong Kong nudibranchs. The first series, still not technically 'nudibranch', are the Pleurobranchomorpha, or 'side-gilled slugs'. They retain a plume-like gill on the right side, beneath the wide-spreading skirt of the 'notum'. Beneath the dorsal surface is embedded a thin, vestigial shell-plate. A few species, such as *Umbraculum japonicum,* carry a large 'chinaman's hat' shell over the whole dorsum. The pleurobranchoid slugs most commony feed on ascidians, boring with the radula into the tests of the simple sea-squirts or grazing over the expanses of zooids of the compound forms. A familiar Hong Kong species is the yellow *Berthellina delicata* (Fig. 12.16). Like most of the local opisthobranchs it is widespread throughout the Indo-Pacific.

The Doridacea are the largest suborder of the true nudibranch slugs and are represented in Hong Kong by 11 families. The family Drorididae tend to be broad and flat, yellowish-brown to reddish in colour, resembling the sponges on which they graze: Hong Kong examples are *Doris verrucosa* and *Homoiodoris japonica*. The Chromodorididae are the best represented in Hong Kong with three genera and 11 species. These are more slender and brightly striped or spotted slugs also feeding on sponges. They are typified by *Chromodoris lineolata* and *Chromodoris festiva,* both illustrated in Fig. 12.16. Like all dorids, these carry two blunt sensory tentacles, the rhinophores, set near the front, with the anus in the posterior mid-line, surrounded by the branched gill-plumes. The cream *Casella atromarginata* has a prominent border of black.

The Asteronotidae are flat and slug-like with retractible rhinophores and plumes. *Halgerda japonica* is illustrated. The only recorded representative of the Polyceridae is *Polycera fujitai* commonly found feeding on the brown bryozoan *Bugula neritina*.

---

[7] Orr, J. 1980. Annotated checklist of Hong Kong nudibranchs. In *The Malacofauna of Hong Kong and Southern China,* ed. B. S. Morton, pp. 109–17. Hong Kong, Hong Kong University Press.

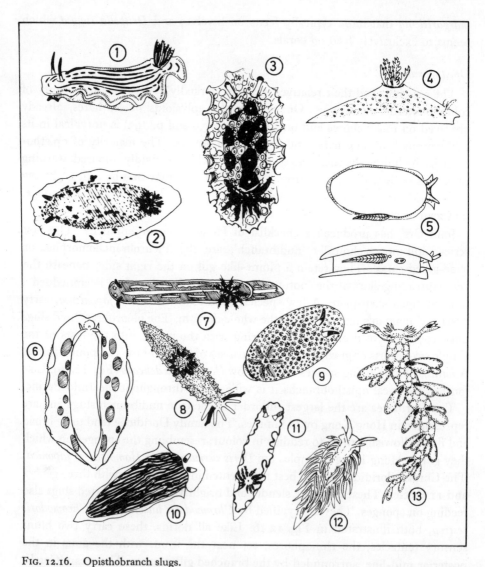

FIG. 12.16. Opisthobranch slugs.
1. *Chromodoris lineata*; 2. *Dendrodoris rubra*; 3. *Dendrodoris denisoni*; 4. *Gymnodoris citrina*; 5. *Berthelina delicata*; 6. *Platydoris speciosa*; 7. *Chromodoris festiva*; 8. *Polycera fujitai*; 9. *Halgerda japonica*; 10. *Armina japonica*; 11. *Casella atromarginata*; 12. *Aeolidiella takanosinensis*; 13. *Bornella digitata*.

The Gymnodoridae have adopted a suctorial diet with reduction or loss of the radula. The yellow or orange *Gymnodoris citrina* has a small radula, and grazes upon polyzoans. In the family Dendrodoridae the radula has disappeared and an eversible proboscis is plunged into the tissues of ascidians or other sessile prey. Common species in Hong Kong are *Dendrodoris rubra*, orange-red, deepest

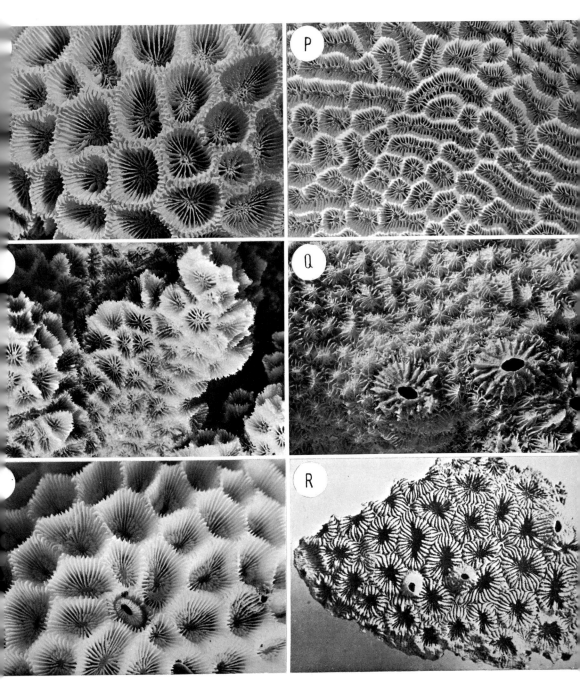

PLATE 25. Some Hong Kong corals: M. *Favia pallida* (Faviidae); N. *Favites pentagona* (Faviidae); O. *Goniastrea aspera* (Faviidae); P. *Platygyra sinensis* (Faviidae); Q. *Hydnophora exesa* (Faviidae); R. *Oulastrea crispata* (Faviidae).

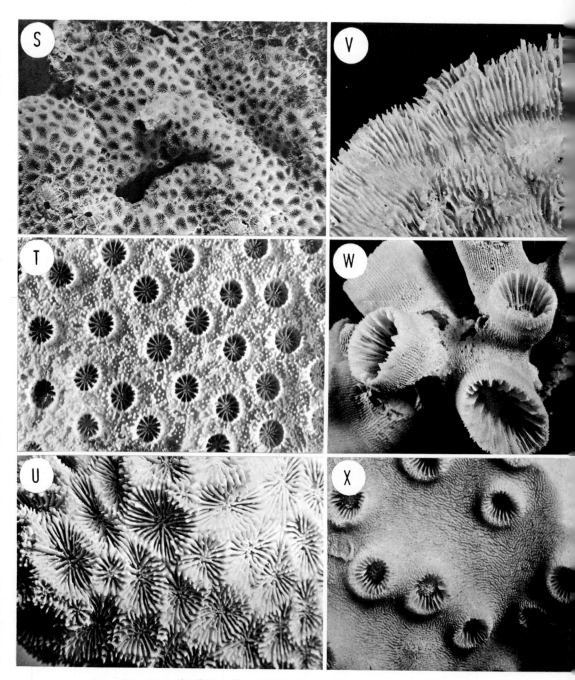

PLATE 26. Some Hong Kong corals: S. *Leptastrea purpurea* (Faviidae); T. *Cyphastrea serailia* (Faviidae); U. *Acanthastrea bowerbanki* (Mussidae); V. *Echinophyllia aspera* (Pectiniidae); W. *Tubastrea diaphana* (Dendrophylliidae); X. *Turbinaria peltata* (Dendrophylliidae).

at the centre and dark-blotched; and the handsome *Dendrodoris denisoni*, with the notum decorated with soft grey-brown folds and mamillae, and smooth areas of deep resplendent blue. The red and golden *Doriopsilla areolata* has been found in the rock pools of Ping Chau.

The remaining families of the Doridacea found in Hong Kong, each with but one or two species, are the Discodoridae *(Discodoris fragilis)*, Rostangidae *(Rostanga arbutus)*, Goniodoridae *(Goniodoris glabrata)*, Archidoridae *(Trippa interciala)* and Phyllididae *(Phyllidia varicosa)*. Three further suborders of Nudibranchs are found—the Aeolidacea, Dendronotacea and Arminacea. The Aeolidacea are represented by four families but have not so far been much collected in the Hong Kong region. They feed on coelenterates. The bright green *Phestilla sibogae* feeds on faviid corals and the bright orange *P. melanobrachia* on dendrophyllid corals, but little is known of the habits of the other recorded species. *Aeolidiella takanosimensis* (Fig. 12.16) is typical of the suborder. *Sakuracolis enoshimensis* is bright red with darker cerata tipped with white. Aeoliid slugs are long and slender, at once identified by the dense clusters of tapered tentacles or 'cerata' into which the notum is produced. These are entered by outgrowths of the digestive gland, and they store at their tips the undischarged nematocysts or stinging cells of their prey athecate hydroids or for some, species of anemones.

The suborder Dendronotacea is solely represented by the single species *Bornella digitata* with a long tapering tan body and widely spaced cerata.

The final suborder Arminacea is similarly only represented by one species—*Armina japonica* with a wide flattened body and a broad head adapted for shallow burrowing. The colour pattern is of narrow longitudinal black and yellow stripes.

The serious study of sub-tidal opisthobranchs and their habits has not yet begun in Hong Kong. Many of the local species are already known from their occurrence in Japan and other parts of the Indo-Pacific; but there is still much of interest to be revealed about their diet and feeding methods, and distribution and habits.

A specialized group of opisthobranchs that should not be overlooked is the order Sacoglossa—these small dark-green or black slugs live, highly specifically, on various sorts of green algae. They lance the cells one by one with highly adapted radular teeth, sucking the cell sap and symbiotically utilising the intact chloroplasts. *Caulerpa* species are host plants among the green algae likely to repay searching; one of their common sacoglossans is frequently a species of *Berthellinia*, primitively a shelled member of the order, uniquely modified into a bivalved gastropod!

Fishes of the Coral Sub-littoral

The fish and their biology lie beyond the scope of this book but a few are mentioned here to illustrate their diversity and importance (Fig. 12.17 and 12.18). A good recent article by W. L. Chan[8], notes that the shallow waters of Hong Kong's eastern sector are widely used as a nursery ground: by scad (Carangidae) with large juvenile stocks in many of the bays in December and January, often

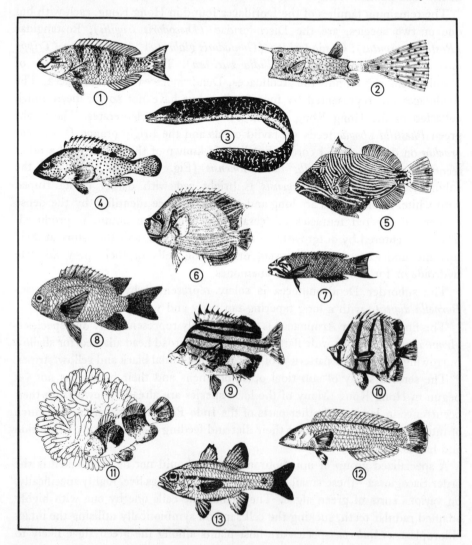

FIG. 12.17. Coral reef fishes from Hong Kong.
1. *Scarus ghobban*; 2. *Lactoria cornutus*; 3. *Gymnothorax undulatus*; 4. *Epinephelus fario*; 5. *Balistapus undulatus*; 6. *Pomacanthus imperator*; 7. *Thalassoma lunare*; 8. *Chromis notatus*; 9. *Microcanthus strigatus*; 10. *Chaetodon modestus*; 11. *Amphiprion percula* living with anemone *Stoichactis kenti*; 12. *Halichoeres tenuispinis*; 13. *Apogon doederleini*.

[8] Chan, W. L. 1976. The fish fauna of Hong Kong. In *The Fauna of Hong Kong*, pp. 1–141. Hong Kong, Hong Kong branch of the Royal Asiatic Society.

The Coral Sub-littoral

swimming with jellyfish, e.g. *Cyanea nozakii*; golden sardine and green pilchard (Clupeidae); chicken grunt (Pomadasyidae) and many others. In Tolo Harbour and the deep inlet of Mirs Bay, with a change of hydrographical regime during

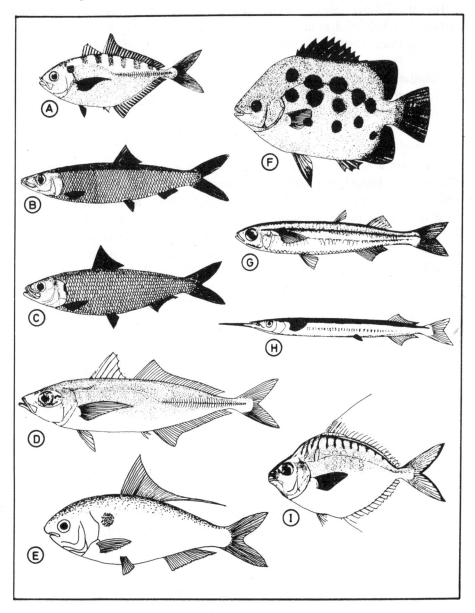

FIG. 12.18. Some inshore pelagic fishes from Hong Kong:
A. *Caranx kalla*; B. *Dussumieria hasseltii*; C. *Sardinella aurita*; D. *Decapterus maruadsi*; E. *Nematolosa nasus*; F. *Scatophagus argus*; G. *Atherina forskali*; H. *Hemiramphus sajori*; I. *Caranx malabaricus*.

monsoons, the fish fauna has a different character. The most abundant forms are Dorosomatidae (gizzard shads), Mugilidae (grey mullets), Leiognathidae (lip-mouths) and Chanidae (milk-fish).

Into these waters come seasonal and spawning migrations of pelagic fish: Atherinidae (silversides) along the west coast of Mirs Bay in May and June, and Scombridae (little tuna) that feed at this time on silversides. Also in summer come the feeding migrations of the large predatory fishes: Carangidae; Scombromorphidae (horse mackerels); Sphyraenidae (barracudas); Istiophoridae (sail-fish); Coryphaenidae (dolphin-fish); Carcharinidae and Sphrynidae (sharks) and Rhincodontidae (whale-sharks).

Coral Reef Fishes

There are a rich and diverse salient of a more tropical fauna, entering the waters of east Hong Kong, with their shallow coral patches, from the reef habitats (Pratas Reef, Paracel Islands, and Macclesfield Bank) to the south. Many of their species are highly territorial, and weak swimmers, intimately specialized for their benthic habit. Their different forms can only be touched upon here; but as interest in diving grows, a firm conservation ethic must be developed towards these beautiful and vulnerable components of the benthic community.

The important families of coral reef fishes, with representative species from Hong Kong, include:

SIGANIDAE (rabbit-fish)   browsing in schools upon algal fronds.

ACANTHURIDAE (surgeon-fish)   browsing and grazing algae.

MULLIDAE (goat-fish, red mullet)   bottom-dwelling carnivores with barbels probing the sand, to search for small worms, echinoderms and crustaceans.

HOLOCENTRIDAE (soldier-fish and squirrel-fish)   cryptic and sheltering in holes and caverns under rubble or coral; foraging at night for gastropods, bivalves, crabs and polychaetes.

SCARIDAE (parrot-fish)   scraping algae and coral polyps from rocks and massive coral heads. (Fig. 12.17. *Scarus ghobban*).

LABRIDAE (wrass)   as typical of coral shores as the corals themselves; abundant and brilliant-coloured, large and small. Feeding on invertebrates (bivalves, gastropods, worms, crabs, hermits, shrimps, urchins and ophiuroids) from sandy and rubble bottoms, crushing these with 'throat teeth' before passage to the stomach. (Fig. 12.17. *Thalassoma lunare; Halichoeres tenuispinis*).

POMACENTRIDAE (damsel fish)   variously feeding, some on algae, some omnivorous, some carnivorous, both from bottom and mid-water.

CHAETODONTIDAE (butterfly fish and angel fish)   with tubular snouts thrusting into interstices and crevices; feeding on worms and polyps, or grazing and browsing algae. Brilliantly coloured, singly or in pairs, wherever

The Coral Sub-littoral 295

coral is growing. (Fig. 12.17. *Pomacanthus imperator*; *Microcanthus strigatus*; *Chromis notatus*; *Chaetodon modestus*).

APOGONIDAE (cardinal fish) small, solitary or schooling fish, in and over patches of branching coral. Voracious carnivores on small fish, crustacea, and other invertebrates. (Fig. 12.17. *Apogon doederleini*).

BLENNIIDAE (blennies) small, active fish living close to, or in refuges in the bottom. Finding food in fine algae and by scraping the bottom for minute animals (Foraminifera, copepods and ostracods).

PEMPHERIDAE (sweep-tails) abundant in surge or under ledges, often schooling; voracious carnivores, taking mainly shrimp-like or crab-like Crustacea.

LUTJANIDAE (reef-snapper) schooling in the deeper waters of coral shores; heterogeneous carnivores.

SERRANIDAE (grouper) voracious carnivores with both large and small species, the latter demersal, hiding in crevices and holes, and under ledges, and making rapid lunges for prey. (Fig. 12.17. *Epinephelus fario*; *Amphiprion percula*).

MURAENIDAE (moray eel) carnivores lurking in crevices to take small fish prey. (Fig. 12.17. *Gymnothorax undulatus*).

SCORPAENIDAE (scorpion fish and their relatives) sluggish, voracious carnivores, often strikingly coloured and bizarrely finned, lurking to take small fish.

BALISTIDAE (trigger-fish) variously omnivorous, making good use of coral as food, and taking molluscs and even spiny urchins, such as *Diadema*. (Fig. 12.17. *Balistapus undulatus*).

TETRAODONTIDAE (puffer and balloon fish) opportunists missing nothing living, but very fond of coral tips and the small fishes living around them.

OSTRACIONIDAE (trunk-fish and box-fish) weak swimmers, cruising round coral heads; small-mouthed, feeding omnivorously from sandy bottoms near coral, especially on polychaetes. (Fig. 12.17. *Lactoria cornutus*).

DIODONTIDAE (spiny puffers) strict carnivores with strong jaws and heavy crushing teeth; feeding on gastropods, Crustacea (especially xanthid crabs) and calcarine Foraminifera.

The Terraces and Pools of Ping Chau

The island of Ping Chau is one of the few places in Hong Kong territories with inter-tidal coral. Its shores, though unusually rich, conform in the essential character with other hard shore zonation. Their pattern is expressed more amply, by the horizontal unrolling pattern of a theme that has elsewhere been vertically compressed on a steep shore.

As elsewhere the geomorphology strongly influences the biology. Metamorphosed sandstones, of the Tolo Formation, form a series of terraces inclined backwards from the sea, with a short steep 'strike' face and a long 'dip'. The perspective view in Fig. 12.18 brings out several structural features. Extensive horizontal and vertical space continues for long distances along the outer (or west-facing) coast of the island. An important feature of these inter-tidal benches is the retention of water, level by level, in shallow pools, seldom seen elsewhere in Hong Kong. Each level has a different ecology. The pools of the bottom terrace are, even at low tide, constantly open to surge. Pools at middle height are isolated and sun-warmed between tides. These pools are the chief places where inter-tidal corals are to be found. The high level pools experience strong fluctuations, between tides, of temperature, pH, salinity and oxygen content; such pools can become physiologically a far more 'difficult' environment than the open rock surface, and animals and plants are correspondingly sparse.

The ecological layout of the Ping Chau terraces is shown in Fig. 12.18. The shore must be classified as moderately exposed (p. 56) with a good length of fetch, generating waves up to several metres high. But with a west-facing outer shore, and the easterly aspect closely protected by the China mainland, Ping Chau does not receive the full effect of the south-easterlies. In exposure/shelter terms (Fig. 4.2) Ping Chau is thus a shore with a strong *Tetraclita squamosa* zone, but with few and only occasional *Balanus tintinnabulum*.

The sequence of terraces was sketched in late February, when the algae had achieved their best growth. Looking down on the shore from behind, thus seeing the long dip-slopes, the golden brown shades of *Sargassum* stand prominently at the bottom. On the next two or three steps higher, on the dip face above pool level, is a belt of *Ulva* and *Enteromorpha intestinalis,* bleaching pale in February by exposure to sun and air.

The zonation is described terrace by terrace, working up the shore from low-water mark.

Terrace 1

The pools are surge-swept at low water, and the rock is pink with a *Lithophyllum* crust. By early March, the pools and open rock are clad with metre-long tresses of *Sargassum hemiphyllum,* with wide-spaced leaves, and mature reproductive branches. Twining about the *Sargassum,* its yellow thongs contrasting with the dark brown is the brown alga *Hypnea musciformis*. It branches repeatedly into clinging tendrils, while its larger branches have small spinules. Conspicuous also among the *Sargassum* are pale brown tufts of a filamentary brown alga, *Ectocarpus*. The bright green alga *Caulerpa peltata* is also common at *Sargassum* level. Of rather stiff, cartilaginous texture, it puts up small, bluntly sub-divided segments from horizontal stolons.

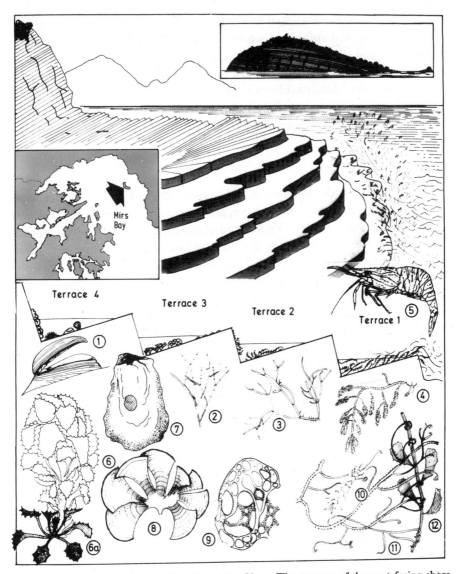

FIG. 12.19. Inter-tidal terraces and pools at Ping Chau. The expanse of the west-facing shore is illustrated, with the arrangements of benches, their pools and algal cover. *(Inset)* South-western shore of Ping Chau from a distance, showing strata of metamorphosed Tolo sandstones. *(Middle)* Sequence of four terraces with pools and emersed rock habitats.

PLANTS AND ANIMALS:
1. *Chama dunkeri*; 2. *Amphiroa valaniodes*; 3. *Gloiopeltis furcata*; 4. *Laurencia japonica*; 5. *Palaemon pacificus*; 6. *Turbinaria ornata* in winter with 6a. reduced summer fronds: 7. *Hyotissa hyotis*; 8. *Padina durvillei*; 9. *Hydroclathrus clathratus*; 10. *Caprella aequilibra*; 11. *Hypnea musciformis* twining on; 12. *Sargassum hemiphyllum*.

Terrace 2

These pools are also breached by surge, and contain *Sargassum*, including the blackened stems and stubble of a former year's growth, and a yellow tangle of *Hypnea*. Beyond the pools the dip slope has small, stiff dichotomous tufts of a dichotomous *Chaetangium* species. On the steeper strike face *Gloiopeltis furcata* is conspicuous and dominant from February onwards. Its dichotomous branches are mucilaginous, first dark reddish-brown and solid, then tubular and light yellow. Above *Gloiopeltis* on the strike face grows the barnacle *Tetraclita squamosa*.

Terrace 3

On the third step, the pools warm up between tides and in summer can become tepid. Sediment collects on the bottom, and turbidity is higher than in the better circulated pools lower down. To seaward of the pools, the dip face has a band of *Gloiopeltis* and the rest of the rock is bare, down to the waterline of the pools, where there is bleached *Ulva* and *Enteromorpha*. Limpets abound on the bare rock. The low-pitched *Cellana toreuma* shows a full range of pattern variants in black rays and gold. The rough coast species, *Cellana grata*, is also moderately common. Three species of pulmonate limpets are abundant: strongly ribbed *Siphonaria japonica*, black-and-white ribbed *S. sirius* and the large, very depressed *S. atra*. Yellow spawn crescents abound in the damper places.

The brown algae of these pools are dense and varied, of quiet water species rather than surge types. Golden brown vesicles of *Colpomenia sinuosa* are abundant everywhere. With them is a tropical relative, *Hydroclathrus clathratus*, essentially like a *Colpomenia* perforated with large holes to form a fleshy network. The brown alga *Padina*, is also common, composed of fan-like branches, fraying into narrower lobes at maturity. Yellow tendrils of *Hypnea musciformis* still appear at this level. A tropical Sargassum-relative, *Turbinaria ornata*, is composed of a ball of modified leaves, flat-topped and with spinulose edges. The reduced summer fronds are fewer and more sparse. The red alga, *Amphiroa valonioides* forms tufts of jointed, dichotomous branches, stiffly calcified. At the bottom of the pools are attached large bivalves, *Chama dunkeri*, mingled with thick-spined and jet-black urchins, *Anthocidaris crassispina*.

Above the water-line, the bare strike face carries crowded ranks of the stalked barnacle *Pollicipes mitella*.

Terrace 4

The terraces of the fourth step are blotched and marbled with black *Hildenbrandtia* and grazed by *Nerita albicilla*. Common also in moist depressions are the nodulose shells of the cerithioid *Ochetoclava sinensis*, and the potamidid *Batillaria sordida*, either living or occupied by hermit crabs.

The Coral Sub-littoral

In the tepid and silty pools, attached animals replace the algae. *Chama dunkeri* continues abundant, along with small pearl oysters, *Pinctada margaritifera*, and large flat coils of the vermetid gastropod *Serpulorbis imbricatus*. Similarly cemented to the flat, sloping bottom of the lower shore pools is a large oyster *Hyotissa hyotis* we have not encountered previously. Belonging to the Gryphaeidae, *Hyotissa* is characteristic of warm tropical waters. There are also snake's locks anemones, *Anemonia sulcata*, and clusters of *Anthocidaris*, some of them in grooves along the open rock. In the deeper and more shaded pools the longer spined *Diadema setosum* is the common urchin. The scarps of these pools possess large numbers of the beautiful prawn *Palaemon pacificus* with a steeply upturned rostrum, clinging to the vertical corals and wandering in search of morsels of detritus.

The stony corals grow in these pools, forming brown or mauve plaques over the rock surfaces. Foremost among these is *Porites lobata*, a species hardy enough throughout the tropics to survive in mid-tidal pools, and to resist moderate turbidity. Clay- or fawn-coloured, it forms light porous crusts and nodules; the small calices have few prominent distinguishing features. Faviid corals are also important in pools, especially the meandrine brain coral *Platygyra sinensis* and nodules of the small-cupped *Cyphastrea serailia*.

The following is a list of scleractinian corals so far recorded from the Ping Chau pools:

Fungiidae	Faviidae
Porites lobata	*Platygyra sinensis*
	Leptastrea pruinosa
	Favites abdita
	Favites pentagona
	Favites flexuosa
	Favia speciosa
	Favia pallida
	Goniastrea aspera
	Oulastrea crispata
	Cyphastrea serailia

Clearly the encrusting Faviidae are here dominant.

The strike face above the pools carries a zone of the rock oyster *Saccostrea cucullata*, sufficiently far back from the surge line to find suitable shelter. This locality must be one of the few places where a rock oyster and a scleractinian live only a few centimetres apart.

By wrenching the flat-based *Anthocidaris crassispina* from the rock, two species of crustacean will often be revealed, sheltering beneath the circum-oral area, where the test is covered only with small spines. The small 'half-crab', *Porcellana ornata*, is greyish-brown, with the legs pink-speckled and the carapace back lightly mottled. The small shrimp, *Athanas dorsalis* has a more intimate, commensal habit. Short-legged and slow in movement, it is usually found in sex pairs, male and slightly larger female together. The body is deep red or black, with a ruby streak down the back. The antennae are short and blunt, the carapace broad and the tail-fan reduced. So closely do they resemble the species to which they cling that they are virtually invisible. *Athanas* poaches the food from the mouth of the echinoid. Like so many other tropical commensals (p. 200), *Athanas* poses for future investigators problems hardly yet touched.

This unique habitat of Ping Chau, protected by its remoteness from urban Hong Kong, both ends and yet starts this book. We began with the simplest of zonation studies of rocky shores and ended with the complex sub-tidal corals. Both come together at Ping Chau and which, now, because of the delicate balance this shore is held in, introduces us to the last chapter—our concern for its continued existence.

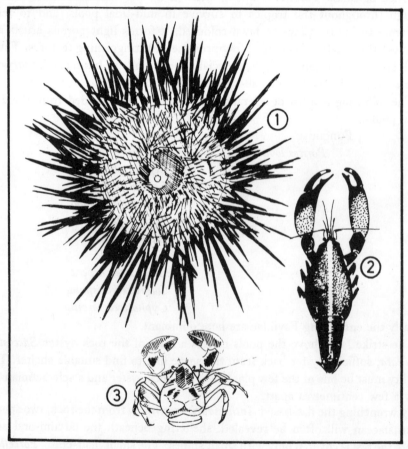

FIG. 12.20. 1. The sea urchin *Anthocidaris crassispina* with, 2. its commensal alpheid shrimp *Athanas dorsalis* (♀) and 3. the crevice-dwelling *Porcellana ornata*.

PART FIVE
The Sea Shore's Future

CHAPTER 13

Epilogue: Pollution and Conservation

The real wonder, perhaps, is that Hong Kong still has many healthy shores. The conservationist must ask how this has come about; and how long it may be maintained.

Round most of Hong Kong's 500 miles of coastline, a diversity of communities is still intact. True, much of the natural coast is already degraded, with a warning of what could rapidly spread. Victoria Harbour has become like a water closet, providentially open-ended and flushed through by the tides. Many smaller inlets are like uncleaned cesspits, with effluent and every sort of rubbish allowed to lodge. Where not already reclaimed, Shatin Harbour is fouled by human and agricultural effluents and other effluvia of industry. The inner reaches of Tolo Harbour are to all intents a dead sea.

But beyond these dead shores, there is a prospect of hope. In many places one can still reach past the plastic drift-line, that, next to effluent pollution, is the greatest blemish of Hong Kong's shores. There are still the splendid marshes of Mai Po; the coral-girded small islands of Mirs Bay; some fine mainland shores like Clearwater Bay, the unspoiled sand-beaches of Lantau and secluded parts of Lamma; and the surf-exposed promontories of these islands, and of Cape D'Aguilar.

Their existence is part of the environmental paradox that is Hong Kong, that has left so much of the once 'Fragrant Harbour' still beautiful. There are terrestrial open spaces, with their own mammals, a large avifauna, and a sub-tropical flora. It is on the sea shore, however, the interface of air, earth and water, that life is most concentrated and diverse, and also most fragile.

Held by many to be the cradle of invertebrate evolution, the sea shore abounds with species of nearly every class and phylum; for many of these it is the only habitat. And as with every interface, it is the critical shore boundary that shows most sensitively where ecological processes on land and in the sea have been adversely affected by man.

Like its society and economy, Hong Kong's shore ecology has survived precariously on a knife-edge. This has happened unplanned, as the benefit from what could be called, in retrospect 'conservation by concentration'. Hong Kong's urban population of over 5 million has been crowded into only a small portion of its 400 square miles. With an estimated half million people to one square mile, Kowloon has some of the highest known human densities.

This sort of conservation, saving the periphery by crowding at the centre, could only have been consciously applied by a ruthless planner's logic, and it has obviously cost Hong Kong dearly in social and economic equality. Today, however, as the population and its expectations still grow, this urban confinement is breaking apart.

Reclamation and Sub-division

New leisure and recreation space will be needed for masses of people; and Government settlement policy in the 1970s has been to develop satellite towns on previously rural areas.

In mountainous Hong Kong, there is so little flat land that quite literally mountains have been moved into the sea. The 'nine hills of Kowloon' have long disappeared; and now other hillsides, in Hong Kong and particularly around Shatin, in Tolo Harbour, are being reclaimed on a grand scale.

The Hong Kong Government's Annual Reports show a total land area for 1976 of 404 square miles as compared with 398 in 1970. This could seem a small fraction, but it must be remembered how concentrated in area is the whole girdle of the inter-tidal shore. And with the slowing down of reclamation within Victoria Harbour, other and less-spoiled sites are being increasingly chosen. Major reclamation projects today include:

  Kwun Tong: housing 548,000: area 913 acres
  Tsuen Wan: housing 850,000: area 2420 acres
  Tuen Mun: housing 547,000: area 2740 acres
  Sha Tin: housing 756,000: area 4450 acres
  Tai Po: housing 236,000: area 37 acres (to date)

To these must be added the reservoirs of Plover Cove (3,600 acres) and High Island (1,700 acres) and a whole spectrum of other reclamations throughout Hong Kong, as at Tai Po in Tolo Harbour, and Sai Kung in Rocky Harbour.

Reclamation spells the death of the natural shore: not merely its relocation by pushing out further to sea. It puts a featureless edge in place of the old diversity of mangroves, saltings, flats and estuaries.

Obviously reclamation will take its chief toll of shallow waters and harbour edges. Often it may be too late to halt what has gone on before, and visual degradation can only be improved by filling in. The affliction of 'water-front blight' can appear wherever more land is needed behind a shore-based activity:

Epilogue: Pollution and Conservation 305

sheds for containers and cargo-handling, oil installations, desalinators, power stations and refuse incinerators. Old water-fronts may fall into disuse, cut off from access with buildings, railways and highways. Thus the city withdraws from the shore, as first began in the 1850s when private-holders made illegal reclamations, first beyond Queen's Road and then outside the newly formed Des Voeux Road.

Such loss of natural shore was inevitable in a thrusting port with little flat land. In newly developing areas it would be ideal to allow no construction on the shore that is not essentially water-related; it is a poor, unplanned makeshift to push out into the sea all those activities, from roads and services, to car-parking and waste disposal, that have been most mis-managed on land.

The second threat to the coast from affluent residential sub-division is directed upon the choicest and most pleasant local shores. It is the threat that hangs over Lantau Island if its seclusion becomes 'opened up' by a road bridge from the New Territories. Worse still could be the private purchase of whole beaches or even small islands at inflated prices, followed by the exclusion of the public. This has already begun on Lantau Island at Sea Ranch. The riparian edge should be for the people's use; and encroachment of boat harbours, marinas and floating villages, should be vigilantly checked.

With development of choice landscapes, there could be opportunities to re-invoke the Chinese tradition of geomancy or 'fung shui', the art of adapting the residences of the living so as to 'co-operate with the local currents of the cosmic breath'. For the living, bad siting was held to cause trouble; good landscape planning to bring health, wealth and happiness. Such features as houses, walls and roads were shaped to the natural landscape so as not to dominate it. Shore design is but an extension of the whole landscape problem to a zone that is also narrow and fragile.

Pollution

If extinction by reclamation is the immediate threat to Hong Kong's shores, the other is the pollution or biological death of those that remain. Human and agricultural wastes and discharges from industry degrade or kill the ecosystem by interfering with its chemical reactions. But paradoxically the worst visual blemish comes from a chemically inert material, plastic. Pieces of toys, containers and plastic flowers abound in the drift-line. Plastic bags come in with every tide, hanging suspended in the wave zone of swimming beaches, or festooned upon mangrove branches at high-water mark.

Chemically active pollution arises from untreated human sewage, the effluents of poultry and piggeries and other farming in the New Territories, and industrial chemical wastes, often resistant to biological degradation. To all this from the land, is added the problem of oil spills and slicks from seaward, and the effects of the detergents used to clean them up.

TABLE 13.1 SIGNIFICANT OIL SPILLS IN HONG KONG

Vessel	Cause	Quantity Spilt
'Columbia Trader' (1968)	Grounding	190 tons of heavy maring diesel oil.
Ap Lei Chau Oil Spill (1973)	Land release	2000–3000 tons of heavy marine diesel oil.
'Eastergate' and 'Circea' (1973)	Collision	200 tons of aviation fuel.
'Korea Hope' (1974)	Grounding	160 tons of aviation fuel.
'Adrian Maersk' (1977)	Grounding	1,100 tons of crude bunkering oil.

The first result of chemical pollution is likely to be de-oxygenation. The oxygen concentration of the water can be lowered by chemical or bacterial oxidation, and also by the presence of sulphite in industrial wastes, and ferrous salts. Deprived of respiratory oxygen, the normal fauna and flora become replaced by large numbers of a few species successful under anaerobic conditions.

Suspended solids are an important constituent of untreated sewage and come also from the dumping of barged wastes and from soil erosion in water-sheds after removal of the natural vegetation. Even though chemically inert, they will cloud the water, causing plant death by reducing photosynthesis and clogging the respiratory and feeding systems of sessile animals, especially corals.

Chemical poisons interfere directly with the metabolism of organisms. To heavy metal poisons, such as lead, cadmium and mercury, new industrial processes are adding many 'un-natural molecules' that organisms have no enzymes to degrade, and which therefore biologically accumulate, as for example DDT insecticides. Their chief effect is to reduce community diversity, leading to the great proliferation of a few resistant species.

It is as easy to disrupt a community by promoting growth as by retarding it. Eutrophication consists of the addition of growth nutrients such as nitrates and phosphates, and sometimes organic carbon compounds and vitamins as well. A few species make rapid preferential growth, disrupting the balance and diversity and consequent stability of the ecosystem. Algae may produce rapid 'blooms' (as are now a permanent feature of Tolo Harbour), until other limiting factors supervene: perhaps the exhaustion of oxygen or the cessation of photosynthesis with reduction of illumination. Massive dying off follows, with the sudden production of anaerobic and often toxic conditions.

A further hazard from sewage pollution will be the high count of pathogenic organisms, generally monitored by counted faecal coliform bacteria though the greatest dangers are from routinely undetectable sewage-borne viruses, as of hepatitis and polyomyelitis. This has been demonstrated for the Deep Bay

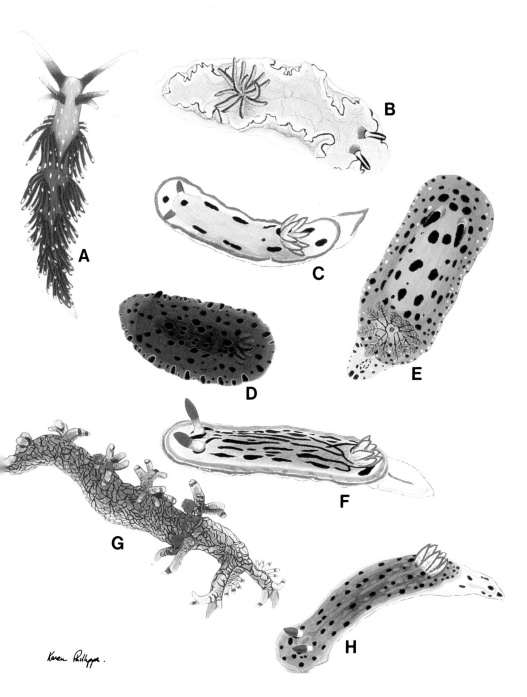

PLATE 27. Hong Kong nudibranch molluscs: A. *Sakuracolis enoshimensis*; B. *Casella atromarginata*; C. *Chromodoris pallescens*; D. *Dendrodoris rubra*; E. *Dendrodoris guttata*; F. *Chromodoris lineolata*; G. *Bornella digitata*; H. *Chromodoris festiva*.

PLATE 28. Some aspects of coastal pollution in Hong Kong: A. Large scale reclamation in Tai Po, Tolo Harbour; B. Luxury buildings on a hitherto public beach, Lantau Island; C. Fish farmers cut coastal mangroves for their ponds, Deep Bay; D. The 'Adrian Maersk' on the rocks of Lamma Island (1977); it released 1,100 tons of crude oil; E. A stinking river at Ho Chung, just prior to its opening onto a beach in the Sai Kung peninsula; F. Garbage floating in the sea.

oysters by Leung and her co-workers[1]. Long accustomed to such pollution, Hong Kong has learned to live with it, until nothing from the sea can be eaten without cooking, and some harbour waters are as dangerous to fall into as to collect food from.

Thermal pollution results from the production of waste heat in the cooling waters of conventional power-generating stations and desalinators. Rise of water temperature can increase the rate of biochemical reactions, with higher energy requirements and respiratory rate while at the same time the solubility of oxygen decreases. Many organisms with narrow thermal ranges may be killed, as their lethal temperatures are reached. Thermal pollution can also have indirect effects, aggravating the effects of poisons and accelerating de-oxygenation. Further, the chlorination of cooling water to prevent clogging growth in the pipes can release dangerous levels of free chlorine in the receiving seawater.

Oil pollution is a difficult hazard for Hong Kong to guard against, with its vital dependence on tanker ships for almost its entire available energy. Even the thinnest films of oil can impede the oxygen uptake of organisms beneath. A gallon of oil covers four acres of water. Oil may originate not only from offshore, but from run-off from used motor oil and malfunctioning marine engines. The ecological impacts of an oilspill upon soft and hard shores of Lamma Island have been estimated by Wormold[2] and Stirling[3].

Detergents used to dispel oil, or carried in sewage, bring their own ecological hazards. High concentrations kill organisms by attacking their cell membranes. At lower concentrations they may act like heat and some poisons in reducing capacity to survive low oxygen tensions.

Carrying Capacity

This is an important characteristic, defined as a habitat's ability to absorb something from outside and still retain its essence. In *physical* terms carrying capacity may relate to visitor-pressure on non-living elements: such a capacity will obviously be lower on a sand-dune than on rock. Limitation of visitor access could be important for fragile habitats so close to urbanized areas. The *ecological* carrying capacity concerns the effects of visitors on the coastal ecosystems. Plant and animal species will be altered if the carrying capacity is exceeded. Shell-fish beds can be polluted from human wastes, or blanketted with sediment from construc-

---

[1] Leung, C., Morton, B. S., Shortridge, K. F. and Wong P. S. 1975. The seasonal incidence of faecal bacteria in the tissues of the commercial oyster *Crassostrea gigas* Thunberg 1793 correlated with the hydrology of Deep Bay, Hong Kong. In *Proceedings of the Pacific Science Association Special Symposium on Marine Sciences, Hong Kong 1973*, ed. B. S. Morton, pp. 114-27.

[2] Wormold, A. P. 1976. Effects of a spell of marine diesel oil on the meiofauna of a sandy beach at Picnic Bay, Hong Kong. *Environmental Pollution* 11: 117-30.

[3] Stirling, H. P. 1977. Effects of a spell of marine diesel oil on the rocky shore fauna of Lamma Island, Hong Kong. *Environmental Pollution* 12: 93-117.

tion sites. Trees and other vegetation can be killed from soil compaction by heavy treading. Nesting and roosting birds can be disturbed until they move elsewhere. Ecological carrying capacity ought to be a constant concern of biologists, as the coast comes to be more intensively observed, especially by large school classes.

Carrying capacity can also be *psychological*: in the degree to which the presence of others affects the experience of being in a particular area. To go from Picnic Bay to the remote granite promontories of Lamma, or from Silvermine Bay to Shiu Hau on Lantau, is to move into a different world; however ironic the mention of 'psychological-carrying capacity' may seem, under the spatial pressures and privations of Hong Kong.

Individualist and business-centred Hong Kong, with the priorities it has borrowed and emphasized from the West, has moved far from the Taoist ideal, that man's first necessity was to understand the nature to which he belonged and that surrounded him: the Tao of Nature that brought everything into being and continues to govern it. In the Third World, and nowhere more than southern Asia, the low priority given to environmental health may be understandable. But to a population as swollen as Hong Kong's, the conservation of open space, even of wilderness, is not something that can be bartered for economic expediency.

Conservation indeed represents ultimate economic sanity, just as with Confucion moderation and good balance. Far more in Hong Kong than in the wild and uncluttered countries of the West, wilderness space could be a prerequisite of psychic health and sanity.

At the 1975 Pacific Science Congress in Vancouver, one of the authors gave a detailed review of the kinds and extent of pollution affecting the shores of Hong Kong[4]; this has been reiterated recently[5] and it is clear that unless pollution in all its forms is controlled now, much of what is described in this book will not survive for later generations.

TOWARDS COASTAL CONSERVATION

The story of Government action to date can be all too shortly related. In Hong Kong, pollution has been recognized as a cause for concern only since about 1970. A series of official Reports, with a growing documentation of aquatic pollution, has stimulated the formation of The Conservancy Association. Not unexpectedly, the first concern has been with aerial and terrestrial pollution,

[4] Morton B. S. 1976. The Hong Kong sea-shore—an environment in crisis. *Environmental Conservation* 3(4): 243–54.

[5] Morton, B. S. 1980. The pollution of Hong Kong territorial waters and coasts—a review. In *Proceedings of the Fourth Symposium for the Cooperative Study of the Kuroshio and Adjacent Regions*, pp. 557–87. Tokyo, Saikon Publishing Co. Ltd. and Hodgkiss, I. J. 1979. The pollution of Hong Kong coastal waters and sea-shore. In *The Future of the Hong Kong Seashore*, ed. B. S. Morton. Hong Kong, Oxford University Press.

Epilogue: Pollution and Conservation

where the evidence is obviously plain to the common citizen.

The 1965 Report of Talbot and Talbot[6] is a landmark in conservation for Hong Kong, though it has had too little Government response. Some mention is made of the coral reefs of the east and the marshes of Mai Po, but its main stress was still on the land. In a recent book, *The Future of the Hong Kong Sea Shore*[7], a strong plea is made for coastal conservation. There is clearly a continuing need for coastal surveys, monitoring the health and diversity of inter-tidal communities. It is hoped the present book will not only direct public feeling towards the coast-line, but provide a background from which future surveys can be made.

At present the Public Works Department of the Hong Kong Government monitors the water quality of the harbours and swimming beaches; but despite growing realization of the local environment crisis, neither Government nor either of the Universities have paid more than lip service to diagnosis and care of environment, with all this entails for the people living in it.

Some planning steps would seem so urgent that their discussion need not wait upon further surveys. In his recent conservation blueprint[7], B. S. Morton has advanced a comprehensive plan, dividing Hong Kong into five coastal areas, worthy of protection.

At the heart of each area, a *Nature Conservation Area* is proposed, being a region of special scientific interest. In these areas, new building would not be prohibited—space in Hong Kong is too valuable—but would be slowed down to retain a simple rural style. Access would be by well-defined routes and patrolled by trained conservation wardens. Once defined, nature conservation areas would be protected, even at high cost, from further interference and pollution.

Surrounding the Nature Conservation area would be a second type of zone, *Nature Education Areas,* to serve as buffers to the scientifically important heart of each system. In each area, a nature conservation centre could be established, to cater for students, teachers and the interested public. These areas should not be isolated from the mainstream of community life, but form an integral part of environment space, with the public aware of them and the facilities they offer.

Surrounding the educational areas a series of wider *Recreational Areas* should be designated, with the provision of open-air facilities for the Hong Kong people. It will be clear from the map that these contain most of Hong Kong's existing and proposed bathing beaches. In areas set aside for recreation, conservationists would have to accept that the natural shore community is in some degree expend-

---

[6] Talbot, L. M. and Talbot, M. H. 1965. *Conservation of the Hong Kong Countryside.* Hong Kong, Government Printer.

[7] Morton B. S. 1979. Future plans for the Hong Kong sea shore: conservation and protection. In *The Future of the Hong Kong Seashore,* ed. B. S. Morton. Hong Kong, Oxford University Press.

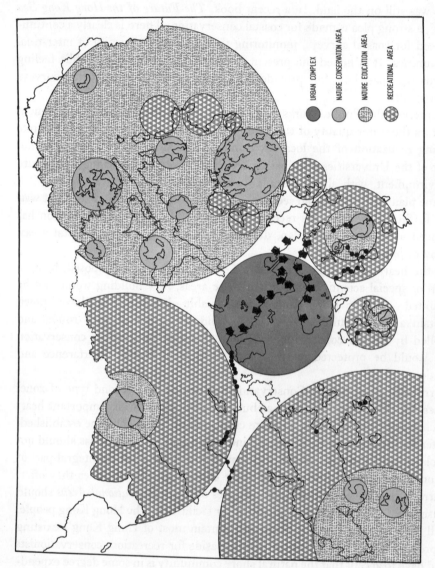

FIG. 13.1. Hong Kong with the urban complex and sewage outfalls arrowed. Conservation, education and recreation area of sea shore are circled, with gazetted recreational beaches shown as small black circles.

Epilogue: Pollution and Conservation

able. But at present, the recreational areas close to urban Hong Kong suffer grossly from pollutants, both chemical and bacterial. To establish safe limits for bathing in Hong Kong is a public health exercise that has never been adequately undertaken.

Fig. 13.1 shows the urban fringed central area that could be said to have permanently lost its amenity value, surrounded by the five proposed protected areas. These are:

1 Deep Bay and the Mai Po area with its marshes.
2 Lantau Island, with its fine sand beaches at Shiu Hau and Pui O.
3 Lamma Island, with its sand beach at Sham Wan, flanked by open promontories.
4 The Cape D'Aguilar Peninsula and Tai Tam Bay, with exposed rocky shores, and soft flats nearest to Victoria.
5 The Sai Kung Peninsula as a whole, the last major land-mass in Hong Kong still relatively unspoiled; with coral formations round the eastern islands and mangrove formations in Tolo Harbour. The existing (gazetted) recreational beaches are identified; and the existing sewage outfalls are shown, in the central harbour area.

For areas (1) to (4), conservation, education and recreation areas are shown. The whole Sai Kung Peninsula (area 5) and its islands are of such high quality, as to be graded a Conservation Area as a whole, with Education and Recreation areas designated within.

To retain Hong Kong's coasts in their beauty and diversity must be worth a continuing price in human and economic forbearance. But with public understanding and commitment, and with due compassion for fellow man—in all senses the most important species on the Hong Kong shore—the preservation of the coasts could be a burden this generation should be willing and able to bear.

GLOSSARY

Adductor	:	a muscle approximating a limb, or a shell valve to another.
Aerobic	:	thriving only in the presence of free oxygen.
Algae	:	marine and fresh-water aquatic plants, some microscopic. others quite large, e.g. sea-weeds.
Alluvial	:	deposits of finely divided soil material laid down by running water.
Ambulacrum	:	the radial zone of an echinoderm, bearing the tube-feet.
Angiosperm	:	flowering plant with seeds enclosed in an ovary which forms a fruit.
Annelida	:	segmented worms, marine representatives of which are the polychaetes.
Antenna(e)	:	the second pair of feelers of a crustacean head.
Antennule(s)	:	the first pair of feelers of a crustacean head.
Anterior	:	pertaining to the front end.
Arachnids	:	eight-limbed arthropods, mostly terrestrial but represented on the shore by the king-crab.
Arthropods	:	largest phylum in the animal kingdom with external jointed skeleton and jointed legs. Includes insects, Crustacea and Arachnida (spiders and mites).
Ascidians	:	simple transparent pre-chordates, attached by the base and filter-feeding via inhalant and exhalant siphons.
Athecate (polyps)	:	hydricid coelenterates devoid of a cup or theca.
Atrium	:	investing fold of the body wall enclosing an outer atrial cavity.
Avifauna	:	the bird fauna of a region.
Axial	:	pertaining to the axis of the body, or (as a shell rib) parallel to the axis.
Axillary	:	arising in the angle of one branch with another.
Backshore	:	portion of a beach above high water.
Backwash	:	the force of water returning after a wave has spent itself.
Barnacles	:	sessile crustaceans belonging to the subclass Cirripedia, found in the sea attached to rocks and other objects, the body enclosed within a series of calcareous skeletal plates.
Beak (bivalve)	:	the oldest or apical part of the shell valve.
Benthic	:	living on the sea bottom.
Berm	:	flat-topped beach terrace above high water.
Biramous	:	having two branches (limbs or appendages).
Bivalve	:	(Mollusca: Bivalvia). Aquatic molluscs characterized by having a shell consisting of two valves connected by an elastic ligament.
Branchial crown	:	a respiratory crown of filamentous tentacles, typically seen in polychaete worms and sea-slugs.

Brachiopod	:	a member of a primitive group of invertebrate animals (Phylum Brachiopoda = lamp-shells) having a superficial resemblance to bivalve molluscs in possessing two valves.
Brackish	:	a mixture of sea (salt) and fresh-water; typical of estuaries.
Break-point	:	the position of wave-break on approaching the shore.
Brood	:	hatch of young birds, or other animals produced from eggs, as in insects, fish, etc.
Bryozoa (ectoproct)	:	a primitive group of animals found in the seas, with a fixed, moss-like appearance on rocks and stones.
Buccal mass	:	the muscular chamber into which the mouth opens (as of a mollusc).
Byssus	:	the horny threads or 'beard' by which some bivalves are anchored to rocks.
Calcareous	:	formed of lime salts, especially calcium carbonate.
Calyx	:	outer whorl of floral leaves; theca of certain hydroids.
Capillarity	:	the tendency of water to rise within narrow spaces.
Capitate	:	with the distal end enlarged and rounded (tentacles).
Carapace	:	the hard covering of the head and thorax of many Crustacea.
Carnivore	:	a flesh-eating animal (occasionally a plant).
Carnivorous	:	feeding on animal flesh.
Catadromous	:	fish which migrate annually from fresh to salt water for breeding purposes.
Ceras (cerata)	:	the club-like appendages carried on the back of certain nudibranchs.
Chaeta(e)	:	the chitinous bristle(s) of the parapodia of polychaete worms (syn. seta).
Chela	:	the claw or pincer of a crustacean.
Chelicerae	:	the biting jaws and poison fangs of spiders and other arachnids.
Cheliped	:	a claw-bearing appendage.
Chitin	:	the horny material of the integument, especially of arthropods.
Chitons	:	coat-of-mail shell. Small, primitive, limpet-like molluscs with eight-valved shells.
Chloroplast	:	a minute granule or plastid containing chlorophyll; responsible for the green colour of plants.
Choanocyte	:	the flagellated collar cell, lining the internal passages in sponges.
Chordate	:	a member of the phylum Chordata which includes the vertebrates.
Cilia	:	microscopic vibratile hair-like processes, keeping up a current over internal or external surfaces.
Circumoral	:	encircling a mouth.
Cirri	:	tentacle-like appendages of limbs (polychaetes) or filamentar limbs themselves (barnacles).
Claspers	:	structures on the body of a male animal used for holding the female during copulation.
Cnidoblast	:	the cell containing a nematocyst (or stinging capsule) (Coelenterata).

Glossary

Coelenterate	:	large primitive phylum including anemones, jelly-fish, hydroids, corals.
Coelenteron	:	the internal body space of a sea-anemone, jelly-fish or other coelenterate.
Coelom	:	the body cavity surrounding the internal organs, as in a vertebrate, echinoderm or annelid.
Coenosarc	:	the common soft tissue, uniting the polyps of a hydroid or coral colony.
Column	:	the cylindrical body of, for example, an anemone or a balanoid barnacle.
Columella	:	the central spire of a gastropod shell giving attachment to the columella muscle.
Commensal	:	an organism living with another and sharing food, both species usually benefitting from the association.
Community	:	a distinct unit of vegetation of different types, in a definite area, with associated animal species.
Conservation	:	protection of natural resources with a view to a sustained yield; preservation of natural habitats.
Continental Shelf	:	area of relatively shallow water down to 200 metres adjacent to coast-lines.
Copepods	:	a very large sub-class of small, sometimes minute crustaceans. Major source food for pelagic fishes.
Coral	:	several orders of the class Anthozoa of marine coelenterates; most build a calcareous exoskeleton.
Corallite	:	cup of a single polyp of coral.
Corallum	:	skeleton of compound coral.
Crenulate	:	with margins minutely scalloped.
Crustacean	:	(Arthropoda: Crustacea). The most primitive class of arthropods, distinguished by their two pairs of antennae, mostly living in the sea.
Crustose lichens	:	lichens which form crusts on the substratum.
Cryptic	:	hidden; concealed.
Dactylozooid	:	a long, tentacle-like polyp with no mouth.
Demersal	:	living on or near the bottom of the sea.
Dendritic	:	branching in tree-like fashion.
Denticulate	:	bearing small teeth.
Dermis	:	the underlayer of the skin or integument.
Detritus	:	material produced by disintegration of plant and animal bodies after death and decomposition.
Diatoms	:	class of microscopic one-celled algae with hard siliceous cell-walls.
Dichotomous	:	regularly branching and rebranching into paired members.
Dicotyledon	:	larger of the two classes of flowering plants whose seedlings have two seed-leaves or cotyledons.
Dimorphic	:	of different shape or size, as applied to the sexes, or the opposite members, as of a pair of appendages.

Dioecious	:	having the sexes separate.
Distal	:	remote from the point of origin (c.f. Proximal).
Diverticula	:	blind-ending tube or sac branching off from a coral or cavity.
Dorsal	:	pertaining to the upper surface of a bilateral animal.
Echinoderm	:	a phylum of wholly marine animals with a radially symmetrical ground plan, moving with the aid of suckers called 'tube-feet'. Star-fish; sea-urchins; sea-lilies; brittle-stars; sea-cucumbers.
Echiuroids	:	unsegmented worms related to the polychaetes. Possessing a proboscis and usually one pair of chetae.
Ecology	:	the relationships between living organisms and their environment, both biotic and physical.
Ecosystem	:	a recognizably self-contained habitat containing its flora and fauna and all the physical and chemical components which make it a separate entity.
Ectoparasite	:	parasite that lives on the surface of its host.
Elytra	:	the scale-like plates covering the back of a polynoid worm.
Endopodite	:	the inner lobe of a crustacean limb.
Entoprocts	:	small colonial animals inhabiting a box and possessing a lophophore feeding organ.
Epifauna	:	the animals living upon the surface of the substrate.
Epiphyte (epiphytic)	:	a plant growing (non-parasitically) upon another plant.
Epithelium	:	layer of cells covering the surface of the body or of living internal surfaces.
Epizoic	:	living on the outer surface of another animal (non-parasitically).
Errant	:	free-moving or wandering (Polychaeta).
Estuary	:	the mouth of a river often experiencing wide fluctuations in salinity.
Eulittoral	:	the lower part of the shore regularly covered and uncovered by the tides.
Euryhaline	:	tolerating a wide range of saline conditions; as opposed to stenohaline.
Eutrophication	:	the addition of excessive nutrients to a water mass resulting in a depletion of oxygen following utilisation of the nutrients by aquatic organisms.
Exopodite	:	the outer lobe of a crustacean limb.
Faecal coliform	:	human or animal bacteria commonly found in the intestine and voided with the faeces.
Family	:	a taxonomic group of animals or plants with certain major characteristics in common.
Faulting	:	cracking of the earth's crust; rocks split and move in different motions so that the strata are discontinuous.
Fauna	:	the animals to be found in a particular area, country, or also a period of time.
Fetch	:	the distance of uninterrupted sea over which a wave may be generated.
Filament	:	a thread-like appendage.

Glossary

Filiform	:	thread-like.
Filter feeding	:	animals that feed by straining out minute plants and animals and also pieces of organic detritus, from water.
Fish-pond	:	a pond, often artificially constructed, for the cultivation of fish.
Flagellate	:	a uni-cellular animal (protozoan) moving by means of a whip-like flagellum.
Flat-worm	:	a flat, unsegmented worm in which creeping motion is by cilia.
Flock	:	a number of birds feeding or travelling together.
Flora	:	the plants peculiar to an area, country or period of time.
Food-chain	:	sequence of organisms in which each is food for a later member in the sequence.
Foraminifera	:	a group of the Protozoa; minute marine animals which form shells of lime.
Foreshore	:	portion of a beach between tidal limits.
Frontal	:	pertaining to the front part.
Fucoid	:	belong to the group of large brown algae containing the genus *Sargassum*.
Gastropod	:	snail; mollusc with a ventral muscular foot adapted for creeping.
Genus	:	a group of closely related species, in the classification of plants and animals.
Geomorphology	:	study of the physical features of the earth's crust and its geological structures.
Girdle	:	the flap of scaly, horny or fleshy mantle surrounding the body (in chitons).
Gnathopod	:	the pincer-bearing limbs of amphipods and some other Crustacea.
Gonad	:	the sex-gland, ovary or testis.
Gonophore	:	sexual zooid produced as a medusoid bud on a hydroid colony, sometimes freed, sometimes kept attached.
Gonotheca	:	the vase-like structure of hydroids, containing the polyp which buds off the medusae.
Gonozooid	:	hydroid polyp modified for reproduction.
Grading (sediments)	:	the distribution of the different particle sizes in a given sample.
Gregarious	:	animals that live together in a herd, flock, etc.
Habitat	:	the type of locality in which a plant or animal lives.
Halophyte	:	a land-plant preferring maritime or salty soils.
Hapteron	:	the attachment branch of a holdfast, in algae.
Herbaceous	:	plants without a woody stem.
Herbivore	:	animal which eats plants.
Herbivorous	:	to eat plants (herbs).
Hermaphrodite	:	producing both male and female sex-cells.
Heteromyarian	:	with adductor muscles unequal in size.
Heterotrophic	:	receiving nourishment from organic substances. All animals and fungi, most bacteria and a few flowering plants.
Hispid	:	having stiff hairs, spines or bristles.

Holothurian	:	class of cylindrical echinoderms with a ring of tentacles around the mouth; sea cucumbers.
Host	:	an organism in or on which another organism (parasite, etc.) spends part of its life history.
Hydranth	:	a nutritive zooid of a hydroid colony.
Hydrography	:	scientific description of the waters of the world.
Hydroids	:	member of the class Hydrozoa; colonial animals forming tuft-like growths on seaweeds etc. Sea fir.
Hydrotheca	:	the small exoskeletal cup containing the polyp of a (thecate) hydroid.
Hypersaline	:	of greater than a given salinity.
Hypofauna	:	the concealed fauna living beneath rocky or other cover.
Hyposaline	:	of less than a given salinity.
Ichthyology	:	scientific study of fishes.
Igneous (rocks)	:	derived from molten materials as distinguished from SEDIMENTARY.
Indigenous	:	native; belonging to the locality.
Infauna	:	animals living within the substrate.
Infest	:	to inhabit the exterior of a host's body.
Insolation (substrate)	:	the accumulation of heat by exposure to the sun.
Integument	:	a covering or coating layer; coat of an ovule.
Interstitial	:	living in small inter-spaces, as in sand or with clusters of large animals.
Intertidal:	:	shore between high- and low-tide marks.
Invertebrate	:	animal which does not possess a bony skeleton and in particular a vertebral column (back-bone).
Iridescent	:	rainbow tinted.
Isopoda	:	order of Crustacea, including wood-lice and sea-skater. Contains terrestrial, fresh-water, brackish and marine species.
Jurassic	:	geological period: lasted from approximately 150 million to 130 million years ago.
Lamella	:	a thin plate scale or sharp upstanding ridge. (adj. lamellose, -ate).
Larva	:	early developmental stage of an animal, often totally unlike the adult form.
Lateral	:	pertaining to the side of a body.
Lenticels	:	small raised ventillating pore developing when, in woody stems, the epidermis is replaced by cork.
Lichen	:	compound plant body formed by a symbiosis between an alga and a fungus, usually small and cryptophytic.
Ligament (bivalve)	:	the uncalcified connecting piece uniting the shell-valves of a bivalve mollusc.
Limpet	:	a marine snail adapted for life on sea-shore rocks with a conical shell and holding on by way of a large sucker-like foot: no operculum.

Glossary

Lithothamnia	:	a group of massive or encrusting calcareous algae.
Littoral	:	the sea shore; that part of the shore that is influenced by the rise and fall of the tides.
Littoral current	:	a local current running along the shore; induced by coastal currents further out.
Littorine	:	organisms living in the littoral zone of the sea shore; sea-shore plants and animals. Also particularly applied to members of the genera *Littorina* and *Nodilittorina*.
Lugs (lugworm)	:	Tube-dwelling marine bristle-worms which form casts or small mounds.
Madreporite	:	a perforated plate in echinoderms by which water is admitted to the water vascular or ambulacral system.
Mamilla	:	nipple-shaped structure.
Mammal	:	a warm-blooded vertebrate animal, with fur or hair, producing living young nourished by milk.
Mandible	:	the biting jaws of an animal.
Mangrove	:	woody swamp vegetation characteristic of tropical estuaries.
Mantle	:	the covering tissue which forms the molluscan shell.
Manubrium	:	the knob-like projection bearing the mouth in a hydroid polyp.
Mariculture	:	the science of cultivating or farming aquatic organisms.
Marine	:	organisms found living in or on the sea.
Maxilla, maxillule	:	the two jaw-like appendages lying behind the mandibles in, for example, the Crustacea.
Maxillipeds	:	an appendage, in one to three pairs, posterior to the maxillae in certain arthropods.
Medusa	:	the free-swimming jellyfish-like stage in the life history of a hydrozoan.
Metamorphosis	:	change in body form during development, such as from larva to adult.
Migrant	:	migrating animal.
Migration	:	change of habitat according to season, climate, etc., shown by various birds and other animals.
Mollusc	:	animal belonging to the Phylum Mollusca, e.g. clams, snails, slugs, squid, etc.
Monocotyledon	:	smaller of two classes of flowering plants (Angiosperms) whose seedlings have only one seed-leaf (cotyledon): includes cereals, palms and orchids.
Monoecious	:	hermaphrodite; both male and female organs borne on same body.
Moult	:	process of periodically shedding the outer body covering of hair, feathers, cuticle, etc.
Nauplius	:	the first planktonic larval stage of many Crustacea.
Nematocyst	:	the explosible stinging capsule of an anemone or other coelenterate, contained in the CNIDOBLAST.
Nemertean	:	the ribbon-worms; long, flat and unsegmented often with simple eyes.

Neuropodium	:	the ventral (lower) lobe of the parapodium in polychaete worms (cf. Notopodium).
Niche	:	the role in a community filled by a given organism.
Nidus	:	a nest; nest-like hollow; cavity for development of spores.
Nocturnal	:	active at night only.
Notopodium	:	the dorsal lobe of the parapodium in polychaete worms (cf. Neuropodium).
Notum	:	the dorsal portion of an insect segment.
Nudibranchs	:	sea-slug with no shell, internal or external.
Ocellus	:	a simple light sensitive organ.
Olfactory	:	pertaining to the sense of smell; applied to stimuli, structures and reactions.
Omnivorous	:	eating both plant and animal material.
Oötheca	:	egg-case.
Operculum	:	a shelly or horny lid closing a gastropod shell, or a worm-tube; also gill covering in teleost fishes.
Oral	:	pertaining to the mouth.
Oral disc	:	the uppermost part of an anemone, bearing the mouth and tentacles.
Oral podia	:	echinoderm tube foot, located around the mouth, and used in the collection of food.
Orbit	:	the space in which the eye or eye-stalk is contained.
Order	:	a classificatory group used in systematics of both animals and plants, e.g. Order Scleractinia—the hard corals, within the Class Anthozoa.
Ornithology	:	the scientific study of birds.
Osculum	:	the large exhalant opening of a sponge.
Osmoregulation	:	the power of an animal to regulate the osmotic pressure of its body fluids, by altering the proportion of dissolved molecules.
Osmotic pressure	:	the pressure required to stop the diffusion of a liquid passing from highest to lowest concentration through a membrane.
Osphradium	:	a sensory organ in the molluscan mantle cavity.
Ostium	:	a small entry, especially one of the inhalant openings of a sponge.
Ovipositor	:	structure used by a female animal in the placing of the egg at oviposition (egg-laying).
Palaearctic	:	a zoogeographical region or sub-region of the holarctic region including Europe, North Africa, Western Asia, Siberia, Northern China and Japan.
Pallial siphon	:	tube for entry or exit of water into the mantle or pallial cavity.
Palmate	:	with branches springing from one centre in palm fashion.
Palp	:	tentacle-like organs, especially tactile (as in polychaetes), or the labial palps of bivalves.
Palpiform	:	resembling a palp or blunt feeler.
Papillose	:	covered with small papillae.
Parasitic	:	an organism living in or on another, to its own advantage in food or shelter and to the disadvantage of the host.

Glossary

Parenchyma	:	plant tissue consisting of living thin-walled cells.
Parietal	:	pertaining to the body wall.
Parapodium	:	the paddle-like limb of a polychaete worm.
Passage Migrant	:	an animal, usually applied to birds, encountered actually during its migrational movement.
Passerine	:	a perching bird, belonging to the Order Passeriformes.
Pathogen	:	a parasite which causes disease.
Paxilla	:	thick plate supporting calcareous pillars, the summit of each covered by a group of small spines in certain sea-stars.
Pedicel	:	the small stalk of a sessile animal.
Pedipalps	:	second head appendage of arachnids; may be locomotor, clawed and used for seizing prey; sensory or modified in the male for fertilization.
Peduncle	:	the stem of a flower or an attached animal body.
Pelagic	:	living in the open waters of a sea or lake, as distinct from on the bottom.
Pereiopod	:	the walking leg attached to the thorax of a crustacean.
Periostracum	:	the external horny layer of a shell.
Periphery	:	the outer-lying parts or circumference.
Perisarc	:	the continuous outer cuticle of a hydroid polyp or colony.
Peristalsis	:	wave-like muscular contractions of a tube or body wall.
Peristome	:	the lip of a spiral shell (Gastropoda) or the skin surrounding the mouth (Echinoidea).
Peristomium (in annelid worms)	:	the second segment, that borders the mouth above and at the sides.
Permian	:	geological period; lasted approximately from 200 to 180 million years ago.
Phoronida	:	worm-like animals possessing a lophophore as a feeding organ.
Photosynthesis	:	the manufacture of carbohydrates by green plants, using the energy of light.
Phylum	:	a major division of the animal or plant kingdom.
Phytophagous	:	plant-eating.
Phytoplankton	:	microscopic plants of the plankton, such as diatoms and flagellates.
Pinnate	:	side-branched in feather-like fashion.
Pinnule	:	a small side branch of an appendage.
Piscivorous	:	fish-eating.
Plankton	:	microscopic aquatic plants and animals floating in the surface layers of the seas or in fresh-water lakes. (Phytoplankton = plants; Zooplankton = animals.)
Plastron	:	a breast-plate or ventral shield.
Pleon	:	the part of a crustacean bearing the swimming limbs or pleopods.
Pleopod	:	the swimming limb attached to the crustacean abdomen.
Pleurobranch	:	gill springing from the lateral walls of the thorax of certain Arthropoda.
Pneumatophore	:	aerating root of certain mangrove plants; air-bladder of certain aquatic plants, etc.
Pollution	:	contamination; destroying the purity of the environment.

Polyculture	:	the cultivation of several species together in the same situation.
Polymorphic	:	individuals of the same species occurring in different forms.
Polyp	:	a simple Actinozoon, or a separate zooid of a colony, with a cylindrical trunk fixed at one end and a mouth surrounded by a ring of tentacles at the other.
Polypide	:	a polyp-like individual, as in Polyzoa.
Polyzoans	:	phylum of sessile colonial organisms with the globular body attached to the substrate by a stalk.
Posterior	:	pertaining to the rear end.
Predaceous	:	being a predator.
Predator	:	a carnivorous animal that kills other animals for food.
Proboscis	:	a trunk-like process projecting from the mouth or anterior part of the head in certain animals.
Propodium	:	the small anterior part of a molluscan foot.
Prostomium (Annelida)	:	the first segment, lying in advance of the mouth.
Protandrous (in hermaphrodites with sex change)	:	developing male characters first (opposite is protogynous).
Protochordate	:	a primitive animal belonging to the Phylum Chordata but lacking the bony skeleton of the vertebrates (including sea-squirts, lancelets, etc.)
Proximal	:	near to the point of origin (cf. Distal).
Pteridophyte	:	ferns; plants with finely divided fronds upon which are born the fruiting bodies. Inhabitants of damp ground.
Pulmonate	:	the fresh-water and land snails and slugs, belonging to the Mollusca, Gastropoda, and breathing with the aid of a lung, not gills.
Pyriform	:	pear-shaped.
Raceme	:	a branching system with a long axis and series of short side pedicels.
Radula	:	the tooth-bearing ribbon used for feeding in snails.
Raptorial	:	being predatory in habits.
Rectum	:	the terminal part of the intestine.
Reptile	:	a cold-blooded vertebrate animal belonging to the Class Reptilia, e.g. snakes, lizards and turtles.
Respiratory pigment	:	pigment concerned with oxidation–reduction processes in living organisms, such as haemoglobin in vertebrates, haemocyanin in crustaceans.
Rhinophore	:	the club-like sensory appendage of a nudibranch.
Rhizoid	:	a small, root-like attachment process.
Rostrum	:	the median process from the front of the head, as in Crustacea.
Scavenger	:	an animal that feeds on dead plants or animals, or on decaying matter, or on animal faeces.
Scutum (pl. scuta)	:	A shield-like plate, horny, bony, or chitinous.
Sea-weed	:	large attached algae, typically of rocky shores.

Glossary

Sedentary	:	not free-living; animals attached by a base to some substrate, as with barnacles.
Sedimentary	:	rocks formed by deposition of material over a period of time.
Septum	:	a wall separating two cavities.
Sere	:	a succession of ecological communities.
Serrate	:	with minute saw-teeth or serrations.
Sessile	:	attached permanently to the substratum.
Seta	:	a minute, bristle-like hair especially of the Polychaeta and Crustacea (Chaeta).
Setigerous	:	bearing fine hair-like bristles (setae).
Sexual dimorphism	:	difference in appearance between sexes.
Siphon	:	a tubular structure in various organisms sub-serving exchange purposes.
Siphonoglyph	:	the ciliated channel running along the stomodaeum of an anemone.
Sipunculid	:	marine sand-dwelling worms with no apparent sign of segmentation.
Somites	:	a mesoblastic segment or compartment; a body segment of an articulate animal.
Sorting (sediments)	:	the segregation of different sizes of particle by waves or currents.
Spawn	:	collection of eggs deposited by fish, frogs, molluscs, etc.; to lay eggs.
Species	:	group of inter-breeding individuals producing fertile young, and showing similar characteristics.
Sponge:	:	a simple encrusting organism, filter-feeding using characteristic cells—choanocytes—with a skeleton of calcareous or silacaceous spicules.
Spongin	:	the fibrous skeletal substance of sponges.
Sporosac	:	a hydrozoan reproductive zooid which does not become medusoid.
Statocyst	:	a small sac containing vibration-detecting sensory hairs, and an often loose concretion or statolith.
Stenohaline	:	tolerating a narrow range of salinity.
Stenothermic	:	organisms adaptable only to slight variations in temperature.
Stenozonal	:	having a narrow zone of distribution.
Sternum	:	ventral surface of insect body; breast-bone.
Stigma	:	an aperture by which water passes through the pharynx wall of ascidians.
Stolon	:	a trailing outgrowth of the body from which new individuals are produced (especially Hydrozoa, Tunicata).
Stratification	:	arrangement in layers; vertical grouping within a community.
Striate	:	marked with striae or narrow lines.
Sub-littoral	:	that part of the sea shore that is never uncovered by the tides.
Substrate	:	underlying basal material.
Swash	:	the impetus of a wave rushing up a shore.
Symbiosis	:	two organisms living intimately together, usually in a beneficial partnership.

Symbion	:	an organism living intimately together with another.
Systematics	:	system of classification of plants and animals.
Taxa	:	a general term used to describe classificatory categories; singular is taxon.
Taxonomy	:	the identification (and naming) of animal and plant species.
Telson	:	the terminal appendage of a crustacean.
Temperate	:	moderate climate, usually without extremes of temperatures, etc.
Terga	:	the chitinous or calcareous plates of a crustacean.
Terrestrial	:	organisms living on land.
Test	:	the firm or hard-walled body covering, as in ascidians (tunic), or echinoids (shelly covering).
Thallus	:	the plant-body of an alga.
Thecate (polyps)	:	provided with cups or thecae.
Thixotrophy	:	the quicksand property of soft water-filled sediments.
Thorax	:	part of the body intermediate between head and abdomen, as especially in the Arthropoda.
Thermocline	:	abrupt change in water temperature in lakes and the sea at a certain depth, due to summer heating.
Tibia	:	the lower leg of mammals, below the thigh, or its equivalent in insects.
Tombolo	:	a small island narrowly tied to the mainland.
Tomentum	:	a fine covering of hair-like processes.
Trabeculae	:	small cross partitions.
Trophozooid	:	the hydroid polyp modified for feeding.
Tropical	:	belonging to the tropical region of the world, that is geographically the areas between the Tropics of Cancer and Capricorn.
Tubercule	:	a small rounded protuberance.
Turbid	:	cloudy, with fine suspended matter.
Umbilicus	:	the cavity at the base of many gastropod shells.
Umbones	:	beak or older part of a bivalve shell.
Uncini	:	small chitinous hooks, in the parapodia of sedentary polychaete worms.
Urban	:	inhabiting, or situated in, a town or city, or associated with man and his dwellings, as opposed to being rural.
Uropod(s)	:	the paired limbs making up the tail-fan in the Crustacea.
Valves (molluscs, barnacles)	:	plates or pieces of the hard shell, e.g. bivalve.
Vegetarian	:	plant-eating.
Veliger	:	a molluscan larva swimming by a velum.
Velum	:	the valve like flap guarding the mantle edge of a scallop; the ridge of long-ciliated cells by which a molluscan larva swims.
Venomous	:	poisonous; having poison-gland and able to inflict a poisonous wound.

Glossary

Ventral	:	pertaining to the lower surface of a bilateral animal.
Vermiform	:	worm-like in shape.
Vertebrate	:	animal belonging to the sub-phylum Vertebrata, and possessing a jointed back-bone and a skull.
Vesicle	:	a bladder-like cavity or sac.
Villi	:	small tag-like appendages, lining an intestine or elsewhere
Viscera	:	soft internal organs, as of the digestive and other systems
Viviparity (viviparous)	:	production of living young as distinct from egg-laying.
Wader	:	bird belonging to the Order Charadriiformes, and typically with long legs for wading in water.
Wetland	:	land with high-water table, and wet surface soil, often with patches of water, marshy.
Whorl	:	a single turn of a spiral, as in a gastropod shell.
Xerophytes	:	a plant growing in desert or alkaline soils.
Zooecium	:	a small compartment holding one of the zooids of a polyzoan colony.
Zoogeography	:	the study of the distribution of animals throughout the world.
Zooid	:	an individual animal of a colony, e.g. of ascidians or polyzoans.
Zoology	:	the scientific study of animals.
Zooplankton	:	the small floating animals of the plankton.
Zooxanthellae	:	uni-cellular plants living in symbiosis with various animals such as corals and giant clams.

TAXONOMIC GUIDE

An outline classification of the plant and animal kingdoms is given here. The classification is highly simplified and deals only with major categories of PHYLUM, CLASS and ORDER, with occasional use of SUB-CLASS, SUB-ORDER and FAMILY. A few examples of local Hong Kong genera are given after the names of the Orders. Certain groups not well represented on the shore have been omitted, or dealt with in very abbreviated form.

I THE PLANT KINGDOM

Phylum PHYCOPHYTA (algae)
  Class CYANOPHYCEAE (or MYXOPHYCEAE) (blue-green algae)
  Class CHLOROPHYCEAE (green algae)
    Order ULVALES *(Ulva, Enteromorpha, Monostroma)*
    Order CLADOPHORALES *(Cladophora, Chaetomorpha)*
    Order CAULERPALES *(Caulerpa, Codium)*
  Class BACILLARIOPHYCEAE (diatoms)
  Class DINOPHYCEAE (dinoflagellates) *(Noctiluca, Gmynodinium)*
  Class PHAEOPHYCEAE (brown algae)
    Sub-class ISOGENERATAE
      Order ECTOCARPALES *(Ectocarpus)*
      Order SPHACELARIALES
      Order DICTYOTALES *(Padina)*
    Sub-class HETEROGENERATAE
      Order CHORDARIALES *(Leathesia)*
      Order DESMARESTIALES
      Order LAMINARIALES *(Petalonia)*
    Sub-class CYCLOSPOREAE
      Order FUCALES *(Sargassum)*
  Class RHODOPHYCEAE (red algae)
    Sub-class BANGIOPHYCIDAE
      Order BANGIALES *(Bangia, Porphyra)*
    Sub-class FLORIDEOPHYCIDAE
      Order NEMALIONALES *(Nemalion)*
      Order GELIDIALES *(Gelidium)*
      Order CRYPTONEMIALES *(Corallina, Hildenbrantia, Amphiroa, Jania)*
      Order GIGARTINALES *(Gigartina)*
      Order RHODYMENIALES *(Rhodymenia)*
      Order CERAMIALES *(Laurencia)*
    LICHENES (symbiotic associations of fungi with algae, rather than separate organisms) *(Caloplaca, Lepraria, Verrucaria, Ramalina)*

Phylum BRYOPHYTA (mosses and liverworts)

Phylum PTERIDOPHYTA (ferns and their allies) *(Acrostichum)*

Phylum SPERMATOPHYTA (highest or 'seed-plants') (Only families with important sea-shore representatives have been here included)
  Class GYMNOSPERMAE
  Class ANGIOSPERMAE
    Sub-class MONOCOTYLEDONAE
      Family Araceae *(Pistia)*
      Family Cyperaceae *(Cyperus, Rhynchospora, Fimbristylis)*
      Family Gramineae *(Zoysia, Sporobolus, Phragmites)*
      Family Hydrocharitaceae *(Halophila)*
      Family Potamogetonaceae *(Zostera)*
      Family Naiadaceae *(Ruppia)*
      Family Pontederiaceae *(Eichornia)*
      Family Liliaceae *(Asparagus)*
      Family Pandanaceae *(Pandanus)*
    Sub-class DICOTYLEDONAE
      Family Chenopodiaceae *(Salicornia, Suaeda)*
      Family Acanthaceae *(Acanthus)*
      Family Euphorbiaceae *(Excoecaria, Macaranga)*
      Family Verbenaceae *(Clerodendron, Avicennia, Vitex)*
      Family Combretaceae *(Lumnitzera)*
      Family Goodeniaceae *(Scaevola)*
      Family Rhizophoraceae *(Bruguiera, Kandelia)*
      Family Sterculiaceae *(Heritiera)*
      Family Aegicerataceae *(Aegiceras)*
      Family Papilionaceae *(Derris)*
      Family Casuarinaceae *(Casuarina)*
      Family Plumbaginaceae *(Limonium)*
      Family Convolvulaceae *(Ipomoea)*
      Family Compositaceae *(Pluchea)*

II The Animal Kingdom

Phylum PROTOZOA (amoeba-like forms, Foraminifera, ciliates and flagellates) *(Noctiluca)*

Phylum PORIFERA (sponges)
  Class CALCAREA (calcareous sponges) *(Scypha, Leucandra)*
    Sub-class CALCINEA
    Sub-class CALCARONEA
  Class DEMOSPONGIA (siliceous and horny sponges) *(Tethya)*
    Sub-class TETRACTINELLIDA
    Sub-class MONAXONIDA
    Sub-class KERATOSA

Taxonomic Guide

Phylum COELENTERATA (= CNIDARIA)
  (hydroids, jelly-fish, sea-anemones, corals)
  Class HYDROZOA
    Order ATHECATA ⎱ (hydroids with polyps dominant) *(Tubularia)*
    Order THECATA ⎰   *(Obelia)*
    Order LIMNOMEDUSAE (hydroids with medusa dominant) *(Gonionemus)*
    Order SIPHONOPHORA (pelagic colonial hydroids) *(Physalia, Velella)*
  Class ANTHOZOA (sea-anemones, corals, soft corals)
    Sub-class CERIANTIPATHARIA *(Cerianthus) (Antipathes)*
    Sub-class OCTOCORALLIA (= ALCYONARIA)
      Order ALCYONACEA (soft corals) *(Nephthya) (Dendronephthya)*
      Order GORGONACEA (horny corals) *(Echinogorgia)*
      Order PENNATULACEA (sea-pens) *(Pteroeides)*
    Sub-class ZOANTHARIA (sea-anemones and corals)
      Order ACTINIARIA (anemones) *(Haliplanella, Epiactis)*
      Order ZOANTHINIARIA *(Palythoa)*
      Order CORALLIMORPHARIA
      Order SCLERACTINIA (true corals) *(Favia, Pavona, Acropora, Goniopora)*
  Class SCYPHOZOA (jellyfish) *(Cyanea, Aurelia)*

Phylum PLATYHELMINTHES (flat-worms)
  Class TURBELLARIA (free-living flat-worms)
    Order POLYCLADIDA *(Stylochus, Notoplana)*

Phylum NEMERTEA (= NEMERTINA)
  Class ANOPLA
  Class ENOPLA *(Nemertopsis)*

Phylum ENTOPROCTA *(Loxosomella)*

Phylum ECTOPROCTA ('Polyzoa' or sea mats)
  Class GYMNOLAEMATA
    Order CYCLOSTOMATA *(Crisia, Tubulipora)*
    Order CHEILOSTOMATA *(Bugula, Schizoporella)*
    Order CTENOSTOMATA *(Flustrella, Amathia)*
  Class PHYLACTOLAEMATA *(Plumatella)*

Phylum PHORONIDA *(Phoronis)*

Phylum BRACHIOPODA
  Class INARTICULATA *(Lingula)*
  Class ARTICULATA (lamp-shells)

Phylum SIPUNCULA *(Sipunculus, Dendrostomum, Siphonosoma)*

Phylum ECHIURA *(Ochetostoma, Thalassema)*

Phylum ANNELIDA
  Class POLYCHAETA (bristle worms)
    'Sub-class ERRANTIA' (errant polychaetes) *(Chloeia, Harmothoe, Diopatra, Eunice, Nereis)*
    'Sub-class SEDENTARIA' (sedentary polychaetes) *(Loimia, Sabellastarte, Lanice, Chaetopterus)*
  Class OLIGOCHAETA (land and fresh-water annelids) (hardly occurring on the shore)
  Class ARCHIANNELIDA

Phylum MOLLUSCA
  Class AMPHINEURA (chitons) *(Liolophura, Onithochiton)*
  Class GASTROPODA (snails and slugs)
    Sub-class PROSOBRANCHIA (shelled marine gastropods)
      Order ARCHAEOGASTROPODA (limpets, top-shells) *(Cellana, Turbo, Monodonta)*
      Order MESOGASTROPODA (numerous herbivorous and carnivorous families) *(Cypraea, Littorina, Cerithium, Batillaria)*
      Order STENOGLOSSA (numerous carnivorous families) *(Conus, Nassarius, Oliva)*
    Sub-class OPISTHOBRANCHIA (sea-slugs and their allies)
      Order CEPHALASPIDEA (= BULLOMORPHA) (bubble-shells) *(Bulla)*
      Order ANASPIDEA (= APLYSIOMORPHA) (sea-hares) *(Aplysia)*
      Order NOTASPIDEA (= PLEUROBRANCHOMORPHA) (side-gilled slugs) *(Umbraculum, Berthelina)*
      Order SACOGLOSSA
      Order NUDIBRANCHIA (true sea-slugs) *(Dendrodoris, Gymnodoris)*
    Sub-class PULMONATA (lung-breathing gastropods)
      Order BASOMMATOPHORA *(Ellobium, Melampus, Siphonaria)*
      Order STYLOMMATOPHORA (land snails and slugs) *(Onchidium)*
  Class SCAPHOPODA (tusk-shells)
  Class BIVALVIA
    Sub-class PROTOBRANCHIA
    Sub-class LAMELLIBRANCHIA
      Order TAXODONTA (ark-shells) *(Arca, Anadara, Barbatia)*
      Order ANISOMYARIA (mussels, scallops, file-shells, oysters) *(Septifer Amussium, Crassostrea)*
      Order HETERODONTA (numerous families, many of them burrowing) *(Tapes, Gafrarium, Geloina)*
      Order ADAPEDONTA *(Hiatella, Martesia, Teredo)*
      Order ANOMALODESMATA *(Laternula)*
  Class CEPHALOPODA (cuttle-fish, squid, octopus)
    Sub-class COLEOIDEA
      Order DECAPODA *(Euprymna, Spirula)*
      Order OCTOPODA *(Octopus)*

Phylum ARTHROPODA
  Class CHILOPODA (centipedes)

Class INSECTA (insects) *(Asclepios, Ceroplastes, Aedes)*
Class CRUSTACEA
　Sub-class BRANCHIOPODA (water-fleas)
　Sub-class OSTRACODA
　Sub-class COPEPODA *(Mytilicola)*
　Sub-class CIRRIPEDIA (barnacles) *(Tetraclita, Balanus, Chthamalus)*
　Sub-class MALACOSTRACA
　　Super-order LEPTOSTRACA
　　Super-order PERACARIDA
　　　Order ISOPODA *(Ligia, Sphaeroma)*
　　　Order MYSIDACEA (opossum shrimps)
　　　Order CUMACEA
　　　Order TANAIPODA
　　　Order AMPHIPODA *(Gammarus, Orchestia)*
　　Super-order HOPLOCARIDA (mantis shrimps) *(Oratosquilla)*
　　Super-order EUCARIDA
　　　Order DECAPODA
　　　　Sub-order NATANTIA (swimming shrimps, prawns) *(Penaeus)*
　　　　Sub-order REPTANTIA (benthal cray-fish, crabs, hermit-crabs) *(Sesarma, Portunus, Grapsus)*
Class ARACHNIDA
　Order PSEUDOSCORPIONES
　Order ARANEAE (spiders)
　Order ACARI (mites)
Class PYCNOGONIDA (sea-spiders) *(Achelis)*
Class TARDIGRADA (bear animalicules)

Phylum ECHINODERMATA
　Class CRINOIDEA (sea-lilies and feather-stars) *(Comatula)*
　Class HOLOTHUROIDEA (sea-cucumbers)
　　Order ASPIDOCHIROTA *(Stichopus, Holothuria)*
　　Order DENDROCHIROTA
　　Order APODA *(Trochodota, Protankyra)*
　Class ECHINOIDEA
　　Order DIADEMATOIDA *(Diadema)*
　　Order ECHINOIDA (sea urchins) *(Temnopleura)*
　　Order CLYPEASTEROIDA (sand dollars, cake urchins) *(Clypeaster)*
　　Order SPATANGOIDA (heart urchins) *(Lovenia, Schizaster)*
　Class ASTEROIDEA (star-fishes)
　　Order PHANEROZONA *(Archaster, Luidia)*
　　Order SPINULOSA
　　Order FORCIPULATA
　Class OPHIUROIDEA (brittle-stars, snake-stars) *(Ophiactis, Macrophiothrix)*

Phylum CHORDATA
　Sub-phylum HEMICHORDATA (acorn-worms) *(Balanoglossus)*
　Sub-phylum UROCHORDATA
　　Class ASCIDIACEA (sea-squirts) *(Ciona, Styela)*

Order ENTEROGONA
Order PLEUROGONA
Sub-phylum CEPHALOCHORDATA (lancelets) *(Branchiostoma)*
Sub-phylum CRANIATA (higher chordates or vertebrates)
  Class AGNATHA (hag-fish)
  Class PLACODERMI
Class CHONDRICHTHYES (sharks & rays)
Class ACTINOPTERYGII (modern or bony fishes) *(Periophthalmus, Scarus, Chaetondon)*
Class AMPHIBIA (frogs, newts, toads)
Class REPTILIA (lizards, snakes) *(Enhydris)*
Class AVES (birds) *(Tringa, Motacilla, Zosterops)*
Class MAMMALIA (otters, porpoises, dolphins) *(Lutra, Neophocaena)*

INDEX OF PLACE NAMES

Aberdeen 3, 12
Aberdeen Harbour 10

Basalt Island 33
Big Wave Bay 137, 138, 139, 140
Bluff Island 33
Boulder Point 33

Canton 10
Cape D'Aguilar 17, 34, 35, 36, 37, 42, 47, 56, 62, 70, 74, 181, 258, 311, Pl.3
Castle Peak 68, 69
Chek Chau (Port Island) 190, 258
Cheun She 211, 212
Cheung Chau 132
Cheung Sha 138
Clear Water Bay 47, 48, 49, 50
Crooked Harbour 32, 250, Pl.13

D'Aguilar Peninsula 58
Deep Bay 9, 14, 18, 68, 108, 214, 217, 221, 230, 231, 234, 238, 240, 241, 242, 244, 249, 250, 311, Pls. 17 & 28
Deep Water Bay 17, 57, 61, 62, 63

East China Sea 6

Hainan Island 264
Ham Tin 147
Hebe Haven 18
High Island 33, 304
Ho Chung Pl.28
Hoi Sing Wan (Starfish Bay) 178, 186, 189, 190, 191, 196, 249
Hole Island 33, Pl.18
Hong Kong Harbour 135
Hong Kong Island 32, 34

Junk Bay 206

Kap Shui Mun 68
Kowloon 3
Kwa Mun 33
Kwun Tong 304

Lai Chi Wo 250
Lamma Island 32, 33, 34, 66, 108, 308, 311, Pls. 7, 8 & 28
Lantau Island 8, 17, 34, 40, 68, 86, 108, 305, 311, Pl.28
Lau Fau Shan 207, 242, 243
Little Sai Wan 3
Long Harbour 178, 190
Luzon Straits 6

Ma Tse Chau 114, 115, 116, 122, 190
Ma Wan 67, 68
Macau 7, 8
Macclesfield Bank 294
Mai Po 233, 234, 235, 236, 237, 309, 311, Pls. 12 & 17
Mirs Bay 32, 257, 278, 283, 293, 294, Pl.18

New Territories 3, 32, 47, 305
Num Shui Wan 32

Pak Sip Lap 131
Paracel Islands 294
Pearl River 7, 8, 9, 10, 12, 13, 20, 66, 68, 135, 242, 257
People's Republic of China 232
Picnic Bay 308
Pillar Point 135
Ping Chau 32, 258, 291, 295, 296, 298, 299, 300, Pl.18
Port Island *see* Chek Chau
Port Shelter 33
Pratas Reef 294
Pui O 135, 136, 137, 138, 208, 311

Repulse Bay 132, 133
Rocky Harbour 33

Sai Kung 218, 304, 311
Sai Kung Peninsula 141, 143, Pl.28
Sai Wan (N.T.) 142
Sandy Hook (New Jersey) 134
Sha Tin 3, 12, 206, 207, 304
Sham Wan 132, 133, 139, 311
Sharp Island 131
Shek-O 42, 132, 137, 250
Shuen Wan Lei Uk 115
Shui Hau 17, 24, 146, 147, 148, 149, 151, 153, 156, 157, 162, 164, 179, 181, 184, 185, 199, 208, 308, 311, Pl.8
Shum Chun River 242
Silvermine Bay 140, 168, 308
South China Sea 6, 14
Stanley Pl.4
Stanley Peninsula 40, 56, 86, 109
Starfish Bay *see* Hoi Sing Wan
Starling Inlet 18
Stonecutter's Island 10, 12

Tai Lam Chung 9, 10, 11, 12, 65, 68
Tai Long Wan 143
Tai Po 3, 304. Pl.28
Tai Po Channel 207
Tai Shang Wai 223
Tai Tam Pl.4
Tai Tam Bay 17, 18, 58, 70, 87, 108, 180, 182, 184, 185, 189, 200, 207, 211, 249, 311, Pl.8
Tai Tam Harbour 86, 99, 147, 158, 166, 178, 180, 181, 182, 183, 243

Tai Tam Reservoir 181
Thai Long Head 34
Three Fathoms Cove 18, 178, 190, 218, 219, 249, Pl.14
Tide Cove 207
Ting Kok 221
Tolo Channel 207
Tolo Harbour 12, 66, 68, 104, 105, 115, 119, 120, 121, 135, 147, 166, 168, 188, 208, 222, 233, 249, 258, 268, 275, 293, 311, Pls. 10, 13 14 & 28
Tong Fuk Pl.7
Tong Fuk Miu Wan 147
Tropic of Cancer 3
Tsim Bei Tsui 10, 11, 12, 65, 68, 244, Pl.17
Tsuen Wan 3, 304
Tuen Mun 3, 304
Tung O 138
Turtle Cove 57, 58, 59, 144

Victoria 3, 75, 76
Victoria Harbour 12, 13, 207, 303, 304

Waglan Island 9, 10, 11, 12, 258
Wang Chau Pl.18
West Lamma 86
White Knights 66, 67
Wu Kwai Sha 178, 190, 208

Yuen Long 233
Yuen Long Creek 242
Yuen Long Plain 10, 112, 237

INDEX OF SCIENTIFIC NAMES
Italicized page numbers refer to illustrations

Abra maxima 248
Acanthastrea 263
Acanthastrea bowerbanki 264, Pl.26
Acanthella 108
Acanthus 208, 218, 222, 233
Acanthus ilicifolius 221, *224*, 234, Pl.16
Acasta dofleini 282
Acasta sulcata 108, 201
Achasmea 176
Achasmea sp. 177, *204*
Achelis superba 52
Acrocephalus arundinaceus 241
Acrocephalus bistrigiceps 241
Acropora 25, 258, 260, 273
Acropora candelabrum 259
Acropora pruinosa 259, Pls. 21 & 23
Acropora tumida 259
Acrostichium aureum 221
Actaea pura 273
Actinia equina 44, *109*, 111
Actinodendron 268
Actitis hypoleucos 233, Pl.20
Actumnus setifer 273
Aedes togoi 51
Aegiceras 102, 208, 220, 221, 222
Aegiceras corniculatus 68, 221, *223*, *224*, 234, Pl.16
Aeolidiella takanosinensis 290, *291*
Agelas 108
Alcedo atthis 240
Alcyonidium 81, 83
Alectryonella plicatula 104, 105, 114, *115*
Alepas pacifica 143, *145*, 203
Alium glaucum 284
Alpheus 27, 118, 271
Alpheus bellulus 185
Alpheus bisincisus 117, 185
Alpheus brevicristatus *183*, 185

Alpheus hippothoe 203, *275*, 276
Alpheus malleodigitus 271
Alpheus parvirostris 271
Alveopora 258, 261, 262, 271
Alveopora irregularis 262, Pls. 22 & 23
Amathia *81*, 83
Amathina tricarinata *165*, 194
Amblygobius albimaculatus 112, *113*
Amblyrhynchotus hypselogeneion 196, *197*
Amphiprion bicintus 203, Pl.24
Amphiprion percula 203, *292*, *295*
Amphiroa ephedraea 40
Amphiroa valonoides *297*, 298
Amphisbetia 79, 80
Amphiura sp. 52, 54
Ammodytes 198
Ammodytes personatus *199*
Amusium japonicum 100, *101*, 280
Amusium pleuronectes 100, 280, *281*
Anadara 179, 200, 207, 234
Anadara granosa 138, 226, 248
Anas acuta 241
Anas clypeata 233, 241
Anas crecca 223, 241
Anas penelope 241
Anas poecilorhyncha 241
Anas querquedula 241
Anchistus custos 202, *205*, 248
Anemonia sulcata 49, 299
Anguilla japonica *199*
Anodontia stearnsiana 248, 249
Anomalocardia 18, 179, 207
Anomalocardia flexuosa 164, *165*, 178
Anomalocardia squamosa 138, 162, 164, *165*, 178, 190
Anomia 101
Anomia achaeus 65, 69, 70, 71, 76, 90
Anomia ephippium 88, 90, 102, 246

Antedon cf. *bifida* 277
Anthenea aspera 277
Anthocidaris crassispina 36, 43, 49, 50, 203, 298, 299, 300
Anthopleura 116
Anthopleura dixoniana 111
Anthopleura japonica 50
Anthopleura pacifica 39, 49
Anthopleura nigrescens 111
Antipathes cf. *densa* 264, 267
Antirrhoea chinensis 87
Aphelasterias japonica 188
Aplysia 190, 192
Aplysia juliana 51
Aplysia sagamiana 51
Aplysilla 108
Apogon doederleini 292, 295
Arca avellana 98, 99, *101*, *121*, *124*, 270, 281, 282
Arcopagia diaphana 168, *169*
Arcopagia inflata 168, *169*
Archaster typicus 178, 188, *191*, 203, 205, Pl.15
Architectonica perspectiva 288
Arcuatula elegans 245, 246
Ardea cinerea 238, *239*
Ardea purpurea 238
Ardeola bacchus 238
Arenicola 196
Armatobalanus allium 203, 205
Armina japonica 290, 291
Asaphis 115
Asaphis dichotoma 114, *115*, 116, 168
Ascidia sydneiensis 76, 77
Asclepios shiranui coreanus 235, *251*
Asparagus lucidus 87
Asparagopsis taxiformis 40
Aspidopholas 245
Aspidopholas obtecta 245, *247*
Assiminea 218
Assiminea brevicula 216
Assiminea dohrniana 235
Assiminea lutea 216
Assiminea subeffusa 216
Assiminea violacea 212, 216, 222
Asthenognathus inaequipes 204
Astraea rhodostoma 49, 50
Astropecten 188

Atactodea 115
Atactodea striata 114, *115*, 143, 166
Atalantia buxifolia 87
Atergatis floridus 272, 273
Athanas dorsalis 203, *300*
Atheleges takanoshimensis 98, *160*
Atherina forskali 293
Atrina 100, 178, 198, 268
Atrina inflata 194, *195*
Atrina pectinata 194, *195*
Atrina penna 194, *195*
Atrina vexillum 165, 190, 194, *195*, 249
Aurelia aurita 143
Auriculastra duplicata 212, 216, 217
Auriculastra subula 216, 217
Avicennia 235
Avicennia marina 181, 220, 221, 222, 223, 224, 234
Axonopus compressus 208, 210, 214

Babylonia areolata 156, 284, *285*
Balanus 104
Balanus albicostatus 67, 87, 116, *117*, 248
Balanus amphitrite 66, 116
Balanus amphitrite amphitrite 70, 75, 76, 234
Balanus pallidus 68
Balanus retriculatus 75, 116
Balanus tintinnabulum 35, 45, 47, 294, 296
Balanus tintinnabulum volcano 36, 38, 43, 70
Balanus trigonus 75, 99, 104, *117*
Balanus variegatus variegatus 70, 116
Balcis kuronamako 203, 205
Balcis shaplandi *191*, 203, 205
Balanoglossus 179, 196, 205
Balanoglossus misakiensis 174, 194, 204
Balistapus undulatus 292, 295
Bangia fusco-purpurea 35, 36, 37, 40, 47
Bankia 145
Bankia carinata 84
Barabattoia mirabilis Pl.22
Barbatia helblingi 270, 281, 282
Barbatia obliquata 270, 282
Barbatia virescens 51, 65, 66, 69, 70, 71, 77, 90, 98, 99, 138
Barentsia 78
Barnea 268

Barnea cf. *dilatata* 121, 124
Batillaria 153, 156, 160, 223, 247
Batillaria cumingii 149, 158, 184
Batillaria multiformis 149, 158, 160, 184, 202
Batillaria sordida 49, 50, 298, Pl.1
Batillaria zonalis 149, 158, 160, 179, 184, 202
Bdella 122
Beania 81, 82
Beroe cucumis 143
Berthellina delicata 289, 290, 291
Biemna 108
Boleophthalmus boddaerti 234, 235, 242
Bolerceroides mcmurrichi 170, 191
Boodleya composita 61
Bornella digitata 290, 291, Pl.27
Bostrychia 234
Bostrychia binderi 40
Bostrychia tenella 40
Botaurus stellaris 238, 239
Botrylloides perspicum 77
Botryllus schlosseri 75, 76, 77
Botula 100
Botula cinnamomea 269
Botula silicula 245, 247, 269
Bougainvillia 80
Bowerbankia 83
Brachidontes 104, 245
Brachidontes atratus 66, 99, 217, 246, 248
Brachidontes variabilis 104
Branchiostoma belcheri 196, 199
Bruguiera conjugata 218, 220, 221, 223, 224, 234, Pl.16
Bryopsis plumosa 42
Bryothinusa 144
Bubulcus ibis 238
Bugula 75, 81
Bugula neritina 82, 289
Bulla ampulla 161, 162, 192
Bursa granularis 96, 99, 103
Bursa rana 104
Bursatella 192
Bursatella leachii 51, 190, 193
Butorides striatus 238

Caecella chinensis 114, 115, 116, 138, 166, 217

Caesalpinea crista 225
Calappa 157, 179
Calappa calappa 158, 187
Calappa hepatica 157, 187
Calappa philargius 187
Calcinus sp. 96, 97
Calidris alpina 238
Calidris ferruginea 229
Calidris subminuta Pl.20
Calliactis japonica 188
Callianassa 179, 185
Callianassa japonica 183, 186, 229, 230
Callianassa petalura 186
Callionymus richardsoni 196, 197
Callyspongia confoederata 107, 108
Caloglossa leprieurii 65, 68
Caloplaca 37, 58
Calothrix scopularum 58
Calyptraea sakaguchi 103, 109
Campanularia 79, 80
Campanulina 80
Canavalia 234
Canavalia maritina 140, 223, 225, Pl.16
Cantellius 205
Cantellius pallidus 203
Capitella 171
Caprella acutifrons 52, 299
Caprella aequilibra 297
Caprella simia 52
Capulus yokoyami 245, 247
Caranx kalla 112, 203, 293
Caranx malabaricus 112, 203, 293
Carcinactis ichikawai 187
Cardita leana 51, 52, 99, 100
Carpilius maculatus 272, 273
Casella atromarginata 289, 290, Pl.27
Cassidula plectorematoides 212, 216, 217, 218
Cassidula schmackeriana 212, 216, 217, 218
Casuarina equisetifolia 140
Catenella nipae 68
Caulacanthus okamurai 42
Caulerpa cupressoides 61
Caulerpa peltata 48, 61, 291, 294, 295, 296, 297
Caulerpa scalpelliformis 48, 61
Caulerpa sertularioides 61
Caulerpa taxifolia 61

Cavernularia obesa 267, 268
Cellana grata 36, 44, 45, 58, 298
Cellana toreuma 62, 63, 90, *109*, 110, 298, Pl.1
Centropus sinensis 241
Ceratodictyon spongiosum 108
Ceratonereis sp. Pl.9
Ceratonereis mirabilis 278, 279
Ceratonereis nuntia var *brevicirrus 115*, 116
Ceratonereis nuntia var. *vallata* 121, 122
Cerianthus 178, 179, 190, 205, 245, 264
Cerianthus filiformis 170, *191*, 192, 203, Pl.9
Cerithidea 156, 158, 179
Cerithidea cingulata 226
Cerithidea ornata 218, 222, 226, Pl.19
Cerithidea rhizophorarum 160, 222, 226, Pl.19
Cerithideopsilla cingulata 246, Pl.19
Cerithideopsilla djadjariensis 222, 245, 246, Pl.19
Ceroplastes rubens 223
Ceryle rudis 240
Chaetangium sp. 300
Chaetodon modestus 292, 295
Chaetomorpha 69
Chaetomorpha antennina 42, *48*, 50, Pl.5
Chaetomorpha brachygona 61
Chaeturichthys hexanema 198
Chama 77
Chama dunkeri 90, 295, 296, 298, 299
Chama reflexa 88, 90, 99
Chamaesipho scutelliformis 68, 119, *121*
Champia parvula 42
Charadrius alexandrinus 191, 233, 238, Pl.20
Charybdis 145
Charybdis bimaculata 94, 201
Charybdis cruciata 94
Charybdis feriatus 143
Charybdis japonica 94
Charybdis miles 94
Charybdis truncata 94, 201
Chasmagnathus 234
Chasmagnathus convexum Pl.19
Chelidonichthys kama 196, *197*, Pl.15
Chelone 245
Chelone mydas 144

Chelura terebrans 83, *84*
Chicoreus microphyllus 99, 104, *105*
Childonias hybrida 240, 241
Childonias leucoptera 240
Chiromanthes bidens 215, 220, 227, *228*, Pl.19
Chiromanthes fasciatum 213, 214, 215, 220
Chiromanthes maipoensis 215, 228
Chironomus sp. 51
Chlamys 100, *101*
Chlamys irregularis 281
Chlamys pyxidatus 280
Chloeia flava 172
Chlorodiella nigra 272, 273
Chlorostoma argyrostoma 109
Chlorostoma nigerrima 102, 110, Pl.6
Chlorostoma rustica 50, 96, 102, 110
Chiorostoma xanthostigmata 109, 110
Chromis notatus 292, 295
Chromodoris festiva 289, Pl.27
Chromodoris lineolata 289, 290, Pl.27
Chromodoris pallescens Pl.27
Chryseulima philippinarum 203
Chrysophrys major 112
Chthamalus malayensis 35, 36, 39, 68
Cicindela 144
Cicindela anchoralis 143
Ciconia nigra 238
Ciona intestinalis 76, 77
Circe scripta 164, *165*
Circus aeruginosus 240
Cirratulus cirratus 118, *123*
Cirrhipathes anguina 264
Cirrhipathes spiralis 265, 267
Cirriformia tentaculata 194
Cladiella digitulata 266
Cladium chinense 37
Cladophora delicatula 61
Clanculus 102
Claudioncha japonica 77, *121*, 124
Clerodendron inerme 37, 140, 181, 221, 225
Clibanarius 96
Clibanarius bimaculatus 50, *96*, 97, 98, 156
Clibanarius infraspinatus 97, 156, *160*, 201
Clibanarius longitarsus 97, 156, *160*
Clibanarius striolatus 96, 97, 201
Clibanarius virescens 96, 97
Cliona cellata 108, 247

Clistocoeloma merguinense 227, 228
Clithon faba 73, 245
Clithon oualaniensis 73, 139, 184, 222, 226, 245
Clithon cf. *retropictus* 72, 73, 245
Clypeaster virescens 189, *191*
Clypeomorus humilis 99, 103, 158, 160, 161, 184, 226
Clypeomorus moniliferum 222, 226
Clytia 79, 80
Codakia 168, 249
Codakia tigrina 165, 166
Codium 76
Codium cylindricum 61, 76, 77
Codium intricatum 61
Coelomactra antiquata 16, 165
Collisella dorsuosa 36, 44, 62
Collisella luchuana 62
Colpomenia 57, 59, 61, 66, 76, 88
Colpomenia sinuosa 49, 50, 67, 77, 89, 298
Comanthus parvicirrus Pl.24
Comatula salaris 277
Conchocoetes artificiosus 95
Conus ebraeus 287
Conus textile 286, 287
Conus sulcatus 286, 287
Corallina (turf) 35, 36, 39, 40, 51, 52, 57, 59, 89, 116
Corallina pilulifera 40, 52
Corallina sessilis 40, 52
Coralliophaga coralliophaga 270, *281*
Coralliophila costularis 288
Corallium 268
Corbula 166, *169*
Corella japonica 99
Coryne 80
Coscinaraea columna 262
Coscinasterias acutispina 43
Crassostrea 101
Crassostrea gigas 14, 108, 202, 242, 245, 246, 247
Crassostrea rivularis 242
Crenatula 200
Crenatula modiolaris 108, 201, 281, 282
Crepidula onyx 103, *109*
Creusia 205
Creusia indica 203
Creusia spinulosa 269, 271

Crinum asiaticum 221
Crisia 81, 82
Cronia margariticola 104, 105, 289
Culcita novaeguinae 277
Culex sitiens 51
Culicea japonica 99
Culicoides 235
Curvemysella paula 156, 200, 201
Cyanea nozakii 143, 203, 293
Cycladicauma cumingii 248, 249
Cyclina orientalis 164, 247
Cyclina sinensis 165
Cyclograpsus intermedius 92, 93, 110, 114
Cymatium echo 99, 103
Cymbium melo 284
Cynodon dactylon 87, 208, 210
Cynoglossus lineolatus 197
Cynoglossus robustus 196, *198*
Cyperus 210, 218, 219
Cyperus javanicus 37
Cyphastrea 263
Cyphastrea microphthalma 262
Cyphastrea serailia 262, 299, Pl.26
Cypraea arabica 99, 103, 283
Cypraea asellus 283, 287
Cypraea erronea 103, 283, 287
Cypraea gracilis 283
Cypraea onyx 283
Cyrtymenia sparsa Pl.5

Daicocus peterseni 196, *197*
Daira perlata 272, *273*
Dardanus haani 188
Dasson japonica 112, *113*
Dasson variabilis 197, *198*
Decapterus maruadsi 293
Dendrodoris denisoni 290, 291
Dendrodoris guttata Pl.27
Dendrodoris rubra 290, Pl.27
Dendronephthya gigantea 266, Pl.23
Dendrophyllia 265
Dendrophyllia gracilis 264
Dendrostomum 119, *121*
Dermonema frappieri 47, *48*, Pl.5
Derris 222, 223, 234
Derris trifoliata 181, 221, 225
Diadema 205, 295
Diadema savignyi 275

Diadema setosum 203, 275, 276, 299, Pl.24
Dictyosphaeria cavernosa 48, 61
Dictyota dichotoma 61
Diminovula punctata 266
Diodora reevei 99, 103
Diogenes edwardsii 97, 98, 156, 160, 200, 201
Diopatra neapolitana 173, *174*
Discodoris fragilis 291
Donax 17, 146, 147, 153, 179
Donax cuneatus 138, *139*, 141, 166, 180
Donax semigranosus 139, 141, 142, 166
Doridium cyaneum 162, *193*
Doriopsilla areolata 291
Dorippe granulata 187, 201
Doris verrucosa 289
Dosinorbis 234
Dosinorbis japonica 248
Dostia violacea 61, 73, 212, 218, 222, 226, 245
Dromia dormia 95, 186, *187*
Drupella rugosa 104, *105*, 289
Dussumieria hasseltii 293

Ebalia scabriuscula 157
Ecdysanthera micrantha 225
Echinaster luzonicus 276, 277
Echineulima mitterei 203
Echinogorgia 267, 268
Echinophyllia 258, 263
Echinophyllia aspera 263, Pl.26
Echinothrix calamaris 275, 276
Ectocarpus 66, 67, 189, 294, 296
Edwardsia japonica 149, *170*
Egretta alba 238
Egretta garzetta 238, 239
Egretta sacra 54, 65, 238, Pl.2
Electra 81
Electra angulata 82
Electroma 101
Electroma japonica 76, 77
Eleocharis afflata 37
Eleusine indica 208, 210
Ellobium 179
Ellobium chinense 212, 216, 217, 218
Ellobium polita 216, 217
Emarginula bicancellata 103
Endarachne binghami 59

Enhydris 242
Enhydris bennetti 236, 241
Enigmonia 101
Enigmonia aenigmatica 102, 227, 248
Ensis 207
Enteromorpha 186, 192, 242, 298
Enteromorpha compressa 58, 66, 67, Pl.5
Enteromorpha intestinalis 294, 296
Enteromorpha linza 60
Enteromorpha prolifera 60, 67, 77
Entomacrodus stellifer 50, 51
Ephippodonta oedipus 203, 275, 276
Epiactis japonica 44
Epinephelus akaara 112
Epinephelus awaora 112
Epinephelus bruneus 112, *113*
Epinephelus fario 292, 295
Epitonium sp. 44
Epitonium scalare 288
Epixanthus frontalis 92, 93, 114, *115*, Pl.6
Ergalatax contractus 104, 105
Erichthonius pugnax 83, 84
Eriocheir sinensis 213, 214, 215
Eriphia laevimana smithii 42, 43
Eriphia sebana 273
Erithacus svecica 241
Etisus laevimanus 272
Euchelus 102
Eucrate alcocki 98
Eudendrium 80
Eudistoma 76
Eulalia viridis 122, *123*
Eunatica papilla 159, *163*
Eunice australis 278, 279
Eunice indica 172
Eunice viridis 122
Euplexaura 268
Euprymna berryi 169, 199
Euraphia withersi 39, 65, 68, 70, 119, *121*, 248
Excoecaria 208, 222, 223, 224
Excoecaria agallocha 181, 183, 218, 221

Falco peregrinus 54
Favia 263, 270
Favia favus 262
Favia lizardensis 262
Favia pallida 262, 299, Pl.25

Index of Scientific Names

Favia speciosa 262, 299, Pl.23
Favites 262, 263
Favites abdita 262, 299
Favites flexuosa 262, 299
Favites pentagona 262, 271, 299, Pl.25
Ficus gracilis 284, 285
Ficus microcarpa 140
Ficus superba 87, 140
Ficus superba japonica 140
Fimbristylis 210
Fimbristylis cymosa 37
Fimbristylis dichotoma 223
Fimbristylis spathacea 218, 219
Flustrella 83
Flustrellidra 81
Fluvia aperta 138
Fronsella 200
Fronsella oshimai 204
Fugu vermicularis 198
Fulica atra 233, 239
Fulvia 207
Fulvia mutica 165, 166, 178
Fungia 261

Gaetice depressa 92, 93, 110, 114, *115*, 246
Gafrarium 179
Gafrarium pectinatum 226
Gafrarium tumidum 164, 202, 207, 226, 248
Galathea elegans 272, 274
Galathea orientalis 274
Galathea subsquamata 95
Galaxaura fastigiata 48, 49, 59
Galaxea astreata 263
Galeomma 200
Galeomma takii 117, 118
Gallinago gallinago 233, 238
Gallinago megala 238
Gallinago stenura 238
Gastrochaena cuneiformis 268
Gastrochaena interrupta 269
Gastrochaena laevigata 269
Gelidiopsis variabilis 42
Gelidium amansii 35, 47, *48*, 49
Gelidium pusillum 56, 58, 59, 89
Gelochelidon nilotica 241
Geloina 227
Geloina erosa 202, 217, 226, 248
Gigartina intermedia 48, 49, 60, Pl.5

Glauconome 234
Glauconome chinense 248
Gloiopeltis furcata 42, 49, 59, 295, 297, 298, Pl.5
Glycera rouxii 171, *172*
Glycymeris 101
Gobius 198
Gomontia sp. 87
Gomphina meleagris 165
Goniastrea 263
Goniastrea aspera 262, 299, Pl.25
Goniodoris glabrata 291
Goniopora 258, 261, 262
Goniopora columna 262, Pls. 22 & 23
Goniopora stutchburyi 262
Gonocephalum pseudopubens 144
Gonodactylus falcatus 272, 273
Gonodactylus mutatus 273
Gonothyrea 79
Gracilaria chorda 193
Grapsus albolineatus 42, 43, 93, 110
Grateloupia filicina 42
Grateloupia imbricata 42
Grateloupia livida 42
Gregariella coralliophaga 270
Gymnodinium microadriaticum 255
Gymnodoris citrina 290
Gymnogongrus flabelliformis 41, Pl.5
Gymnothorax undulatum 292, 295

Halcium 80
Halcyon pileata 222, 240
Halcyon smyrnensis 240
Halgerda japonica 289, 290
Haliaetus leucogaster 54, 57, Pl.2
Halichoeres nigrescens 112
Halichoeres tenuispinis 292, 294
Halichondria okadai 107
Haliclona permollis 107, 111
Haliotis 102, 159
Haliplanella luciae 56, 76, 77, 88, 90, 117, 170, 192
Haloa japonica 192, *193*
Halophila beccarii 223, 245, 250, *251*, Pl.16
Halophila ovata 190, 207, 223, 245, 249, 251
Halosydna brevisetosa 277

Harmothoe 246
Harmothoe imbricata 117, 118, *123*
Harpiosquilla japonica 230, *231*
Hebella 80
Helice 234
Helice tridens 227, *228*
Hemifusus ternatana 156, 284, *285*
Hemigrapsus penicillatus 245, *246*
Hemigrapsus sanguineus 92, *93*, 110
Hemiramphus sajori 293
Herdmania momus 76, *77*
Heritiera 223
Heritiera littoralis 220, 221, *224*
Heteromacoma 234
Heteromacoma irus 248
Heteropilumnus ciliatus 92, *273*
Hiatella flaccida 99
Hiatella orientalis 121, *124*
Hibiscus 88, *115*
Hibiscus tiliaceus 37, 87, 181, *183*, 221
Hildenbrandtia 56, 57, *58*, 63, 89, 298
Hippa pacifica 139, *143*
Hippocampus cf. *kuda* 46
Hippothoa 81
Hippothoa hyalina 82
Hircinia 108
Hirundo rustica 241
Histrio histrix 45, *46*, Pl.1
Holometopus dehaani 213, *215*
Holometopus serenei 213, *214*, 215, 220
Holothuria arenicola 111, *275*
Holothuria leucospilota 109, 111, 203, 205, *275*, Pl.6
Homoiodoris japonica 289
Huenia proteus 46, *47*
Hydnophora 271
Hydnophora exesa 262, *263*, Pls. 23 & 25
Hydractinia 80
Hydroclathrus clathratus 61, 297, *298*
Hydrogamasus 122
Hydroides 88, 122
Hydroides dirampha 278
Hydroides elegans 77, *90*, 121
Hydrophasianus chirurgus 239
Hydrophis cyanocinctus 241
Hydroprogne caspia 55
Hypnea cervicornis 41
Hypnea japonica 41

Hypnea musciformis 41, 294, *295*, 296, 297, 298
Hyotissa hyotis 297, *299*

Ianthina 145
Ianthina globosa 143
Ibacus ciliatus 236, *237*
Ibla 119, 121, *124*
Ibla cumingii 121
Ilyoplax tenella 185
Iphione muricata 275, 276, *277*
Ipomoea 35
Ipomoea pes-caprae 37, *139*, 140
Iodictyum axillare 76, *82*
Ircinis sp. *107*
Irus mitus 56
Ischaemum indicum 218, 219, *233*
Ischnochiton coreanicus 91
Ischnochiton lepidus 91
Isognomon 101
Isognomon acutirostis 63, *64*
Isognomon legumen 121, 124, *281*, 282
Isognomon isognomum 100
Istiblennius tanagashimae 51
Ixeris repens 140
Ixobrychus cinnamomeus 238
Ixobrychus eurhythmus 238
Ixobrychus sinensis 238, *239*

Jania 61
Jania undulata 40, 51, *52*
Junceella 268
Junceella juncea 267

Kandelia 208, 218, 220
Kandelia candel 68, 221, 223, 224, 233, *234*, Pls. 11 & 16

Labroides dimidiatus 112
Lactoria cornutus 292, *295*
Laemodonta 218
Laemodonta exaratum 216, *217*
Laemodonta punctatostriata 109, 216, *217*
Laemodonta punctigera 216, *217*
Lambis lambis 285, *287*
Lambrus validus 187
Lanice conchilega 174, *194*
Lantana 183

Lantana camara 221
Laomedia astacina 229, *231*
Larus argentatus 55, Pl.2
Larus ridibundus 55, 241
Larus saundersi 241
Lateolabrax japonicus 237
Laternula 207
Laternula truncata 202, *248*, 249
Lates calcarifer 236, 237
Launaea sarmentosa 140
Laurencia japonica 42, *48*, 50
Laurencia undulata 43, 50
Lecanora 37
Lecidia 37
Leonnates nipponicus 278
Lepas antifera 38, 143, *145*
Lepas fascicularis 143
Lepidonotus squamatus 277, *279*
Lepidozona coreanicus 88
Leptastrea pruinosa 262, 299
Leptastrea purpurea 262, Pl.26
Leptomithrax sp. 272
Leptoseris mycetoseroides 262, Pl.21
Leptosynapta ooplax 89, 111
Leucandra sp. 107
Leuconia solida 106
Leucosia craniolaris *155*, 157
Liagore rubromaculata 272, 273
Lichenopora *81*, 83
Ligia 97
Ligia exotica 38, 87, 93, *109*, 110
Lima 101
Lima lima 280, *281*
Limaria fragilis 280, *281*, 282
Limaria hongkongensis 282, Pl.24
Limnoria 97
Limnoria tripunctata 83, *84*
Limonium sinense 212
Limosa lapponica 230, 233, 238, Pl.20
Lineus 177
Lingula 179, 196
Lingula lingua 151, *152*, 190, 207, Pl.15
Liolophura japonica 36, 45, 202, Pl.1
Liomera venosa 272, 273
Lipura 122
Liriope tetraphylla 143
Lithophaga 101, 247, 269
Lithophaga antillarum 270

Lithophaga hanleyana 270
Lithophaga lima 270
Lithophaga malaccana 56, 100, 270
Lithophaga mucronata 270
Lithophaga simplex 269, 270
Lithophaga teres 269, 270
Lithophyllon 258
Lithophyllon edwardsi 261, Pl.22
Lithophyllum 35, 36, 45, *50*, 294, 296
Litsea glutinosa 181
Littorina 63
Littorina brevicula 62, 66, 72, 87
Littorina melanostoma 62, 68, 72, 212, 214, 222, 226
Littorina scabra 62, 65, 66, 68, 69, 72, *88*, 214, 222, 226
Lobophytum 266
Loimia medusa 174, *194*, 204
Lonchura malacca 241
Lophiotoma leucotropis 286
Lovenia elongata 189, *191*, Pl.15
Loxosomella 78
Luidia hardwicki 188
Luidia longispina 188
Luidia maculosa 191
Lumnitzera 208, 218, 222
Lumnitzera racemosa 221, 223, 224, Pl.16
Lunella 51
Lunella coronata 63, 64, 76, 91, 96, Pl.6
Lutjanus argentimaculatus 112
Lutjanus johni 112
Lutjanus russelli 112, 113
Lutjanus sanguineus 112, 113
Lyrodus 145
Lyrodus medilobatus 84
Lyrodus pedicellatus 84
Lyrodus tristi 84
Lysidice ninetta 278, 279

Macaranga tanarius 140
Macoma 207
Macrobrachium nipponense 236, 237
Macrophiothrix 277
Macrophiothrix longipeda 203, 273, 275, 276, Pl.24
Macrophiothrix variabilis 275
Macrophthalmus 207, 234
Macrophthalmus boteltobagoe 157, 229

Macrophthalmus convexus 156, 201, 229
Macrophthalmus definitus 157, 223, 227
Macrophthalmus latreille 157, 227, *231*
Macropipus corrugatus 155, 158, 186
Mactra veneriformis 165, 166
Magilus striatus 269, 286
Maldanella harai 173, *174*
Malleus 101, 268
Malleus malleus 100, 282
Malleus regulus 282
Malosydna brevisetosa 279
Malthopsis annulifera 197, 198
Mancinella echinata 105
Marcia hiantina 166, *167*
Marcia japonica 166, *167*, 202
Marcia marmorata 166, *167*
Marphysa 173
Marphysa adenensis 270
Marphysa sanguinea 116, *121*, *122*, *123*, 270
Martesia striata 83, *84*, 124
Mastigias papua
Matuta 179
Matuta lunata 155, 157, 178, 187
Melampus triticens 212, 216, 217
Melo melo 285
Membranipora 81, 82
Menaethius monoceros 46, 47
Meropesta nicobarica 165, 166
Meretrix 17, 18, 144, 179
Meretrix meretrix 138, *149*, 150, 151, 164
Mesochaetopterus japonicus 173, *174*, 204
Metapenaeus intermedius 236, 237
Metopograpsus frontalis 92, 93, 110
Metopograpsus messor 227, *228*, 246
Micippa philyra 47, 274
Microcanthus strigatus 292, 295
Mictyris 157, 178, 185, 190, 207
Mictyris longicarpus 153, 154
Milvus migrans 54, 55
Modiolus 51
Modiolus agripetus 52, 76, 77
Modiolus elongatus 65
Modiolus metcalfei 245, 246
Moerella juvenilis 207, 248
Monodonta 51
Monodonta australis 63, 64, 90, 96, *109*, 110, 214, Pl.6

Monodonta neritoides 109, 110
Monostroma 189, 192
Monostroma nitidum 67
Montacutona 200
Montacutona olivacea 192
Montastrea curta 262
Montipora 258, 260, 270
Montipora informis 259
Montipora striata 259
Montipora venosa 259, Pl.21
Morinda umbellata 87
Mortensella forceps 176, 177, 204
Morula margaratifera 64, 99
Morula marginata 99
Morula musica 36, 45, 64, 68, 69, 91, 96, 105
Morula spinosa 105
Motacilla flava 241
Mucronalia fulvescens 191, 203, 205
Mugil cephalus 112, *113*, 236, 237
Mugil engeli 198
Murex torrefactus 160
Musculista senhausia 77, 100, 138, 245, 246
Mycale 108
Mylio berda 112
Mylio latius 112, 236, 237
Mylio macrocephalus 237
Myra fugax 187
Mytilicola intestinalis 202
Myxicola infundibulum 278, 279

Nanosesarma minutum 92, 93, *154*, 158, 215
Nassarius 247
Nassarius festivus 159, *161*
Nassarius nodiferus 159, *161*
Nassarius pullus 149, 159, *161*, 170, 202
Nassarius teretiusculus 159, *161*
Natica 162
Natica alapapilionis 159, *163*
Natica gautteriana 159, *163*, 180
Natica lineata 159, *163*
Natica maculosa 159, *163*
Natica tigrina 247
Natica vitellus 159, *163*
Naxoides hystrix 187
Neanthes unifasciata 278, 279
Neanthes oxypoda 279

Nemertopsis 177
Nemertopsis gracilis 39, *121*, 201
Neofibularia 108
Neosarmatium punctatum 213, 214, 215
Nephthya 266, 271
Nephtys ciliata 171, *172*
Nereis succinea 171, *172*
Nerita albicilla 63, 64, 68, 71, 73, 91, 96, 109, 110, 298, Pl.6
Nerita chamaeleon 64, 65, 73, *109*
Nerita costata 64, 73
Nerita lineata 72, 73, 218, *222*, 226
Nerita plicata 64, 73
Nerita polita 73, *109*, 110, 116
Nerita undata 63, 68, 69, 73, 91, *109*, 110, *160*, 222, 246
Nerita yoldii 212, 214, *222*, 226
Nerocila phaeopleura 97
Neyraudia reynaudiana 140
Nipponomysella 200
Nipponomysella subtruncata 204
Nobia 205
Nobia conjugatum 203
Noctiluca scintillans 206, 255
Nodilittorina sp. 35, 36, 57, 58
Nodilittorina exigua 37
Nodilittorina millegrana 35, 37, 56, 62, 63, 66, 69, 72, 87
Nodilittorina pyramidalis 35, 37, 56, 62, 63, 66, 69, 72, 87
Notoacmea concinna 62, 63, *88*, 90
Notoplana delicata 121
Notosinister granulata 52, 53
Numenius arquata 233, 238
Numenius phaeopus 233, 238, Pl.20

Obelia 79, 80
Obisium 122
Ochetostoma 177
Ochetostoma erythrogrammon 175, *176*, *191*, 204, Pl.9
Ochetoclava sinensis 99, 103, 298
Octolasmis 205
Octolasmis neptuni 201
Octolasmis tridens 201
Octolasmis warwicki 201
Ocypode 140, 141, 147, 178, 179, 181
Ocypode ceratophthalma 141, 142, 153, *154*

Ocypode cordimana 141, 142, 153
Ocypode mortoni *139*, 141, 142, 153
Octopus dollfusi 200
Octopus membranaceous *169*, 200
Oligometra serripinna serripinna 277
Oliva annulata *161*, 162
Omobranchus loxozonus *121*, 125
Omobranchus uckii 245
Onchidium 212
Onchidium verraculatum 68, 217
Onithochiton hirasei 36, 45
Ophiactis savignyi 275, 276
Ophiarachnella gorgonia 275, 276
Ophichthys apicalis 199
Ophidiaster cribrarius 43
Ophiomastix mixta 111
Ophioplocus japonica 275
Ophiotrichoides gratilla 275
Opisthoplatia orientalis 143
Opuntia dillenii 140
Oratosquilla oratoria 200, 201, 230, *231*
Orbinia 171
Orchestia platensis 52
Orthopyxis 79, 80
Oudemansia esakii 218
Oulastrea crispata 262, 299, Pl.25
Ovula ovum 283
Ovalipes 179, 186
Ovalipes punctatus *155*, *158*

Pachycheles pisum 274
Pachycheles sculptus 272, 274
Padina australis 60
Padina durvillei 297, 298
Paederia scandens 87
Pagurus samuelis 188
Pagurus trigonocheirus 96, 97
Palaemon pacificus 297, 299
Palaemon serrifer 229, *231*, Pl.15
Palaemonetes 229
Palola siciliensis 270
Palythoa 264, 266
Pandanus 222
Pandanus tectorius 37, 140, 181, 221
Pandion haliaetus 240
Panicum repens 218, 219
Panulirus stimpsoni 201

Paphia cf. *philippinarum* 116
Paphia undulata 166, *167*
Paracentropogon indicus 112, *113*
Paracondylactis hertwigi 149, 170
Paralepida takii 117
Paralepidonotus ampulliferus 172
Paranthus sociatus 149, 170, 202
Parapilumnus trispinosus 273
Paraplagusia blocki 196
Parasalenia gratiosa 275, 276
Parasesarma affinis 215
Parasesarma pictum 87, *92, 93*, 110, Pl.6
Parasesarma plicatum 183, 215
Parasicyonis actinostoloides 203, Pl.24
Paspalum vaginatum 210
Patella flexuosa 35, 45
Patelloida pygmaea 63, 68, *88*, 90, 110, *121*, 158
Patelloida lampanicola 158, 202
Patelloida saccharina 36, 44, 58, 90, Pl.1
Pavona 261
Pavona decussata 258, 262. Pls. 21 & 23
Peasiella sp. 62, *63, 72*
Pecten 100, 280
Pectinaria hyperborea 173, *174*
Pedicellina 78
Pedum spondyloideum 280
Pegasus volitans 197, 198
Pelagia panopyra 143
Pelecanus crispus 241
Pelecanus philippensis 239, 241
Penaeus 27
Penaeus japonica 236
Penaeus latisulcatus 236
Penaeus monodon 236, *237*
Penaeus semisulcatus 236, *237*
Pennaria 80
Pentaceraster regulus 277
Percnon 93
Percnon affinis 42, *43*, Pl.1
Periclimenes demani 250, *251*
Perinereis cultrifera 123
Periophthalmus cantonensis 234, 244, Pl.19
Perna 101
Perna viridis 51, 69, *71, 76*, 77, 99
Peronella lesueuri 189
Petalonia 59
Petalonia fascia 57, 58, 60, Pl.5

Petrolisthes coccineus 42, *43*, 274
Petrolisthes japonicus 92, 95, 109, 110
Petrospongium 59
Petrospongium rugosum 60
Phalacrocorax carbo 54, *55*, Pl.2
Phalium glaucum 285
Pharaonella perna 138, 168, *169*
Pharella 234
Pharella acuminata 248
Phascolosoma scolops 245, 246, 269, 270
Phestilla melanobrachia 291
Phestilla sibogae 291
Philine orientalis 161, 162, 192
Philyra pisum 155, 157
Phoronis australis 191, 192, 203, 205
Phragmites 132, 233, 234, 241
Phyllanthus cochinchinensis 87
Phyllidia varicosa 291
Physcia 37
Pilumnus longicornis 273
Pilumnus minutus 43, *52*, 273
Pilumnus verspertilio 186
Pinctada 101
Pinctada furcata 99, 100
Pinctada margaratifera 100, 299
Pinctada martensii 99, 100, *138*
Pinctada maxima 100
Pinna 100, 179, 192, 205
Pinna atropurpurea 194, *195*, 202, 249
Pinna attenuata 194, *195*
Pinna muricata 194, *195*
Pinnixa 205
Pinnixa balanoglossana 174, 204
Pinnixa rathbuni 78, 204
Pinnotheres 205
Pinnotheres affinis 202
Pinnotheres cyclinus 165, 202
Pinnotheres dilatatus 202
Pinnotheres sinensis 202, 245
Pisaster 43
Pitar pellucidium 165
Placamen tiara 164, *165*
Placuna 101, 179, 207, 234, 280
Placuna placenta 102, 248, 249
Plagusia 93
Plagusia dentipes 42
Plagusia depressa tuberculata 42, 43
Planaxis 51

Planaxis sulcatus 63, 64, 68, 88, 90, 96, 110, Pl.6
Platalea leucorodia 239
Platalea minor 239
Platydoris speciosa 290
Platygyra 263, 271
Platygyra daedalea 262
Platygyra pini 262
Platygyra sinensis 262, 299, Pl.25
Plesiastrea versipora 262, 263, Pl.22
Pleurobrachia globosa 143, 145
Plicatula 101
Plicatula plicata 102, 114, 115
Plotosus anguillaris 236, 237
Plumularia 80
Pluvialis dominicus Pl.20
Pocillopora 273
Pocockiella variegata 61
Podarke angustifrons 203
Podiceps ruficollis 239
Polinices 162
Polinices didyma 159, 163
Polinices melanostomus 149, 159, 163
Polinices peselephanti 159, 163
Polinices tumidus 159, 163
Pollicepes 56, 68
Pollicepes mitella 35, 38, 39, 47, 58, 66, 67, 69, 70, 109, 110, 119, 121, 177, 201, 299, Pl.1
Polycera fujitai 82, 289, 290
Polycheira 245
Polycheira rufescens 89, 109, 111, 114, Pl.6
Polydora 245, 247
Polymastia 108
Polysiphonia spp. 42, 68
Pomacanthus imperator 292, 295
Pomatoceros triqueter 77
Pomatoleios kraussi 88, 89
Porcellana ornata 272, 274, 298, 300
Porcellana picta 203, 267, 268
Porites 258, 259, 261, 270
Porites lobata 262, 280, 299, Pl.22
Porphyra 56, 58, 68
Porphyra dentata Pl.5
Porphyra suborbiculata 35, 36, 38, 40, 44, 47, 49, 67, 109
Porpita 145
Porpita porpita 143

Portunus gladiator 94
Portunus pelagicus 94, 201
Portunus sanguinolentus 94, 186
Portunus trituberculata 94
Potamon hongkongensis 213, 214, 215, 220
Prinia flaviventis 241
Prinia subflava 241
Proclava kochi 99
Protankyra bidentata 174, 189
Protobonella 278
Protoeces ostreae 202
Psammocora 262
Psammocora contigua 260
Psammocora haimeana 260
Psammocora superficialis 260, Pl.21
Pseudochama retroversa 90, 99
Pseudionella pyriforma 98, 160
Pseudopythina subsinuata 200, 201
Pseudosesarma patshuni 213, 214, 215
Pseudostegias setoensis 96, 98, 201
Pteria 101
Pteria brevialata 281
Pteria penguin 100
Pteroeides sparmanni 203, 267, 268
Pueraria phaseoloidea 140
Pycnonotus sinensis 241
Pylaiella 66, 67
Pyrene punctata 50
Pyrgoma 205
Pyrgoma sp. 269, 271
Pyrgoma cancellatum 203
Pyrogopsella 205
Pyrogopsella stellula 108, 201
Pythia cecella 216, 217, 218
Pythia fimbriosa 216, 217, 218

Ralfsia verrucosa 59, 60, 89
Rallus aquaticus 233, 239
Rallus striatus 239
Rapa rapa 286, 288
Reimarochloa oligostachya 140
Rhadosarga sarda 112
Rhizoclonium 67, 69
Rhizoclonium hookeri 61
Rhizoclonium riparium 77
Rhizophora 223, 230
Rostanga arbatus 291
Ruppia maritima 250

Ryenella cuprea 76, 77, 282

Sabellastarte indica 88, 90
Sabellastarte japonica 174, 194, Pl.9
Saccharum arundinaceum 140
Saccostrea 57, 104
Saccostrea cucullata 56, 59, 70, 71, 89, 90, 242, 245, 246, 248, 299
Sacculina 98, 205
Sagaretia theezans 87
Sakuracolis enoshimensis 291, Pl.27
Salmacis 178
Salmacis sphaeroides 188, 190, 191
Salmonella 207
Sanhaliotis planata 36, 45
Sarcophyton 283
Sarcophyton elegans 266
Sardinella aurita 293
Sargassum 35, 36, 40, 41, 42, 45, 47, 57, 58, 68, 257, 294, 296, 298, 300
Sargassum hemiphyllum 41, 50, 59, 60, 294, 295, 296, 297
Sargassum horneri 41, 45, 59, 60
Sargassum patens 41, 59, 60
Sarmatium germaini 215, 227, 228
Savignium crenatum 203, 205
Scaevola 35, 115
Scaevola frutescens 37, 181
Scaevola hainanensis 212
Scaevola sericea 139, 140
Scapharca 179, 207, 234
Scapharca cornea 226, 248
Scarus ghobban 292, 294
Scatophagus argus 293
Schizaster lacunosus 189, 191
Schizophrys aspera 274
Schizoporella 81
Schizoporella unicornis 82
Schoenus 218, 219
Scintilla vitrea 117, 118
Scirpus 218, 219
Scoloplos armiger 171, 712
Scopimera 18, 141, 142, 156, 178, 179, 183, 190
Scopimera globosa 183, 185, 207
Scopimera intermedia 153, 154, 185
Scutus unguis 99, 103
Scylla 98

Scylla serrata 201, 228, 229, 231
Scypha ciliata 106, 107
Scytosiphon 59
Scytosiphon lomentaria 60, 67, 68, Pl.5
Sebasticus marmoratus 112
Semicassis persimilis 284, 285
Sepia lycidas 199
Sepia pharaonis 145, 199
Septifer 101
Septifer bilocularis 36, 39, 51, 66, 69, 70, 71, 77, 99, Pl.1
Septifer virgatus 99, 281, 282
Sermyla tornatella 237
Serpula vermicularis 278, 279
Serpula watsoni 278
Serpulorbis imbricatus 65, 98, 99, 106, 299
Sertularella 79
Sertularia 79, 80
Sesarma 179, 207
Sesarmops sinensis 154, 158, 215, 220, 228
Siganus fuscescens 113
Siganus oramin 112
Sigaretotornus plana 176, 177, 204
Sigmadocia symbiotica 108, 201, 281, 282
Sillago sihama 196
Sinularia polydactyla 266
Sinum japonicum 159, 163
Siphonaria atra 62, 88, 90, 298
Siphonaria japonica 45, 58, 62, 63, 65, 68, 69, 298
Siphonaria sirius 36, 44, 45, 58, 298, Pl.1
Siphonochalina truncata 107, 108
Siphonosoma cumanense 175, 191, 204, Pl.9
Sipunculus nudus 175, 191, 204, Pl.9
Solecurtus divaricatus 165, 170
Solen 207
Solen corneus 248, 249
Solen strictus 149, 168
Soletellina diphos 168, 169, 178, 202
Soletellina olivacea 168, 169
Sphaeroma walkeri 83
Sphaerozius nitidus 92, 93
Sphyraena jello 112, 113
Spinifex littoreus 140
Spirastrella vagabunda 108, 201
Spirobranchus tricornis 278, 279
Spirorbis 104, 105, 122
Spirorbis foraminosus 77, 88, 90

Spondylus 101
Spondylus barbatus 102, 280, *281*
Spongia reticulata 108
Sporobolus fertilis 210
Sporobolus virginicus 37, 181, 223
Squilla 179
Stachyptilum dofleini 267
Stegocephalus inflatus 52, 53
Stenopus 275
Stenopus cf. *hispidus* 203, 273, 276
Sterna albifrons 241
Sterna hirundo Pl.2
Stichaster 43
Stichopus 275, 276
Stoichactis kenti 203, 292
Stomolophus meleagris 143, 145
Striarca afra 77, 98, 99
Strombus isabella 156
Strombus luhuanus 160, 285, *287*
Strombus vittatus 285, *287*
Struvea delicatula 61
Styela canopus 77
Styela clava 76
Styela plicata 76, 77
Stylariodes parmata 269, *271*
Stylocheilus longicauda 50, *51*
Stylochoplana pusilla 202
Stylochus ijimai 117, 118
Stylocoeniella 260
Stylocoeniella guentheri 259, Pl.21
Stylophora 258
Suaeda australis 212, 223, Pl.16
Suberites 108, 201, 281, 282
Sycon 106
Sycon okadai 107
Sygnathus acus 46
Symplectoscyphus 79
Synalpheus coutierei 271, 272
Synalpheus gravieri 271
Synthecium 79, 80
Syphopatella walshi 103, 156, *160*, 201

Tachypleus 27, 179
Tachypleus gigas 150
Tachypleus tridentatus 149, 150, 201
Tadorna tadorna 241
Talorchestia brito 117
Tapes 115

Tapes dorsatus 166, 167
Tapes literatus 166, 167
Tapes philippinarum 89, 114, *115*, 138, 164, 167
Tapes variegatus 166, *167*
Tectus pyramis 99, 102
Tedania 108
Tellinella 178
Tellinella virgata 168, *169*
Tellina vestalis 168, *169*
Tellinides 178
Temnaspis amygdalum 201, 205
Temnopleurus reevesi 178, 188, 190, Pl.15
Temnopleurus toreumaticus 188, *191*
Terebella ehrenbergii 123
Terebralia 158
Terebralia sulcata 160, 212, 214, 222, 226, Pl.19
Teredo 145
Teredo furcifera 84
Teredo navalis 84
Tethya aurantia 107
Tethya robusta 108
Tetraclita 67, 89
Tetraclita squamosa 35, 36, 38, 39, 44, 45, 47, 49, 56, 57, 58, *63*, 66, 68, 69, 70, 75, 294, 296, 298, Pl.1
Tetraclitella purpurascens 109, 110
Tetralia 273
Tetrodon spadiceus 198
Thais carinifera 245, 246, 247
Thais clavigera 45, 69, *105*, Pl.1
Thais luteostoma 36, 45, *63*, 96, 104, 105
Thais tissoti 105
Thalamita 98
Thalamita picta 94
Thalamita prymna 92, 94, 118, 158, 186
Thalassema 190
Thalassema fuscum 175, *176*
Thalassina anomala 229
Thalassoma lunare 292, 294
Thelepus setosus 117, 118
Thelidiopsis 37
Therapon jarbua 112, 198
Threskiornis melanocephalus 239
Thyone sp. Pl.24
Tilapia mossambica 237
Tmethypocoelis ceratophora 183, *185*

Tonna fasciata 284, 285
Tonna tessellata 156
Trapezia 273
Trapezium 270
Trapezium liratum 77, 100, *121*, 124, 245, 246
Trapezium sublaevigatum 66, 99, 100
Tricellaria 81
Tricellaria occidentalis 82
Trichocanace sinensis 235
Tridacna 282
Tridacna maxima 281, 283
Tridentiger 198
Tridentiger trigonocephalus 112, *113*
Tringa hypoleucos 54, 238
Tringa nebularia 233, 238, Pl.20
Tringa totanus 233, 238
Tripneustes gratilla 275
Trippa interciala 291
Trisidos semitorta 165, 170, 226
Tritodynamia 205
Tritodynamia horvanthi 174, 204
Tritodynamia rathbuni 174, 204
Tropiometra afra macrodiscus 276, 277
Truncatella sp. 214, *222*, 226
Trypauchen vagina 198, *199*
Tubastrea aurea 264
Tubastrea diaphana 264, Pl.26
Tubularia 80
Tubulipora 81, 82
Turbinaria mesenteria 264, 265
Turbinaria ornata 295, 297, 298
Turbinaria peltata 264, 265, Pl.26
Turbinaria trialata 41, 258
Turbo argyrostoma 99, 102
Turricula nelliae spurius 286
Turritella 234
Turritella terebra cerea 245, 247

Uca 18, 184
Uca arcuata 184, 185, *228*, 229, 234, Pl.19
Uca chlorophthalmus 183, 184, 185
Uca lactea 153, 178, 179, 181, 183, 185, 207
Uca vocans 184, 185, *228*, 229

Ulva 189
Ulva conglobata 58, 66, 67, 77, 294, Pl.5
Ulva fasciata 58, 67
Ulva lactuca 42, 58, 67, 68, 298
Ulva reticulata 60, *193*
Ulothrix sp. 51
Umbonium 14, 18, 144, 150, 151, 179
Umbonium costatum 148, *149*
Umbonium japonicum 289
Umbonium moniliferum 148, *149*
Upeneus tragula 112, *113*
Upogebia 179, 229
Upogebia major *183*, 185, 186
Urechis caupo 177

Varuna litterata 227, 228
Velella velella 143
Venericardia 100
Vepricardium sinense 166
Verrucaria 57
Verrucaria maura 35, 37, 58
Villogorgia 268
Virgularia 267, 268
Viriola tricincta 52, 53
Vitex rotundifolia 37, 87, 140
Volva 283
Volva brevirostris rosea 288
Vulsella 200
Vulsella vulsella 201, *281*, 282

Watersipora 81

Xantho reynaudi 272, 273

Zebina tridentata 109, 110
Zebrias zebra 198
Zonaria coriacea 61
Zostera nana 250, *251*
Zosterops japonica 222, 241
Zoysia 18, 88, 114, 115, 179, *183*, 184, 208, *210*, 214, 217, 218, 219
Zoysia sinica 37, 68, 87, 140, 153, 181, 210, *212*, 223
Zygometra comata 277

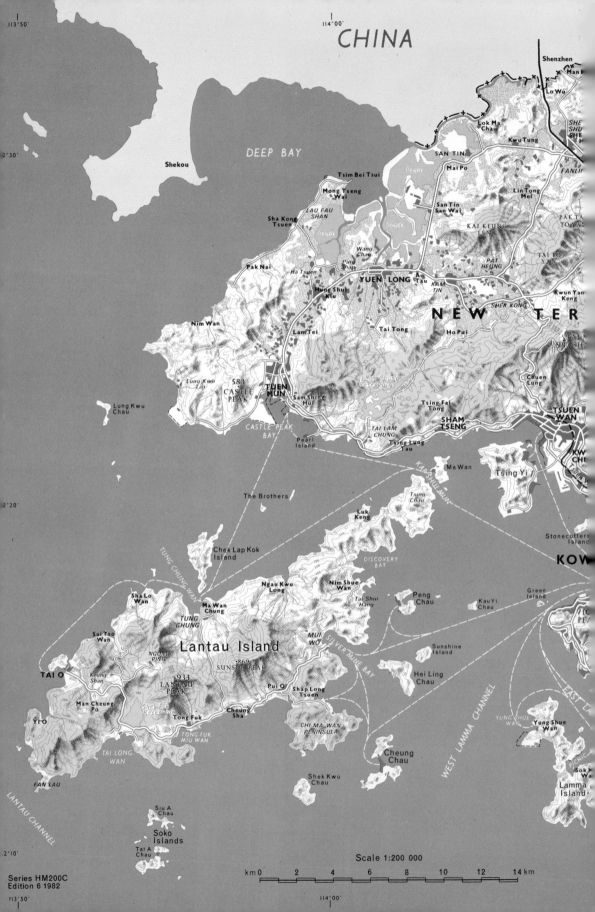